Supercooled Liquids

ACS SYMPOSIUM SERIES **676**

Supercooled Liquids

Advances and Novel Applications

John T. Fourkas, EDITOR
Boston College

Daniel Kivelson, EDITOR
University of California—Los Angeles

Udayan Mohanty, EDITOR
Boston College

Keith A. Nelson, EDITOR
Massachusetts Institute of Technology

Developed from a symposium sponsored by the
Division of Physical Chemistry at the
212th National Meeting of the American Chemical Society,
Orlando, Florida,
August 25–29, 1996

American Chemical Society, Washington, DC

Library of Congress Cataloging-in-Publication Data

Supercooled liquids: advances and novel applications / John T. Fourkas, editor . . . [et al.]

p. cm.—(ACS symposium series, ISSN 0097–6156; 676)

"Developed from a symposium sponsored by the Division of Physical Chemistry at the 212th National Meeting of the American Chemical Society, Orlando, Florida, August 25–29, 1996."

Includes bibliographical references and indexes.

ISBN 0–8412–3531–7

1. Supercooled liquids—Congresses. 2. Supercooled liquids—Industrial applications—Congresses.

I. Fourkas, John T., 1964– . II. American Chemical Society. Division of Physical Chemistry. III. American Chemical Society. Meeting (212th: 1996: Orlando, Fla.) IV. Series.

QC145.48.S9S86 1997
530.4′2—dc21 97–29912
CIP

This book is printed on acid-free, recycled paper.

PRINTED IN THE UNITED STATES OF AMERICA

Advisory Board

ACS Symposium Series

Foreword

THE ACS SYMPOSIUM SERIES was first published in 1974 to provide a mechanism for publishing symposia quickly in book form. The purpose of the series is to publish timely, comprehensive books developed from ACS sponsored symposia based on current scientific research. Occasionally, books are developed from symposia sponsored by other organizations when the topic is of keen interest to the chemistry audience.

Before agreeing to publish a book, the proposed table of contents is reviewed for appropriate and comprehensive coverage and for interest to the audience. Some papers may be excluded in order to better focus the book; others may be added to provide comprehensiveness. When appropriate, overview or introductory chapters are added. Drafts of chapters are peer-reviewed prior to final acceptance or rejection, and manuscripts are prepared in camera-ready format.

As a rule, only original research papers and original review papers are included in the volumes. Verbatim reproductions of previously published papers are not accepted.

Contents

EXPERIMENTAL ADVANCES IN SUPERCOOLED LIQUIDS

POLYAMORPHISM AND SUPERCOOLED WATER

NOVEL APPLICATIONS: SUPERCOOLED LIQUIDS AND PROTEIN DYNAMICS

INDEXES

Preface

NEITHER THE WINDOW IN YOUR OFFICE nor the Hawaiian Islands could have been formed without the assistance of supercooled liquids. Indeed, supercooled liquids are omnipresent in the physical and natural sciences and in engineering and technology. Because of this broad interest, supercooled liquids have long been the subject of considerable scientific scrutiny (and will continue to be so until their innermost secrets are revealed).

This volume is the result of a symposium presented at the 212th National Meeting of the American Chemical Society, titled "Experimental and Theoretical Approaches to Supercooled Liquids: Advances and Novel Applications", sponsored by the ACS Division of Physical Chemistry, in Orlando, Florida, August 25–29, 1996. This symposium brought together scientists from a wide range of disciplines to allow the cross-fertilization of ideas from these different areas, to assess the state of the field, to chart new experimental and theoretical directions for the field, and to identify other areas of science in which the tools developed for understanding supercooled liquids might provide some illumination. Those attending the symposium included chemists, physicists, biologists, biophysicists, geologists, nuclear engineers, and polymer engineers with expertise in experiments, theory, and simulations. Special attention was devoted to the properties of supercooled water and to the connections between supercooled liquids and the conformational dynamics of proteins.

The chapters of this volume provide a representative sampling of the science discussed at the symposium. We have endeavored, however, to make this book more than a proceedings of the symposium. Together with the authors of the chapters, we have prepared what we hope is a graduate-level primer on the state of the art of supercooled liquids and the techniques that are used to understand them. To that end, Chapter 1 is intended as a brief introduction to the field of supercooled liquids for the neophyte. Austen Angell presents a detailed scientific overview of this volume in Chapter 2. This overview is followed by sections devoted to theoretical aspects of supercooled liquids, to recent experimental advances in the field, to polyamorphism and supercooled water, and to the connections between supercooled liquids and protein dynamics.

Acknowledgments

We are grateful to the authors and other participants in the symposium, the Petroleum Research Fund, the College of Arts and Sciences at Boston College, and the ACS Division of Physical Chemistry for helping to make this symposium and this volume possible.

JOHN T. FOURKAS
UDAYAN MOHANTY
Eugene F. Merkert Chemistry Center
Boston College
Chestnut Hill, MA 02167

DANIEL KIVELSON
Department of Chemistry and Biochemistry
University of California
Los Angeles, CA 90095

KEITH A. NELSON
Department of Chemistry
Massachusetts Institute of Technology
Cambridge, MA 02139

June 18, 1997

INTRODUCTION

Chapter 1

A Brief Introduction to Supercooled Liquids

John T. Fourkas[1], Daniel Kivelson[2], Udayan Mohanty[1], and Keith A. Nelson[3]

[1]Eugene F. Merkert Chemistry Center, Boston College, Chestnut Hill, MA 02167
[2]Department of Chemistry and Biochemistry, University of California, Los Angeles, CA 90095
[3]Department of Chemistry, Massachusetts Institute of Technology, Cambridge, MA 02139

In nature, supercooled liquids play an important role in areas as diverse as volcanoes and hibernating animals. From a technological standpoint, supercooled liquids are linked intimately to processes ranging from the development of new materials to the storage of nuclear waste. It is also becoming increasingly evident that the dynamics of supercooled liquids can shed light on numerous problems that, upon a superficial examination, appear unrelated. As a result of this broad importance, the study of supercooled liquids has received considerable scientific attention over the past century. Supercooled liquids have yet to yield their deepest secrets to our scrutiny, however.

This introduction is devoted to the newcomer in the field, for whom we outline the current hot topics in the field and provide the background necessary to understand the contemporary experiments and theory on supercooled liquids contained within this volume and in the literature in general. In the following chapter Austen Angell presents a detailed overview of the field and of this book.

Current Issues

We begin with a description of the phenomenology that characterizes the field of supercooled liquids, which we define as metastable liquids that have been cooled below the melting point (T_m) but not below the temperature at which a dynamic arrest occurs (i.e., the glass-transition temperature, T_g). Although it is presumably only the crystalline phase that is stable below T_m, the first-order crystallization process may be blocked dynamically by a high free energy of activation, so that the crystalline phase need not be considered except when it makes an unwanted heterogeneous entry. The supercooled liquid may therefore be considered "equilibrated," and the feature that should concern us first and foremost is the drastic increase of its viscosity (η) and structural-relaxation (generally denoted α–relaxation) time with decreasing temperature. The more drastic this temperature dependence, the more "fragile" the supercooled liquid is said to be. Fragile liquids are typified by van der Waals organic liquids with weak intermolecular interactions and widely varying intermolecular geometries with similar energies. At the opposite extreme are "strong" liquids, which are typified by silicate and other network materials with a smaller number of strongly favored local interionic geometries.

Below T_g the relaxation time becomes so long that the liquid can no longer structurally equilibrate. The glass-transition temperature can be specified in numerous ways, since it represents a crossover between relaxation and experimental times. Unlike the melting point, which is well defined thermodynamically, the glass transition temperature is generally defined on the basis of an empirical standard such as the viscosity reaching a value of 10^{13} Poise. The time scale on which "dynamic arrest" occurs is defined by essentially the same limits of human patience, not by any fundamental parameter. In this sense, we should not expect T_g to provide meaningful theoretical insights.

Characteristic Temperatures. This is not to say that characteristic temperatures are not important to the understanding of supercooled liquids; indeed, the question of the characteristic temperature about which to scale or expand functional forms describing the properties of viscous liquids is fundamental to the field. Thus we look with some degree of envy at the description of phase transitions and critical phenomena, in which a critical temperature T_{cr} can generally be defined such that many static and dynamics quantities (including the heat capacity and the cooperative relaxation time) follow the form $|T-T_{cr}|^a$, where a is a critical exponent. Although some approaches focus on the melting point, we would suggest that crystallization is a distinct and separate phenomenon from supercooling. There is an inherent dynamic insulation from crystallization in the supercooled state, and T_m consequently has little connection with the theory of supercooled liquids.

There are currently three schools of thought on the subject of choosing a characteristic temperature. One school focuses on a possible critical point (T_o) that lies below the glass transition temperature. Extrapolation of the rapidly increasing viscosity and α–relaxation time to temperatures below T_g suggests that there may indeed be a divergence at T_o. Furthermore, extrapolation of the rapidly decreasing entropy to temperatures below T_g suggests that the entropy of the liquid is equal to that of the crystal at a temperature (T_k) that lies in the vicinity of T_o; this is known as the Kauzmann paradox. At some temperature above 0 K but below T_k the extrapolated entropy of the liquid becomes negative. This apparent violation of the third law of thermodynamics is merely a consequence of the extrapolation, and might be avoided through a phase transition in the vicinity of $T_o \approx T_k$, where both the dynamic relaxation time and the thermodynamic entropy appear to exhibit anomalies.

Another school has focused on the opposite end of the temperature scale, postulating a cross-over temperature T^* above which the liquid is molecular in character and below which it exhibits enhanced collective behavior. T^* lies near or above the melting point, so this picture is vastly different from the one above.

A third school considers the possibility of a critical or characteristic temperature T_c that lies between T_g and T_m. In this region, many properties of supercooled liquids exhibit what might be call cross-over behavior; for instance, a plot of $\log(\eta)$ vs. T^{-1} exhibits a considerable elbow and the time scales of various relaxations become distinct in this temperature region. One must keep in mind that all of these temperatures (T_g, T_o, T_k, T^* and T_c) are floating about in this volume and in the literature of supercooled liquids in general. If the newcomer begins to feel a sense of confusion upon seeing such a wide range of descriptions for what is ostensibly the same phenomenon, then it is probable that we have provided a realistic introduction to the field!

Phenomenology. We should now identify those properties that are generally considered to define supercooled liquids, insofar as these properties cannot be derived from a simple extrapolation of the properties of liquids at temperatures well above T_m. We have already mentioned the dramatic increase in the α-relaxation time and the rapid decrease in entropy with decreasing T. The temporal behavior of the relaxation function describing the response of a supercooled liquid to a perturbation (and the subsequent return of the liquid to its non-crystalline "equilibrium" state) is also of interest. At long times this relaxation appears to be described by a stretched exponential function $\exp[-(t/\tau)^{\beta}]$, where τ is a characteristic relaxation time and the stretching exponent β lies between zero and one. The relaxation becomes slower (i.e., τ increases) and often increasingly stretched (i.e., β becomes smaller) as the temperature is lowered. At shorter times, the α-relaxation appears to be described by a power law of the general form $a\text{-}b(t/\tau)^{\alpha}$, where τ is a characteristic relaxation time and $0<\alpha<1$ (this is sometimes known as von Schweidler relaxation). A different power-law relaxation (known as fast-β relaxation) occurs at yet shorter times, in addition to what appears to be purely molecular relaxation processes at the shortest time scale. Complicating this picture is the fact that within the von Schweidler range one often encounters an apparently distinct process (slow-β or Johari-Goldstein-β relaxation) and within the fast-β region one finds additional dynamic features associated with so-called "Boson peaks" in the light-scattering spectrum. As a result, it is not always clear in the literature what authors means when they refer to α- and β-relaxations.

This overall picture, with power laws extending over limited ranges, is difficult to depict unambiguously and gives rise to a full spectrum of uncertainty and controversy. These assorted problems are apt to remain controversial unless (or until) there is a well-accepted theory that yields sufficiently good quantitative descriptions of these relaxations. Closely associated phenomena that are the subject of current scrutiny include the decoupling of translational and rotational diffusion (i.e., the breakdown of the Stokes-Einstein relation for translational diffusion) and the temperature dependence of the amplitude of the stretched exponential decay (a quantity known as the non-ergodicity factor).

Homogeneity vs. Heterogeneity. A current question of great interest in supercooled liquids is whether structure and dynamics can be described as essentially homogeneous or heterogeneous. Heterogeneity implies correlations over supermolecular distance scales, which is to say that the liquid can be thought of as being comprised of domains or supermolecular structures. Such domains may be treated as long-lived entities or, alternately, as slowly-relaxing fluctuations with a supermolecular correlation length. The picture is clearest if the domain lifetime is long compared to the measured relaxations, which may or may not be the case in reality; certainly if the domain lifetime is very short, the system should be considered homogeneous. There is recent evidence, some of which is discussed in this volume, that lends support to the heterogeneous picture. The resolution of the question of which type of model is preferable remains a joint challenge to the experimentalist and the theorist.

Polyamorphism. Another topic that is attracting considerable attention currently and is touched upon throughout this volume is that of polyamorphism (which is also known as polymorphism). The fact that a number of glass-forming materials may transform to apparently amorphous phases that differ both from the supercooled liquid and from the glass is of great interest; the possibility that this phenomenon may be inherently related to the properties of supercooled liquids makes polyamorphism all the more intriguing. Studies of potential polyamorphism in water have been going on for some time, and more recently have been extended to other tetrahedrally-bonded systems and non-tetrahedrally-bonded systems as well; all of these types of systems are represented here in contributions by Kivelson *et al.* and Yarger *et al.*

Water. Water is not exactly the "prototypical" supercooled liquid. There is a limit to the amount of supercooling the bulk liquid can withstand, and even approaching this limit is an experimental challenge; as such, much of the behavior described above cannot be observed in water. Yet, as many of the chapters in this volume attest, supercooled water has received at least as much scrutiny as any other supercooled liquid one cares to name. This interest stems not only from the ubiquity and universal importance of this liquid, but also from the richness and complexity of its phase behavior. In one way or another, polyamorphism is an underlying theme in the experimental and the theoretical studies of water by Stanley *et al.*, Tanaka, Chen *et al.* and Walrafen *et al.* contained herein.

Proteins. Techniques that have been used to help understand supercooled liquids are increasingly finding application in other (often apparently unrelated) areas of science. One area in which this cross-fertilization of ideas is somewhat more mature is the conformational dynamics of proteins. Although proteins and liquids at first glance bear little resemblance to one another, there exists a similarity between the complex conformational energy landscape of proteins and that of supercooled liquids. Furthermore, the conformational dynamics of a protein is linked inextricably to the dynamics of the liquid both surrounding and within that protein. In these senses, it should not be surprising that a connection exists between the dynamics of proteins and supercooled liquids. Such a connection is explored from different perspectives in chapters by Jonas and by Fayer and Dlott.

Experimental Techniques

The subject of supercooled liquids, represented as it is by a wide range of materials including polymers, organic molecular liquids, ionic melts, aqueous solutions, proteins, colloidal solutions and others, has of course been approached with an equally wide range of experimental techniques. A comprehensive survey is impossible here. Instead, the main classes of experimental measurements will be summarized and some of the primary challenges faced in the experimental characterization of glass-forming liquids will be emphasized. The experimental investigations described in this volume will be discussed in this context.

The Main Challenge: Every Time Scale! The single deepest challenge to experimental elucidation of supercooled liquid behavior is that this behavior spans nearly all time scales over which measurements can be made. At a single temperature, relaxation dynamics may be observed over a substantial range of time scales (three or more decades is typical), and as the temperature is varied the entire range shifts from picoseconds or faster at high T to seconds (or even years) at low T. By restricting the discussion to "liquids", we can exclude the very slowest time scales, but even this exclusion is somewhat capricious since several models posit strong connections between dynamics in the supercooled liquid and glassy states.

One could argue readily that every time scale is in fact accessible experimentally. For example, optical spectroscopy provides access to very fast time scales (e.g. femtosecond spectroscopy in the case of time-resolved measurements or IR absorption and Raman spectroscopies in the case of frequency-resolved measurements) as well as very slow ones (as nearly any optical measurement can be conducted while sample evolution is under way on slow time scales). However, one soon discovers that connections among different types of experimental observables are often problematic, so that while measurement of fast dynamics through light or neutron scattering and measurement of slow dynamics through mechanical methods may cover many time scales, these measurements may produce results that are difficult to connect to each other in a rigorous way. In some cases even the same experimental observable is mediated by different phenomena on different time scales. For example, molecular vibrations, molecular orientational motions, sound waves, and other types of dynamics all contribute to the light-scattering spectrum of a supercooled liquid. The spectrum may cover an enviable frequency range, but the dynamics of these different types of motions are not related to each other in ways that permit a comprehensive elucidation of supercooled liquid relaxation dynamics.

Classification of Measurements. One could attempt a classification of the experimental techniques used to study supercooled liquids based on the type of measurement (e.g. scattering, absorption, etc.) within which the material properties responsible for the measured responses under various conditions could be elaborated. Alternatively one can base the classification on the material property examined, and the methods that may be used for its measurement may be elaborated. We will use the second approach here.

Probes of Collective Behavior

Structural Relaxation Dynamics. The dynamics of macroscopic structural change usually can be subdivided into density and shear, or longitudinal and shear, relaxation. On slow time scales (milliseconds and longer), mechanical methods of various sorts may be used. Structural relaxation dynamics on faster time scales can be deduced from acoustic properties, in particular from the characteristically high acoustic damping rates (and dispersive acoustic velocities) at acoustic frequencies that overlap structural relaxation rates. Traditional methods for measurement of acoustic responses include ultrasonics (mainly in the 1-100 MHz frequency range), whose utility has been limited by strong damping rates and other features; inelastic (Brillouin) light scattering, mainly in the 1-10 GHz range; and inelastic neutron scattering in the 100-GHz range. Time-domain variants of light scattering have also been used for examination of acoustic responses, mainly in the 10 MHz -10 GHz range. Finally, structural relaxation contributes directly to light and neutron scattering in the form of a quasielastic peak (labeled the "Mountain" mode in the light-scattering spectrum) whose width provides a measure of the average relaxation time and whose non-Lorentzian shape is due to the nonexponential form of the relaxation dynamics. Of particular recent importance has been the use of depolarized light scattering for elucidation of structural relaxation dynamics over a wide dynamic range from molecular vibrational frequencies down to about 1 GHz. Also important, especially for slower time scales, have been time-domain variations on the light scattering theme. Coherent scattering methods have permitted observation of structural relaxation dynamics on nanosecond through microsecond time scales. Photon correlation spectroscopy has also been used for measurements, primarily in the microsecond range.

The elucidation of structural relaxation dynamics from the measurements mentioned above may or may not be straightforward. In the case of mechanical or acoustic measurements, including inelastic light or neutron scattering from acoustic waves, the observed quantity is related directly to the density-density correlation function, and so density relaxation dynamics (or in some cases, longitudinal or shear relaxation dynamics) are determined. However, there are large frequency gaps between neutron scattering, light scattering, ultrasonics, and mechanical measurements, even in the favorable cases where all these methods have been applied effectively. Quasielastic light scattering including depolarized light scattering, coherent scattering, and photon correlation methods can fill in much of the dynamic range, but the analysis must be conducted carefully because in general it is not only the density-density correlation function, but some combination of it and others, that contribute to the signals. Some of these issues are discussed in the article by Yang and Nelson.

Structural Relaxation Dynamics at High Wavevectors

Dielectric Relaxation Dynamics. The complex dielectric constant can be measured over an extraordinary range of time scales, using methods that are reviewed briefly in the contribution by Lunkenheimer *et al.* From a macroscopic point of view, the results always can be analyzed in terms of the time-correlation function of the dielectric constant, whose relaxation dynamics are elucidated over the entire dynamic range. Dielectric relaxation measurements have played a major role in the formulation of empirical relaxation dynamics in polymers and other glass formers, as well as in the testing of models and fundamental theories of supercooled liquids. Of course, the microscopic interpretation of dielectric relaxation on widely varying time scales requires consideration of different sorts of responses including molecular rotation, motion of polar side groups on polymers, and so on. In addition, there are some motions that have little influence on the dipole moment and so are not probed through dielectric measurements. In this sense comparisons among dielectric, light scattering, neutron scattering, and other results are of interest in revealing the dynamics of different contributions, but in general such results should not be expected to show detailed agreement. The article by Lunkenheimer *et al.* illustrates nicely the similarities and differences among these techniques.

Thermodynamic Measurements. Measurement of heat capacities and other thermodynamic quantities usually are conducted on a static basis, that is, with the measurement made on a time scale that is slow compared to any significant relaxation times (or at least with that intent, within the limits of human patience), although dynamic aspects have been explored as well. The entropy of supercooled liquids has been of fundamental interest since early in their study, and the problem of its potential reduction below that of the crystalline state at a temperature below T_g remains a current issue. The contribution to the present volume by Kivelson *et al.* provides an illuminating example of calorimetric measurements whose results indicate that a single substance may enter distinct amorphous phases depending on its thermal history.

Local Probes. There exists a wide range of local properties that are mediated by collective relaxation dynamics and whose elucidation therefore can shed light on collective behavior. These properties include spin relaxation, measured through nuclear magnetic resonance (NMR), molecular vibrations and electronic levels, and a host of others. Since there are far more such properties than can be discussed here, we restrict the present comments almost exclusively to those represented by contributions to this volume.

NMR has particularly powerful capabilities, in that it can be used to determine time-correlation functions involving more than a single time interval. This capability is crucial in the evaluation of whether the unique properties of supercooled liquids are of homogeneous or inhomogeneous origin. Results presented in this volume by Diezemann *et al.* address this issue directly. An indication of significant site-to-site variation in the local spin dynamics suggests that the nonexponential relaxation dynamics characteristic of all supercooled liquids may be due at least in part to local inhomogeneities rather than intrinsically complex behavior of individual degrees of freedom. NMR can also be conducted on samples under applied pressure, if a nonmagnetic pressure cell is used. In the contribution by Jonas, information about denatured and partially folded protein structures was deduced on the basis of NMR measurements during denaturation which, at elevated pressures, occurs at low temperatures.

Nonlinear optical measurements, like multiple-pulse NMR measurements, can provide information about high-order time-correlation functions. This has been used to advantage in the work presented by Ma *et al.* in which the dynamics of molecular solvation are examined. The results reveal complex relaxation including nonexponential dynamics whose form and temperature dependence show the characteristics normally associated with collective structural relaxation. Thus an incisive probe of structural relaxation dynamics with capabilities for extremely high time resolution is provided.

Nonlinear optical methods such as photon echoes have been exploited extensively in the study of molecular electronic states, which resemble the two-level or finite-level systems characteristic of nuclear spins. Only recently have these techniques been extended to molecular vibrations, which pose far more difficulties due to their many quantum levels and relatively weak anharmonicities. The value of this capability is made clear in the contribution by Fayer and Dlott. Vibrational relaxation and dephasing dynamics are unmasked from beneath inhomogeneously broadened IR absorption bands, revealing the supercooled-liquid-like, complex potential energy landscape of the protein.

As mentioned earlier, light scattering measurements may be sensitive to local molecular motions, especially molecular vibrations, as well as to collective motions characterized by lower frequencies. In the contribution by Walrafen et al., Raman spectra of supercooled water OH stretching vibrations are used to deduce the intermolecular H-bonded network geometries through comparison to behavior exhibited by clathrate crystals that form similar networks.

Not included in this volume, but of great current interest, are light scattering and other probes of what is commonly called the "Boson peak", a broad feature typically in the 0.1-1 THz frequency range. This is attributable mainly to intermolecular or interionic vibrations whose classification as "local" could be debated, but that should be considered either local or "mesoscopic" (that is, intermediate between molecular and macroscopic). Probes of this broad feature have led to intensive debate concerning the extent to which the Boson peak is homogeneous or inhomogenous; is predictable on the basis of current models; is revealing of local vibrational anharmonicity; and so on. We conclude this section with the suggestion that this region of the spectrum should be the next one to be explored by nonlinear optical methods, which will permit its origins to be revealed more incisively.

Theoretical Perspectives

The theories that have been advanced over the past century to explain the properties of supercooled liquids number in the hundreds (if not thousands). In this limited space, we cannot hope to do justice to all of these models, much less to the subset of models that are currently popular. We shall therefore confine our discussion primarily to the theories and theoretical tools that are touched upon within the chapters of this volume.

We should begin by pointing out that the long time scales of so many of the phenomena that are characteristic of supercooled liquids had for many years all but restricted theorists to pencil-and-paper treatments at a time when molecular dynamics and Monte Carlo simulations were opening new vistas into the study of normal liquids. That this situation has evolved in recent years is readily apparent from the theoretical contributions in this book, many of which draw upon computer simulations to varying degrees. Simulations provide the ability to test theoretical concepts in an idealized environment in which the role of molecular details in determining the dynamics of the system can be minimized (or at least readily distinguished). As such, computer simulations will take on an increasingly important role in the study of supercooled liquids.

Mode-Coupling Theory. Without a doubt, mode-coupling theory (MCT) enjoys the highest visibility of any theory of supercooled liquids in the current literature, as will be evident from many of the chapters in this volume. MCT is based upon differential equations that arise from an approximation to the equation of motion for density fluctuations in a supercooled liquid. Given this approximation, the behavior of the liquid can be described out to the time scale on which diffusion begins to take place. According to this so-called idealized version of MCT, at the aforementioned temperature

T_C (which lies well above the traditional glass-transition temperature) structural arrest occurs and ergodicity is lost in the liquid. This phenomenon is moderated in the extended version of MCT, in which an activated hopping process is introduced to permit relaxation to occur below T_C, thus restoring ergodicity. However, the underlying singularity at T_C persists, and its ostensible observation in a number of model glass-forming liquids has stimulated intense interest in the theory.

Neither version of MCT includes hydrodynamic interactions, which may be crucial for describing systems such as colloidal glasses. MCT does not incorporate specific molecular detail either. On the one hand, this means that MCT cannot predict on an absolute time scale the quantitative time dependence of any dynamic variable. On the other hand, MCT does make specific and testable predictions about how the dynamics of supercooled liquids depend on temperature. MCT predicts that certain features of any dynamic variable should be universal (such as a power-law behavior in the temperature dependence of relaxation times). These predictions are discussed in detail in Kob's chapter on MCT.

MCT has many vocal advocates and equally many outspoken critics. However, all would agree that MCT has had a profound impact on the field, particularly in causing the supercooled liquids community to scrutinize data from both experiments and simulations in new (and often more stringent) ways. A desire to develop better tests of the predictions of MCT has also been one of the driving forces behind the current trend of attempting to obtain dynamic data on supercooled liquids over as large a frequency or time range as possible. Although other theories may well supplant MCT in the future, it is incontrovertible that MCT has stimulated a great deal of constructive thought and activity and has focused efforts into significant experiments.

Domain Theories. As discussed above, many models treat supercooled liquids as being composed of heterogeneous domains. Many of the theoretical contributions in this volume, while distinctly different in formulation and predictions, can be grouped loosely into this category of domain models. One of the foundations of such models is the intuitively appealing idea that the strong temperature dependence of the behavior of supercooled liquids is due in some way to a change in the distance scale that characterizes cooperativity. In this sense, the "cooperative rearranging region" model of Adam and Gibbs might be said to be the intellectual grandfather of contemporary domain theories. We should hasten to point out, however, that despite intriguing parallels between the Adam-Gibbs model and the models in this volume, except for that of Mohanty the models herein are not to be understood as extensions or generalizations of the Adam-Gibbs model.

Adam-Gibbs theory postulates that a supercooled liquid is composed of subunits that are essentially independent of one another, in the sense that each region can rearrange itself through some sort of cooperative motion that is unconstrained by the other regions. These cooperative regions have a distribution of sizes, each of which is larger than some minimum size that is necessary for cooperativity. This minimum size is linked closely to the configurational entropy of the system, from which many thermodynamic and dynamic properties of supercooled liquids can be calculated.

The domain concept is approached from a number of different viewpoints in this volume. Mohanty uses concepts from the theory of critical phenomena to generalize the Adam-Gibbs model and to make a connection with the Kauzmann temperature. Stillinger employs a theory of "fluidized domains" to investigate the decoupling of translational and rotational diffusion. This phenomenon and other apparent anomalies in supercooled liquids are attacked by Tarjus, Kivelson and Kivelson with a theory of domains whose growth is limited by the inherent structural frustration of the liquid. Mountain presents intriguing simulations that suggest, at least in two dimensions, that there is indeed a characteristic length scale that grows with increased supercooling.

Topographic Concepts. The potential energy of a supercooled liquid composed of N molecules that are each composed of n atoms in turn can be viewed as a $3nN$-dimensional hypersurface. The coordinates of this surface that are intramolecular may have relatively little effect on the properties of the liquid. The remaining coordinates comprise the complex energy landscape of the supercooled liquid.

Many theories of supercooled liquids are concerned with the topographic features of this landscape. Obviously, the deepest minima in the landscape correspond to crystalline structures. The very fact that a liquid is supercooled implies that it is trapped in a portion of the landscape from which the nearest crystalline basin is not readily accessible. It is the features of this local region that determine the behavior of the supercooled liquid. For instance, the frequency and depth of local minima is related to the fragility of the liquid.

At any given moment, it is likely that the liquid does not lie in a local minimum of the potential energy hypersurface. Some theories focus on the local basins into which the liquid will fall when quenched, which are called the "inherent structures" of the supercooled liquid. Mohanty, for instance, discusses the connection between inherent structures and the Adam-Gibbs model. Other models concentrate on how the liquid will behave at any given point along the hypersurface. For instance, instantaneous normal mode (INM) theory treats any given configuration of the liquid as a set of harmonic normal coordinates, which can then be used to approximate the short-time behavior of the system. These coordinates are derived by diagonalizing the Hessian matrix (the matrix of second derivatives of the potential energy with respect to any two atomic coordinates) of the liquid. The resultant matrix will have both positive and negative eigenvalues, the former of which correspond to minima along that coordinate and the latter of which correspond to barriers. Zürcher and Keyes incorporate INM ideas into an extension of the two-level system model of glasses to successfully reproduce many of the temperature-dependent properties of supercooled liquids.

Spin Glasses. Spin glasses are a theoretical construct that was originated to understand the behavior of magnets. Dilute magnetic alloys are prototypical materials that would be described as spin glasses. In these alloys, the non-magnetic ions are randomly substituted by magnetic ions. Similarly, in a spin glass quantum-mechanical spins are placed at lattice points and some form of coupling is introduced such that the value of any given spin is influenced by those around it.

While simple in concept, spin glasses exhibit a wealth of fascinating behavior. The random couplings between the spins are usually "frustrated," and there exists a characteristic temperature below which the spins freeze without long-range order. The magnetic susceptibility displays cusp-like behavior near this temperature, and irreversible behavior of other magnetic properties is observed below this temperature.

Due to frustration, the number of ground states of a spin glass of N spins is proportional to $\exp(N)$, and hence many order parameters are required to describe such a system. The configuration space thus consists of many valleys, each valley representing a possible state of the system. The valleys are separated by free energy barriers. In addition to quenched disorder and randomness, aging and a slowly-decaying relaxation function are also features of spin glasses.

One feature that distinguishes spin glasses from real glasses (which are often called structural glasses in this context) is that the latter do not exhibit quenched disorder. Nevertheless, there is a long-standing belief that there is a fundamental connection between spin glasses and structural glasses. This belief is bolstered by the fact that there is an intimate connection between the equations governing mean-field spin-glass models and MCT. The relationship between spin and structural glasses remains to be understood fully, and some of the connections are explored in this volume. In particular, Parisi describes how a statistical mechanics technique called the replica method, which has been highly successful in the study of spin glasses, can be applied to the study of structural glasses.

Other Models. We now touch briefly upon two other theoretical models that appear in this book. We begin with the Coupling Model (CM), which postulates that on short enough time scales, the fundamental dynamic units of a supercooled liquid exhibit exponential relaxation, but that at longer times the relaxation becomes stretched-exponential as cooperativity becomes important. Ngai and Rendell present evidence in support of the CM in their contribution to this volume.

Corrales considers the phase behavior of binary silicates, which are of practical importance as materials for the disposal of radioactive waste. The statistical mechanics technique used in this system is Flory-Huggins theory, which uses a sort of combinatorics to calculate entropies of mixing between species.

Concluding Remarks

The following chapter contains a detailed overview of this volume, but we hope our whirlwind tour of contemporary research in supercooled liquids will make the rest of this volume accessible to the newcomer to the field, and that this book in general will assist in advancing our understanding of supercooled liquids.

Overview

Chapter 2

The Viscous-Liquid–Glassy-Solid Problem

C. A. Angell

Department of Chemistry, Arizona State University, Tempe, AZ 85287–1604

The different aspects of liquid state behavior currently in focus are classified in terms of their place in the range of relaxation times separating the microscopic (phonon) time from the structural relaxation time at the normal glass transition T_g. The place within this scheme of each of the contributions to this symposium is identified. Special attention is given to the crossover between "free diffusion" and "landscape-dominated" regimes. Thermodynamic estimates of the "height of the landscape" are given for strong and fragile liquids and a correspondence is found between the thermodynamic estimate and characteristic temperatures T_c of mode coupling theory and T_x of Rössler scaling obtained from different analyses of the dynamic properties. Fragile liquids have all their configurational microstates (configurons) packed into a small energy band between kT_K and the upper limit at kT_u i.e. T_u is relatively close to T_K. In the landscape-dominated regime, clustering occurs. We outline a cooperative cluster model which can generate high fragility and which in extreme cases can provoke a first-order phase transition to the fully clustered state.

It is now becoming widely appreciated that systems which slow down with decreasing temperature (or increasing pressure) to the point where internal equilibrium is lost and only solid-like characteristics remain, are very widespread in nature and comprise some very important systems, both in physics and biology. It has further become evident that the problem of adequately describing and accounting for the ergodicity breaking process is a profound one. The contributions to this volume, which records many of the contributions to the challenging symposium on the subject organized by John Fourkas, Udayan Mohanty, Keith Nelson and Daniel Kivelson for the Fall meeting of the ACS, provide a well-rounded description of the key elements of this currently exciting problem area for a particular but very important case of this phenomenology. This is the case in which the elements which "get stuck" in the glassy state are atoms, molecules, or "beads" on a polymer or biopolymer chain.

The most challenging aspects of the problem may be thought of as occurring in three very different domains of time. The first (and, to many, the most challenging) concerns the very short time domain in which these condensed systems first resolve the very chaotic motions characteristic of the dense gas and hot liquid states into separable components. The slow and (increasingly) temperature-dependent diffusive modes separate from fast oscillatory modes and the gap in characteristic time scales opens up increasingly as temperature decreases or as pressure increases – a scenario which is revealed by many neutron scattering and computer simulation studies. This is generally well accounted for by the mode coupling theory of Götze and co-workers discussed below.

The second is found at intermediate times, of order ns-ms depending on the system under consideration. The mode coupling theory in its idealized form predicts dynamical jamming, but this is averted by the onset of activated processes (basin-hopping) in configuration space, which we may regard as a crossover to "landscape domination" of the relaxation processes. In at least some well-known cases, this is also where the so-called secondary ("β-slow") relaxations split off from the main relaxation. In real space terms, this may be related to some initial stage of the clustering phenomenon which is being observed in a later stage in the third domain under current investigation.

The third domain is found at the other end of the accessible time scale range, *i.e.* at very long times where experiments can no longer record the complete response functions, and the systems therefore become non-ergodic. It is in the latter range (just before ergodicity-breaking occurs), that the amorphous structures appear to be developing a measurable degree of structural non-uniformity. This is evidently the result of some sort of clustering phenomenon which then would appear to have much to do with the specially provocative relaxational and thermodynamic behavior exhibited by certain of the systems which vitrify – the so-called "fragile" systems.

While the above comments refer to the standard viscous liquid problem, a fourth component to the overall problem has been recognized by the symposium organizers. This is significant because the systems involved are some of the most important of all, for instance water and proteins in the life sciences, and silicon and silica in the inorganic world. This latter component involves an increased level of complexity in the overall problem, wherein the last-mentioned tendency of fragile liquids to resolve their structures into distinct components differing in structural organization, goes to the limit of splitting the systems into two distinct liquid (or, at least, amorphous) phases. This twist, classified under the name of polyamorphism, is of recent recognition, and is attracting a lot of attention at the moment.

Hot Liquids, Liquid "Structure," and Relation to Dynamics

In the dense gas and hot liquid phases, except near the critical point, most of the physics can be understood in terms of isolated binary collisions. With increasing density, usually brought about by decrease of temperature, the behavior becomes more complicated as the particle motions become increasingly oscillatory with increasing "caging". At some point in density the structure reaches the point where a shear rigidity can be detected for high enough frequency probes and the liquid can be said to have a "structure." In the simplest types of fluids based on hard spheres, this state is related to a vibrationally expanded version of Bernal's "loose random packing" of hard spheres *(1)*. Instantaneous removal of all thermal energy from such a structure [by conjugate gradient quenching*(2)*] leaves the most open of Stillinger's "inherent structures," a "house of cards" structure which would collapse to something denser under the slightest agitation. Between this structure and the lowest energy most densely packed structure conceivably consistent with amorphous packing lies an immense number, of order e^N, *(2,3)* alternative mechanically stable structures which together make up the structures of any conceivable "glassy" state that could be made by the most extreme quenching or geologically slow cooling process that could be imposed on the system (in the absence of crystallization). These collectively make up the energy "landscape" *(2,4)* for the system. Later in this volume, the energy landscape is specifically invoked to account for the observations of the multidimensional NMR relaxation study of Diezemann *et al.*

The dynamic behavior of liquids seems to be sensitive to the arrival at the densities characteristic of these coherent structures. As mentioned above, for lower densities (higher temperatures), the dynamical behavior seems to be described rather well by the mode coupling theory developed by Götze *et al (5)*. There is some uncertainty as to whether some of the experiments which support the detailed predictions of the theory have been looking at the right sort of fluctuations *(6)*. Certainly a lot of viscosity and diffusivity data, starting at the highest temperatures, can be well represented over a limited range of temperatures by the power laws that the theory predicts *(5,7,8)*. However, then, instead of jamming into a rigid solid at the power law viscosity divergence temperature, there is a crossover to a slower divergence, heralded by breakdowns of the Stokes-Einstein viscosity-diffusion law and the Einstein-Debye diffusivity-rotational diffusivity law *(9,10)*.

The short time scale measurements of Berg, and of Nelson and co-workers, in the present volume give new tests of the mode coupling description of the short time dynamics and new data on the crossover temperature. The accord with theory shown by more ideally simple systems is tested by computer simulation studies of LJ mixtures in the paper of Kob. Similar systems are studied in the paper by Zurcher and Keyes in their alternative description of the dynamics in terms of real and imaginary normal modes of quenched structures. Mountain also uses simulations of similar systems in an attempt to define a characteristic length scale which grows with increasing density and decreasing diffusivity.

In about the same range of densities, the relaxation function seen by experimental probes bifurcates into a fast Arrhenius non-exponential process with a strongly temperature dependent halfwidth in frequency, and a slower also non exponential process with a slowly increasing halfwidth in frequency. To distinguish this fast process from the fast process invoked in MCT, the MCT process is now designated "fast β-" and the process splitting off from the a-relaxation is called the "slow β-" process. The latter, but not the former, is characterized by a maximum in the susceptibility for the measurement (dielectric or calorimetric) used to detect it.

Scaling Relations and Clustering. Different ways of scaling the existing viscosity and relaxation time data help to bring out the relation between structure and dynamic characteristics. In the recent Fujimori-Oguni scaling of dynamic data *(11)*, the temperature of the β-glass transition is used, in place of the usual $T_{g(\alpha)}$, to scale the relaxation time data (Figure 1). This scaling brings out the idea that in fragile liquids, the separation into α and β processes occurs at lower reduced temperatures, hence longer relaxation times, for the less fragile liquids. According to Fujimori and Oguni *(11)*, this is because in fragile liquids the process leading to the clustering which determines the α-relaxation time is more cooperative. On the other hand, Perrera and Harrowell *(12)* argue it is because, in the less fragile liquids, the fast process (which they say provides the actual mechanism whereby all particle rearrangements occur) has a larger activation energy than for the more fragile liquids. The consequence is that the temperature at which the dynamics becomes dominated by the cluster growth, revealing the α-relaxation, is postponed to lower values. The presence of both dynamically fast and slow domains, the latter being long-lived, has been documented in the photobleaching experiments of Ediger and co-workers *(13)* and is supported in some detail by the multidimensional NMR measurements of Diezemann *et al* given in this volume.

The origin of the clusters themselves, in frustration limited growth process, is the subject of the paper by Tarjus *et al* in this volume. This model leads to a quite different scaling representation of the viscosity or relaxation data than that of Fujimori and Oguni since it does not specifically recognize the occurrence of a (slow) β-process. On the other hand the nature of the slow β-process, as a primitive (Arrhenius and exponential) relaxation which then becomes the α–relaxation by cooperative slowing down, is the emphasis of the coupling model whose description of the short time dynamics in relaxing liquids is presented by Ngai in this volume.

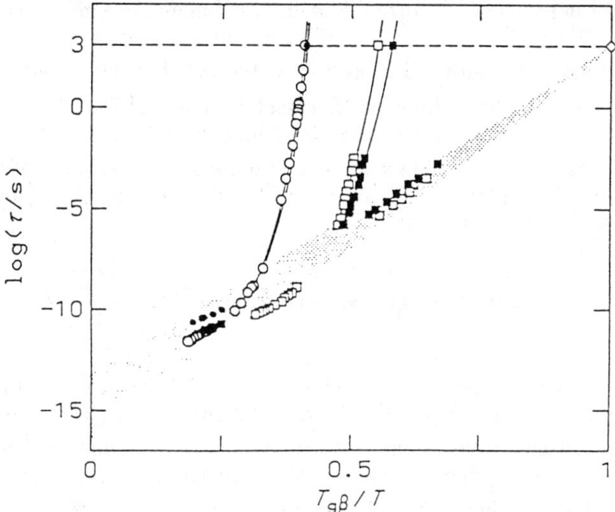

Figure 1. Representation of the relation between primary and secondary relaxation processes in liquids of different fragility given by the scaling principle suggested by Fujimori and Oguni *(11)*. Here the most fragile liquids, triphenyl phosphite and PMS (a diphenylated disiloxane) are those in which the bifurcation into α– and β- process occurs on the shortest relaxation times. This is interpreted *(11)* in terms of these liquids having the highest cooperativities.

In the other recent scaling proposal, that of Rössler and co-workers *(14)*, the α-glass transition temperature is retained as a scaling parameter but a second characteristic temperature, T_c, is introduced which serves to collapse all liquid viscosities on to the same Vogel-Fulcher curve. This second characteristic temperature is higher relative to the first (T_g) for the stronger liquids, and indeed lies close to the crossover temperature of mode coupling theory according to Rössler and Sokolov *(14,15)*. It would appear to correlate with the temperature at which the energy landscape begins to dominate the dynamics *(14)*, but says nothing about onset of decoupled fast modes (slow β-relaxations), which would then have to be associated with inherent kinetic characteristics within the energy minima of the landscape. Somewhere in this domain lies the source of the decoupling of diffusivity from viscosity first described by Sillescu and coworkers *(10)*. An account of this phenomenolgy in terms of "fluidized domains" within an inert matrix is given in this volume by Stillinger. The relation of Stillinger's fluidized domains to the fast regions held to account for the (slow) β-relaxation is unclear, but they would appear to be unrelated since the fluidized domains are essential to the vagaries of the α-relaxation. Thus they are more closely related to excitations across the landscape than to the complexity of fluctuations within a single minimum as is descriptive of the (slow) β-processes. The "height" of the landscape in kT can be estimated *(16)* by calculating the temperature T_u in the expression,

$$S_c = k_B \ln W = {}_{\sim} k \ln \exp(N) = R/\text{mole} = \int_{T_K}^{T_u} \frac{\Delta C_p}{T} dT \qquad (1)$$

where ΔC_p is the heat capacity increment associated with exploration of the landscape (*i.e.* the jump in C_p as $T > T_g$). Of course the height is then greater for liquids with small ΔC_p (*i.e.* strong liquids, in accord with the higher T_c found by Rössler *et al* for the strong liquids. The value of T_u turns out to depend only weakly on the functional form assumed for ΔC_p. For both of the simple choices,

$$\Delta C_p = \text{constant}, \quad \text{and} \quad \Delta C_p = \text{constant}/T \qquad (2)$$

T_u found to be the same for values of the constants given by experiment on the simplest fragile liquids *(17)*. For liquids like S_2Cl_2, and also Lennard-Jones argon *(18)*, ΔC_p at T_g is found to be about 17J/K per mole of heavy atoms, and T_u is then $1.59T_K$ which is about $1.27T_g$. Interestingly enough, this is almost the same as the value of T_c/T_g found by Rössler and co-workers for the ratio of upper scaling temperature T_c to T_g which means it is the same as the ratio T_c/T_g where T_c is the mode coupling theory T_c. The identification of T_c with the temperature characterizing the top of the landscape is consistent with the idea that MCT fails at low temperature because of the crossover to landscape-dominated dynamics. Our estimate of T_u is probably a minimum value because a part of the ΔC_p at T_g (of unknown magnitude, but thought to be small) is vibrational in nature. A value of T_u somewhat above T_c would seem appropriate because T_c presumably represents a crossover in dominant relaxation mechanism rather than a real end-point in structural character. In some model systems for which data are available, e.g o-terphenyl, T_c is also found to correspond with the bifurcation temperature into distinguishable α- and β-relaxations and with the lower limit temperature for accurate data-fitting by the "high temperature Vogel-Fulcher law" in the Stickel-plot analysis *(19)*. The reason that a V-F law (with unphysical parameters) should fit in the free diffusion domain is not at all clear. This is the domain in which a power law fit of the same data yields the T_c found by the other criteria, so the V-F fit may be a trivial consequence of its relation to the power law through the Bardeen singularity *(20)*.

The relation between entropy, relaxation time, and the position of kT on the energy surface is summarized in the three part ensemble of Figure 2 which shows the regime of free diffusion above the highest features of the landscape but below the boiling point. T_c falls somewhat below the highest energy features of the landscape.

An alternative approach to accounting for the difference between strong and fragile behavior in liquids has been advanced in this volume by Mohanty, stimulated by the correlations noted by Sokolov *et al (21)* between the strength of the so-called boson peak and the fragility of the liquid in question. Some evidence for the onset of a strong boson peak as a consequence of rapid increase in the harmonic character of the system is reported for the particular case of water on the basis of MD studies by Chen *et al.* Water is seen here both as an ideal mode-coupling glassformer, and as a system which at the lowest temperatures has transformed to a strong liquid. Indeed, in this volume there is a group of papers which focus on the unusual character of water in this overall scheme. This brings up the third component of the subject mentioned in our introduction.

Phase Changes in Liquids, and Polyamorphism in Glasses

Water is of course famous for its strange behavior. What is not generally appreciated is how water-like certain other liquids, like liquid silicon for instance, can be. The reason seems to lie in the similar inversion of the usual energy/volume correlation in these cases.

Contributions from Walrafen, Stanley and co-workers, and Tanaka, in this volume give different slants on the phenomenology. The latter two contributions are both MD simulations using well known pair potentials for water, and both agree on the important conclusion that water is capable of undergoing a first order phase transition between liquid states of different density. There are only differences in the estimates of the (critical) pressure which is needed for a first order transition, rather than a continuous transition, to be manifested. The two types of transition are separated by a critical point, the existence of which was previously unsuspected. On either side of the first order transition line, and radiating out from the critical point, are spinodal lines (in the mean field sense) at which fluctuations would diverge if kinetics did not precipitate a prior nucleation of the new liquid phase. This scenario is a theoretical consequence of a bond modified van der Waals model by Poole *et al (22)* and also of a two-state model by Moynihan *(23)* which is parameterized using data from the experimental observations of Mishima on polyamorphic forms of vitreous water *(24)*. Even the simple bond lattice model of the author gives a phase transition when the lattice excitations are allowed to be cooperative. On the other hand, Debenedetti and co-workers in a series of papers using different theoretical models have shown that the strange properties of water can be accounted for by models in which no singularities exist.

The diverging heat capacities and plunging diffusivities of supercooling water, which imply super-fragile character in the liquid, are also observed preceding a first order phase transition in the case of computer-simulated liquid silicon *(25)*. The transformation in the laboratory substance is from metallic liquid to tetrahedral semi-conductor. In this case, it is found to be first order at ambient pressure both in the simulations and in laboratory flash heating and quenching experiments *(26)*. As in the case of water, the transition is preceded during cooling by a density maximum *(25)*. A number of other liquids (of substances known to have open packed tetrahedral crystals (like InSb) also show density maxima, and might be expected to have comparable metastable state transitions *(27)*. Indeed there is a considerable literature on phase transitions in such systems described in terms of two-fluid models *(28)*. Even liquid silica has a density maximum, and there is speculation about the

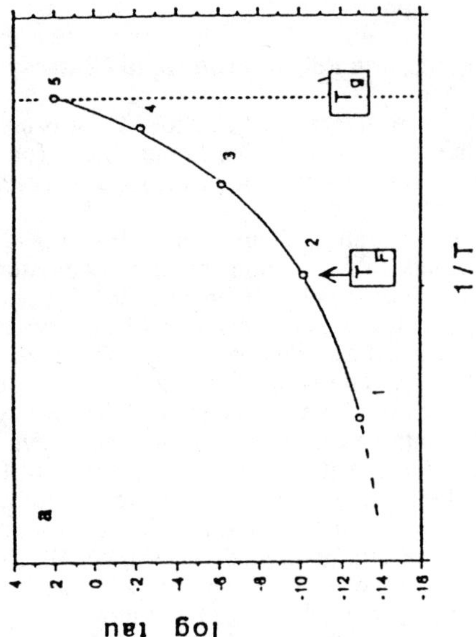
log tau
1 / T
4
2
0
-2
-4
-6
-8
-10
-12
-14
-18
a
1
2
3
4
5
T F
T g

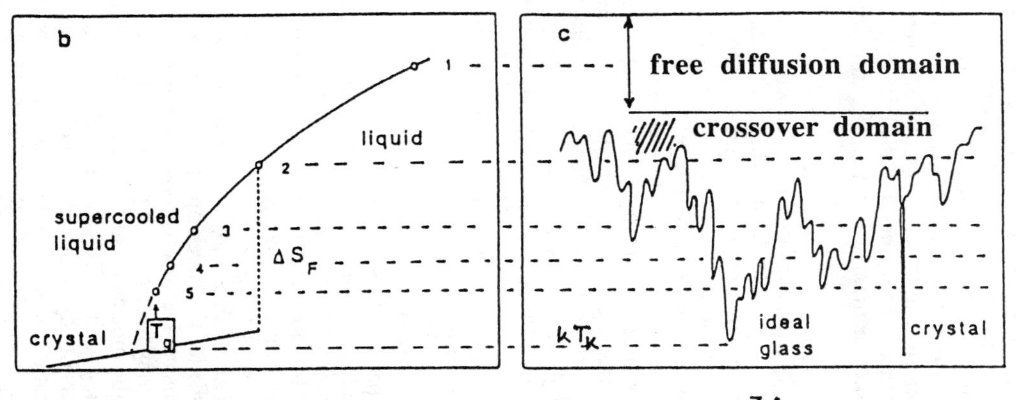

Figure 2. Summary of phenomenology of glassformers showing diverging relaxation times (in part (a)) being related, by points 1, 2, 3. ... on the plot, to vanishing excess entropy (in part (b)) and finally, in part (c), to the level of energy minima on the potential energy hypersurface (or "landscape"). The temperature T_K corresponds with the energy of the lowest minimum on the amorphous phase hypersurface. Many vertical spikes, corresponding to configurations in which particle core coordinates overlap, are excluded from this diagram for clarity.
The number of minima on the surface is of order exp(N), where N is the number of particles. The height of the landscape relative to kT_K is a measure of the "strength" of the liquid.

role that a submerged transition (below T_g) may play in the phenomenology of immiscibility in alkali metal silicates *(25,29)*.

Transitions between species, if not phases, in liquid silicates is the subject of the contribution of Corrales *et al* in this volume. This is an important issue in deciding on the applicability of the term "polyamorphism" in the case of transformation between glassy phases, which is where the whole subject started. The matter of discontinuous vs. continuous transitions between glassy phases is taken up in the contribution of Yarger *et al*, where the critical importance of the narrowness of bond angle distributions in the amorphous phase is brought into focus. The pressure-induced transformation between high density and low density glass phases in the case of silica is only weakly hysteritic compared with the case of water and even GeO_2, and this is attributed to major differences in the intertetrahedral angles in the respective low density glasses and the resultant differences in transformation cooperativities.

Finally there is the mysterious case of triphenyl phosphite and its transformation to the "glacial phase", purportedly a polyamorphic form of the supercooled liquid, reported by Kivelson *et al* in this volume. While the possibility that this highly disordered phase is simply a crystal polymorph of low lattice energy (with thermodynamic stability always less than the normal crystal – a not-uncommon finding with glassformers), has not yet been eliminated, its interpretation as a product of a liquid-liquid phase transition is well within the possibilities encompassed by the above-mentioned models. There is, for instance, nothing in the two species water model of Moynihan *(23)*, or the cooperative cluster model described below, that requires that the system have a density maximum preceding the phase transition. And triphenyl phosphite does exhibit the high fragility which should always precede a liquid-liquid phase transition. Thermodynamic treatment of the exchange of molecules between the slow and fast regions of the earlier-discussed microheterogeneous model of fragile liquids (in a cooperative two-state approximation) leads directly to a phase transition to the cluster phase in the limit of high fragility. Indeed such a model can provide a simple-minded bridge between all the different aspects of the phenomenology discussed in this chapter, and so we outline it in the last section of this paper. First, however, we need to put in perspective another fascinating aspect of the polyamorphism phenomenology.

We start by noting the unfortunate circumstance that all the above liquid-liquid transitions occur under conditions which are highly metastable with respect to crystallization of the liquid, so the events of greatest interest cannot be studied in the laboratory (the possibility of exceptions is suggested by the case described by Aasland and McMillan *(30)*. Fortunately, an analogous phenomenon, *viz.* the occurrence of reversible first order transitions between aperiodic structures, occurs under thermodynamically stable conditions in mesoscopic systems of great biological significance. We refer here to the denaturing transitions of proteins and the analogous unfolding transitions in other complex biomolecules such as the RNAs. There are three contributions to this volume dealing with different aspects of the dynamics of relaxation within proteins or with the thermodynamics of the unfolding transitions themselves. Dlott and colleagues look at the short time dynamics within the folded protein of the CO recombination process which has been the subject of much longer time dynamics studies *(31)*, and of the first application of the landscape paradigm to biomolecules *(32)*. Jonas examines the unfolding transition in the low temperature domain using pressure both to stabilize the liquid state and also to move the temperature of the unfolding transition itself.

Cooperative Clusters, Fragility, Ideal Glasses, and Phase Transitions

We give here a brief version of a model which could be used to explain how fragility and phase transitions can arise in a dense molecular system in which molecules can exist either in clustered form or in loose intercluster regions, as in the Fujimori-Oguni *(11)* and Perrera-Harrowell *(12)* scenarios. It is a two state model in the sense of the author's bond-lattice model *(33)* and the more recent two-species regular solution model for water polyamorphism of Moynihan *(23)* but one in which the non-ideality is introduced in a more physical manner. Details will be given elsewhere *(34)*.

We suppose in first approximation that the molecules in a cluster have a lower enthalpy H_1 per mole and a lower entropy S_1 per mole than molecules in the intercluster region where they are H_2 and S_2. Then the distribution of molecules between clusters and "tissue" (intercluster) regions will vary according to the usual two state thermodynamic relations

STATE I <=======> STATE II
(cluster) <=======> (intercluster)
A <=======> B

$$K = [A]/[B] = X_B/(1 - X_B) \tag{1}$$

$$\Delta G = \Delta H - T\Delta S = RT\ln K_{eq} \tag{2}$$

from which the mole fraction of B can be shown to be

$$X_B = [1 + \exp(\Delta H - T\Delta S)/RT]^{-1} \tag{3}$$

and the associated heat capacity is

$$C_p = (\partial H/\partial T)_p = R(\Delta H/RT)^2 \cdot X_B(1 - X_B) \tag{4}$$

This heat capacity is of the familiar Schottky anomaly form *(35)*, a dome with onset commencing at a temperature determined by the molar enthalpy increment per bond break, ΔH, and rapidity of increase and maximum value determined by the associated entropy change, ΔS. The general behavior for different ΔH, ΔS combinations may be seen in Figures 2 and 3 of ref. 30., and two cases (for ΔS finite and zero) are given in Figure 3. This is just the Angell-Rao bond-lattice model for the glass transition *(36)* with the excited state ("off-bond" *(36)*) now identified with a non-cluster molecule.

Now as the size of a cluster increases, a smaller percentage of its molecules are on the surface, and correspondingly the average Debye temperature for the cluster will decrease and its free energy per mole will be lowered. Let us suppose that clusters grow larger with decreasing T and hence decreasing X_B, so that the free energy difference between cluster and non-cluster molecules decreases with increasing X_B. Thus in second approximation, we assume the simple one-parameter form

$$\Delta G = \Delta G_o/(X_B + \delta) \tag{5}$$

in which small δ, the only additional parameter introduced, means high cooperativity. According to Eq. (5) the free energy accompanying excitation from cluster to tissue crosses over from a $\Delta G_o/\delta$ at the low temperature extreme, where X_B tends to zero, to the smaller value $\Delta G°/(1+ \delta)$ as the temperature (and state of excitation) increases. Then the mole fraction of intercluster molecules increases cooperatively at low temperatures according to

$$X_B = [1 + \exp(\Delta G_o/RT(X_B+ \delta))]^{-1} \tag{6}$$

which is a continued fraction whose value can be obtained by iteration. In Figure 3 we compare results for this model with $\Delta H_o = 750$ cal/mole, $\Delta S_o = 2.0$ cal/mole, K and three different degrees of cooperativity $\delta = 0.5$, 0.1, and 0.05, with results for two non-cooperative cases. In the first, ΔS_o is kept at 2.0 cal/mole K and ΔH in Eq. (3) is set to the high temperature limit of the first case ($\Delta H^o/1.1$) to give correspondence at higher temperatures. In the second, ΔS_o is set to zero, meaning that the excited (inter-cluster) state is non-degenerate, and we have the normal Schottky anomaly.

It is seen that the introduction of the cooperative cluster effect can initially sharpen the rate at which the fully clustered state is approached, i.e. increase the fragility, and at sufficient cooperativity, it can turn the crossover from partly clustered to fully clustered state into a first order transition (as some single cluster goes critical). Thus the Kivelson glacial phase in triphenyl phosphite (TPP) described in this volume and elsewhere *(37,38)* could be rationalised as a case of extreme cooperativity and linked to the extreme fragility observed in the normal supercooled liquid state of this substance. It is notable however, that while both Fan *(39)* and Fujimori and Oguni *(11)* found an unusually large ΔC_p for TPP, neither study found any sign of a high temperature heat capacity anomaly which is expected *(37,38)* for the case where there is a narrowly avoided critical point at temperatures above T_c (and even above T_m) (though it seems reasonable to expect such an anomaly to be most pronounced in the most fragile liquids). This weakens the case for the frustrated cluster interpretation of the putative phase transition and for the associated liquid viscosity model *(37,38)*.

It takes only a minor change of wording for the model described here to account for events observed in water. The fact that single molecules are either in clusters or not could account for the finding by Moynihan *(23)* that the active unit in his two species mixtures model of water was an $(H_2O)_1$ species, eliminating the broken H-bond as the active unit. A distribution of cluster sizes would be enough to account for the failure of the Prigogine Defay ratio to be unity as the simple two-state model requires.

A final observation on the utility of the cooperative cluster model is the following. Since, per mole, there are N_A distinct energy states between unclustered and fully clustered limits, one can crudely consider each one to represent a minimum on the potential energy hypersurface, and each excitation to require a shift from one minimum to another. Since this shift requires the displacement of a molecule from a cluster to an intercluster position the excitation is intrinsically diffusive. Thus the intimate connection between relaxation and configurational entropy seen in Fig 2 and almost quantitatively accounted for by the Adam-Gibbs equation *(40,41)*, can be qualitatively understood.

While the above ideas are quite simplistic, and are simplistically dealt with in the above model, they are physically based and the results are easily understood. Thus they may be suggestive of the origin of some of the more provocative features of the viscous liquid state that are currently being revealed.

Conclusions

As the contents of this symposium volume show, the viscous liquid state, dying as the lifeblood entropy drains away with decreasing temperature, provides in its manner of dying an inversely lively source of discussion, not to say argumentation. It is a problem which has resisted solution for more than a century, despite having "flowered" into high activity states at regular 30 year intervals (vitrearchs: Simon (1900s), Tammann (1930s), Gibbs, Goldstein, Litovitz (1960s) and now many contenders in the 1990s). It will of course be interesting to see if the problem yields in the present cycle. This volume pushes the solution a little closer, but most conferees will agree it is still a long way off.

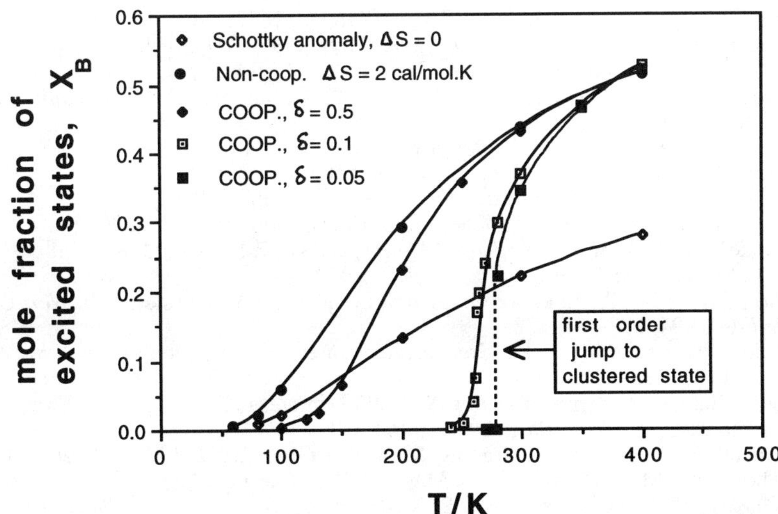

Figure 3. Variation of the mole fraction of unclustered molecules,with temperature in the cooperative and non-cooperative cluster models of a glassforming liquid in the lansdscape-dominated regime of viscous liquid behavior. Curve a shows simple Schottky behavior; Curve b shows behavior for degenerate excited state but non-cooperative two state model, while curves c and d and e show three degrees of cooperative behavior. The third case e shows how, above a certain cooperativity level, a first order phase change to the fully clustered state can occur.

Acknowledgments

This work was supported by the National Science Foundation under Grant Nos. DMR 9108028-002 and 9614531, and INT.9602884. The author is indebted to Peter Harrowell for many invigorating discussions of the subject matter of this overview during a sabbatical leave at Sydney University, N.S.W., Australia.

Literature Cited

1. Bernal, J. D. *Nature* **1960,** *183,* 141.
2. Stillinger, F. H.; Weber, T. A. *Science* **1984,** *225,* 983.
3. Speedy, R. J.; Debenedetti, P. G. *Mol. Phys.* **1996**, *88*, 1293.
4. Goldstein, M. *J. Chem. Phys.* **1969,** *51,* 3728
5. Götze, W. *Liquids, Freezing, and Glass Formation,* ed. J.vP. Hansen and D. Levesque; Les Houches, 1989.
6. Cummins, H. Z.; Li, G.; Du, W.; Pick, R. M.; Dreyfus, C. *Phys. Rev. E* **1997**, *53*, 896.
7. Taborek, P.; Kleiman, R. N.; Bishop, D. J. *Phys. Rev. B*, **1986**, *34*, 1835.
8. Angell, C. A. *J. Phys. Chem. Sol.* **1988,** *49,* 863.
9. Rössler, E. *Phys. Rev. Lett.* **1990**, *65,* 1595.
10. Fujara, F.; Geil, B.; Sillescu, H.; Fleischer, G. *Z. Phys. B* **1992**, *88*, 195.
11. Fujimori, H.; Oguni, M. *Solid State Commun.* **1995,** *94,* 157; Oguni, M. *J. Non-Cryst. Solids* **1996.**

12. Perrera, D.; Harrowell, P. *Phys. Rev. E* **1996**, *52*, 1652.
13. Cicerone, M. T.; Ediger, M. D. *J. Chem. Phys.* **1995**, *103*, 5684.
14. Rössler, E.; Sokolov, A. P. *Chem. Geol.* **1996**, *128*, 143.
15. Novikov, V. N.; Rössler, E.; Malinovsky, V. K.; Surovtev, N. V. *Europhys. Lett.* **1996**, *35*, 289.
16. Angell, C. A. *17th Sitges Conference* (in press).
17. Angell, C. A.; Tucker, J. C. (to be published).
18. Clarke, J. H. R. *Trans. Faraday Soc. 2* **1976**, *76*, 1667.
19. Stickel, F.; Fischer, E. W.; Richert, R. *J. Chem. Phys.* ***1996***, *104*, 2043.
20. Anderson, P. W. In *Ill-Condensed Matter*; Balian, R.; Maynard, R.; Toulouse, G., Ed.; Les Houches: North Holland, 1979; p 171.
21. Sokolov, A. P.; Kisliuk, A.; Quitmann, D.; Kudlik, A.; Rössler, E. *J. Non-Cryst. Solids* **1994**, *172-174*, 138.
22. Poole, P. H.; Sciortino, F.; Grande, T.; Stanley, H. E.; Angell, C. A. *Phys. Rev. Lett.* **1994**, *73*, 1632.
23. Moynihan, C. T. *Symp. Mat. Res. Soc.* **1997** (in press).
24. Mishima, O. *J. Chem. Phys.* **1994**, *100*, 5910.
25. Angell, C. A.; Borick, S.; Grabow, M. *J. Non-Cryst. Solids* **1996**, *205-207*, 463.
26. Thompson, M. O.; Galvin, G. J.; Mayer, J. W. *Phys. Rev. Lett.* **1984**, *52* 2360.
27. Glazov, V. M.; Chizhevskaya, S. N.; Evgen'ev, S. B. *J. Phys. Chem.* **1969**, *43*, 201.
28. Ponyatovsky, E. G.; Barkalov, O. I. *Mat. Sci. Rep.* **1992**, *8*, 147.
29. Angell, C. A.; Poole, P. H.; Hemmati, M., *Proc. 12th Conf. Glass and Cermaics*, Varna, Sept. 1996 (in press).
30. Aasland, S.; McMillan, P. F. *Nature* **1994**, *369*, 633.
31. Austin, R. H.; Beeson, K. W.; Eisenstein, L.; Frauenfelder, H. *Biochem* **1975**, *14*, 5355.
32. Frauenfelder, H.; Sligar, S. G.; Wolynes, P. G. *Science* **1991**, *254*, 1598.
33. Angell, C. A.; *J. Phys. Chem.* **1971**, *75*, 3698.
34. Angell, C. A.; to be published.
35. Kittel, C. *Introduction to Solid State Physics;* 2nd ed.; Wiley: New York, 1956; Appendix A.
36. Rao, K. J.; Angell, C. A. In *Amorphous Solids*; Douglas; Ellis, Ed.; Wiley and Sons: 1971.
37. Kivelson, S. A.; Zhao, X.; Kivelson, D.; Fischer, T. M.; Knobler, C. M. *J. Chem. Phys.* **1994**, *101*, 2391.
38. Kivelson, S. A.; Zhao, X.; Nussinov, Z.; Tarjus, G. *Physica A* **1995**, *219*, 27.
39. Fan, J. Ph.D. thesis, Arizona State University, 1995.
40. Adam, G.; Gibbs, J. H. *J. Chem. Phys.* **1965**, *43*, 139.
41. Angell, C. A. *J. Res. NIST* (in press).

THEORETICAL PERSPECTIVES ON SUPERCOOLED LIQUIDS

Chapter 3

The Mode-Coupling Theory of the Glass Transition

Walter Kob[1]

Institute of Physics, Johannes Gutenberg Universität, Staudinger Weg 7, D–55099 Mainz, Germany

We give a brief introduction to the mode-coupling theory of the glass transition, a theory which was proposed a while ago to describe the dynamics of supercooled liquids. After presenting the basic equations of the theory, we review some of its predictions and compare these with results of experiments and computer simulations. We conclude that the theory is able to describe the dynamics of supercooled liquids in remarkably great detail.

The dynamics of supercooled liquids and the related phenomenon of the glass transition has been the focus of interest for a long time [1]. The reason for this is the fact, that if a glass former is cooled from its melting temperature to its glass transition temperature T_g, it shows an increase of its relaxation time by some 14 decades without a significant change in its structural properties, which in turn poses a formidable and exciting challenge to investigate such a system experimentally as well as theoretically. Despite the large efforts that have been undertaken to study this dramatic growth in relaxation time, even today there is still intense dispute on what the underlying mechanisms for this increase actually is. Over the course of time many different theories have been put forward, such as, to name a few popular ones, the entropy theory by Adams, Gibbs and Di Marzio [2], the coupling-model proposed by Ngai [3], or the mode-coupling theory (MCT) by Götze and Sjögren [4]. The approach by which these theories explain the slowing down of the supercooled liquid with decreasing temperature differs radically from case to case. In the entropy theory it is assumed, e.g., that the slowing down can be understood essentially from the thermodynamics of the system, whereas MCT puts forward the idea that at low temperatures the nonlinear feedback mechanisms in the microscopic dynamics of the particles become so strong that they lead to the structural arrest of the system.

[1]Electronic mail: kob@moses.physik.uni-mainz.de
http://www.cond-mat.physik.uni-mainz.de/~kob/home_kob.html

One of the most outstanding advantages of MCT over the other theories of the glass transition is the fact that it offers a wealth of predictions, some of which are discussed below, that can be tested in experiments or computer simulations. This, and its noticeable success, are probably the main reason why this theory has attracted so much attention in the last ten years. This is in contrast to most other theories which make far fewer predictions and for which it is therefore much harder to be put on a solid experimental foundation. This abundance of theoretical predictions has, of course, its price, in that MCT is a relatively complex theory. Therefore it is not surprising that doing *quantitative* calculations within the framework of MCT is quite complicated, although it is remarkable that the theory gives well defined prescriptions how such calculations have to be carried out and for simple models such computations have actually been done.

The goal of this article is to give a brief introduction to the physical background of MCT, then to review some of the main predictions of the theory and to illustrate these by means of results from experiments and computer simulations. In the final section we will discuss a few recent developments of the theory and offer our view on what aspect of the dynamics of supercooled liquids can be understood with the help of MCT.

Mode-Coupling Theory: Background and Basic Equations

In this section we give some historical background of the work that led to the so-called mode-coupling equations, the starting point of MCT. Then we present these equations and some important special cases of them, the so-called schematic models.

In the seventies a considerable theoretical effort was undertaken in order to find a correct quantitative description of the dynamics of dense simple liquids. By using mode-coupling approximations [5], equations of motion for density correlation functions, described in more detail below, were derived and it was shown that their solutions give at least a semi-quantitative description of the dynamics of simple liquids in the vicinity of the triple point. In particular it was shown that these equations give a qualitatively correct description of the so-called cage effect, i.e. the phenomenon that in a dense liquid a tagged particle is temporarily trapped by its neighbors and that it takes the particle some time to escape this cage. For more details the reader is referred to Refs. [6] and references therein.

A few years later Bengtzelius, Götze and Sjölander (BGS) simplified these equations by neglecting some terms which they argued were irrelevant at low temperatures [7]. They showed that the time dependence of the solution of these simplified equations changes discontinuously if the temperature falls below a critical value T_c. Since this discontinuity was accompanied by a diverging relaxation time of the time correlation functions, this singularity was tentatively identified with the glass transition.

Let us be more specific: The dynamics of liquids is usually described by means of $F(\vec{q},t)$, the density autocorrelation function for wave vector $\vec{q}$, which is defined as:

$$F(\vec{q},t)=\frac{1}{N}\langle\delta\rho^*(\vec{q},t)\delta\rho(\vec{q},0)\rangle \qquad \text{with} \qquad \rho(\vec{q},t)=\sum_{j=1}^{N}\exp(i\vec{q}\cdot\vec{r}_j(t)), \qquad (1)$$

where N is the number of particles and $\vec{r}_j(t)$ is the position of particle j at time t. The function $F(\vec{q},t)$, which is also called intermediate scattering function, can be measured in scattering experiments or in computer simulations and is therefore of practical relevance.

For an isotropic system the equations of motion for $F(\vec{q},t)$ can be written as

$$\ddot{F}(q,t)+\Omega^2(q)F(q,t)+\int_0^t\left[M^0(q,t-t')+\Omega^2(q)m(q,t-t')\right]\dot{F}(q,t')dt'=0 \quad . \quad (2)$$

Here $\Omega(q)$ is a microscopic frequency, which can be computed from the structure factor $S(q)$ via $\Omega^2(q)=q^2k_BT/(mS(q))$ (m is the mass of the particles and k_B Boltzmann's constant), the kernel $M^0(q,t)$ describes the dynamics at short times and gives the only relevant contribution to the integral at temperatures in the vicinity of the triple point, whereas the kernel $m(q,t)$ becomes important at temperatures where the system is strongly supercooled. If we assume that $M^0(q,t)$ is sharply peaked at $t=0$, and thus can be approximated by a δ-function, $M^0(q,t)=\nu(q)\delta(t)$, we recognize from Eq. (2) that the equation of motion for $F(q,t)$ is the same as the one of a damped harmonic oscillator, but with the additional complication of a retarded friction which is proportional to $m(q,t)$.

It has to be emphasized that these equations of motion are *exact*, since the kernels $M^0(q,t)$ and $m(q,t)$ have not been specified yet. In the approximations of the *idealized* version of mode-coupling theory, the kernel $m(q,t)$ is expressed as a quadratic form of the correlation functions $F(q,t)$, i.e. $m(q,t)=\sum_{\vec{k}+\vec{p}=\vec{q}}V(\vec{q};\vec{k},\vec{p})F(k,t)F(p,t)$, where the vertices $V(\vec{q};\vec{k},\vec{p})$ can be computed from $S(q)$. With this approximation one therefore arrives at a closed set of coupled equations for $F(q,t)$, the so-called mode-coupling equations, whose solutions thus give the full time dependence of the intermediate scattering function. These are the above mentioned equations that were proposed and studied by BGS [7]. It is believed that they give a correct (self-consistent) description of the dynamics of a particle at short times, i.e. when it is still in the cage that is formed by its neighbors at time zero, and of the breaking up of this cage at long times, i.e. up to the time scales when the particle finally shows a diffusive behavior.

We also mention that in Eq. (2) the quantities $\Omega^2(q)$, $M^0(q,t)$ and $V(\vec{q};\vec{k},\vec{p})$ depend on temperature. This temperature dependence is assumed to be smooth throughout the whole temperature range, an assumption which is supported by experiments and computer simulations. Thus any singular behavior in the solution of the equations of motion are due to their nonlinearity and not due to a singularity in the input parameters. Since the strength of this nonlinearity is related to the (temperature dependent) structure factor, we thus see that the relevant temperature dependence of the solution of Eq. (2) comes from the one of $S(q)$.

Due to the complexity of the mode-coupling equations their solutions can, unfortunately, be obtained only numerically. Therefore BGS made the approximation, which was proposed independently also by Leutheusser [8], that the structure factor is given by a δ-function at the wave vector q_0, the location of the main peak in $S(q)$ [7]. With this approximation Eq. (2) is transformed into a single equation for the correlation function for q_0, all the other equations vanish identically. By writing $\Phi(t) = F(q_0, t)/S(q_0)$ we thus obtain

$$\ddot{\Phi}(t) + \Omega^2\Phi(t) + \nu\dot{\Phi}(t) + \Omega^2 \int_0^t m[\Phi(t-t')]\dot{\Phi}(t')dt' = 0, \qquad (3)$$

where $m[\Phi]$ is a low order polynomial in Φ. Such an equation for a single (or at most very few) correlation function is called a *schematic model.* Originally it was believed that such schematic models reproduce some essential, i.e. universal, features of the full theory. In the meantime it is understood what these universal features of the full theory are. Therefore we know now, how to construct schematic models, i.e. memory kernels $m[\Phi]$, so that they reproduce the desired features of the full MCT equations.

We mentioned above that in the full mode-coupling equations the relevant temperature dependence of the equations is given by the one of the structure factor, which enters the coefficients of the quadratic form of the memory kernel $m(q, t)$. In analogy to this one therefore assumes that in the schematic models the coefficients of the polynomial $m[\Phi]$ are also temperature dependent.

Since in these simplified models the details of all the microscopic information has been eliminated they cannot be used to understand experimental data *quantitatively.* However, their greatly reduced complexity, as compared to the full equations, make them amenable to analytic investigations from which many qualitative properties of their solutions can be obtained. Thus many of the predictions of MCT stem from studying the solutions of the schematic models, work that was done in the last ten years mainly by Götze, Sjögren and coworkers. We will discuss these predictions in more detail in the next section. For the moment we just mention briefly that the analysis of the schematic models shows that if the nonlinearity, given by the memory kernel $m[\Phi(t)]$, exceeds a certain threshold, the solution of the equation does not decay to zero even at infinite times. This means that a density fluctuation that was present at time zero does not disappear even at long times, i.e. the system is no longer ergodic. Thus in this (ideal) case this dynamic transition can be identified with the glass transition.

We mentioned earlier, that when BGS wrote down for the first time what today are called the mode-coupling equations [Eq. (2), with $m(q, t)$ given as a quadratic function of $F(q, t)$], they neglected in these equations certain terms. Later it was found that at very low temperatures these terms do become important, since they lead to a qualitatively different behavior of the time dependence of the solution of the equations of motion [9]. In particular it was shown that the above mentioned singularity in this solution disappears, i.e. that even at low temperatures all the correlation functions decay to zero at long times and that thus the system is always ergodic. Since one of the mechanisms that can lead to the relaxation of the system, and that is not taken into account by the idealized mode-coupling equations, is a process in which one particle overcomes the walls of its cage in an activated way, these processes are usually called *hopping processes.* The version of MCT in which the effects of such hopping processes are taken into account is called the *extended version* of MCT. So far, however, the investigations of such processes has been restricted to discuss solutions of mode-coupling equations in which the hopping processes have been taken into account only in a crude way, since even in these relatively simple cases the addition of one additional parameter, the strength of the hopping processes, makes the discussion of the solution quite a bit more cumbersome, as compared to the case where hopping is absent. Nevertheless, it is of course important to gain insight whether the presence of such hopping processes changes *all* the predictions of the idealized MCT or whether there are only a few which are modified, since, after all, in real materials such processes are always present. We will see in the next section to what extend the solutions of the mode-coupling equations of the idealzed theory differ from the ones in which hopping processes are taken into account.

Mode-Coupling Theory: Predictions and Tests

In this section we will present some of the main predictions of MCT and compare these with the results of experiments and computer simulations of supercooled liquids. The results presented here are by no means comprehensive in that the theory makes significantly more predictions, quite a few of which have been tested in experiments or computer simulations, which we do not discuss here. For a better overview the reader is referred to the review articles [4] and the collection of articles in Ref. [10].

One of the main predictions of the idealized version of MCT is that there exists a critical temperature T_c at which the self diffusion constant D, or the inverse of the α-relaxation times τ of the correlation functions, vanishes with a power-law, i.e.

$$D \propto \tau^{-1} \propto (T - T_c)^{\gamma} \quad , \tag{4}$$

where the exponent γ is larger than 1.5. Thus one can attempt to see whether for a given system the low temperature dependence of these quantities is given by such a law. In Fig. 1 we show that for a binary Lennard-Jones system such a temperature dependence can indeed be found and that the critical temperature T_c is, in accordance with MCT, independent of the quantity investigated. Furthermore MCT predicts that the exponent γ should be independent of the quantity investigated. From the figure we recognize that this is reasonably well fulfilled for this system if one compares the two relaxation times or the two diffusion constants with each other, that, however, the exponents for D and τ^{-1} are definitely different. Thus this prediction of the theory does not seem to hold for this particular system.

It is very instructive to study the full time and temperature dependence of the solution of schematic models of the form given in Eq. (3), since they can be compared with the relaxation dynamics of real systems and thus allow to perform a more stringent test of the theory. In Fig. 2 we show the correlation functions for a model with $m(\Phi) = \lambda_1\Phi + \lambda_2\Phi^2$, with $\lambda_i > 0$, that were computed by Götze and Sjögren [12] (with no hopping processes). The different curves correspond to different values of the coupling parameters λ_i and are chosen such that their distance to the critical values decreases like $0.2/2^n$ for $n = 0, 1, \ldots$ (liquid, curves A,B,C ...) and increases like $0.2/2^n$ for $n = 0, 1, \ldots$ (glass, curves F', D', ...). Note that what in real systems corresponds to a change in temperature corresponds in these schematic models to a change in the coupling constants λ_i. In order to simplify the language we will, however, in the following always use temperature as the quantity that is changed.

From Fig. 2 we recognize that at short times the correlation functions show a quadratic dependence on time, which is due to the ballistic motion of the particles on these time scales. For high temperatures, curve A, this relaxation behavior crosses directly over to one which can be described well by an exponential decay. If the temperature is decreased, curve D, there is an intermediate time regime, the so called β-relaxation regime, where the correlation function decays only very slowly, i.e. the Φ versus log(t) plot exhibits an inflection point. For even lower temperatures, curve G, the correlation functions show in this regime a plateau (log(t) $\approx$ 5). Only for even longer times the correlation function enters the so-called α-relaxation regime, the time window in which it finally decays to zero.

The reason for the existence of the plateau is the following: At very short times the motion of a particle is essentially ballistic. After a microscopic time the particle starts to realize that it is trapped by the cage formed by its nearest neighbors and thus the correlation function does not decay any more. Only for much longer times, towards the end of the β-relaxation, this cage starts to break up and the particle begins to explore a larger and larger volume of space. This means that the correlation function enters the time scale of the α-relaxation and resumes its decay.

The closer the temperatures is to the critical temperature T_c, the more this β-relaxation region stretches out in time and the time scale of the α-relaxation diverges with the power-law given by Eq. (4), until at T_c the correlation function does not decay to zero any more. Upon a further lowering of the temperature the height of the plateau increases and the time scale for which it can be observed moves to *shorter* times. MCT predicts that for temperatures below T_c, this increase in the height of the plateau is proportional to $\sqrt{T_c - T}$, which was indeed confirmed by, e.g., neutron scattering experiments on a polymer glass former [13].

Apart from the existence of a critical temperature T_c, one of the most important predictions of the theory is the existence of three different relaxation processes, which we have already seen in curve G of Fig. 2. The first one is just the trivial relaxation on the microscopic time scale. Since on this time scale the microscopic

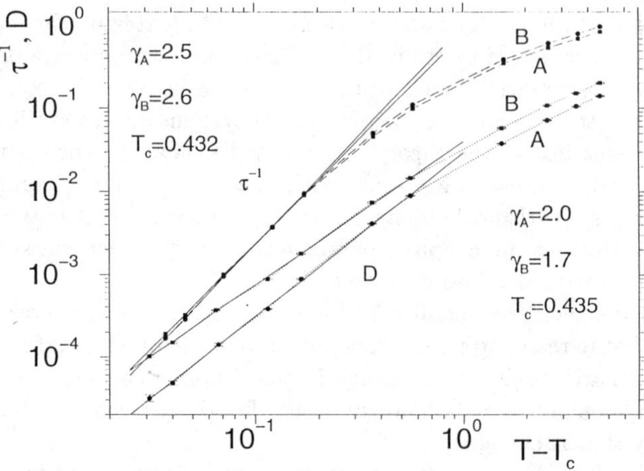

Figure 1: Temperature dependence of the diffusion constant D and the inverse relaxation times τ for the intermediate scattering function for the two types of particles (type A and type B) in a binary Lennard-Jones system. The straight lines are fits with a power-law. Reprinted with permission from reference [11]. Copyright 1994 The American Physical Society. All rights reserved.

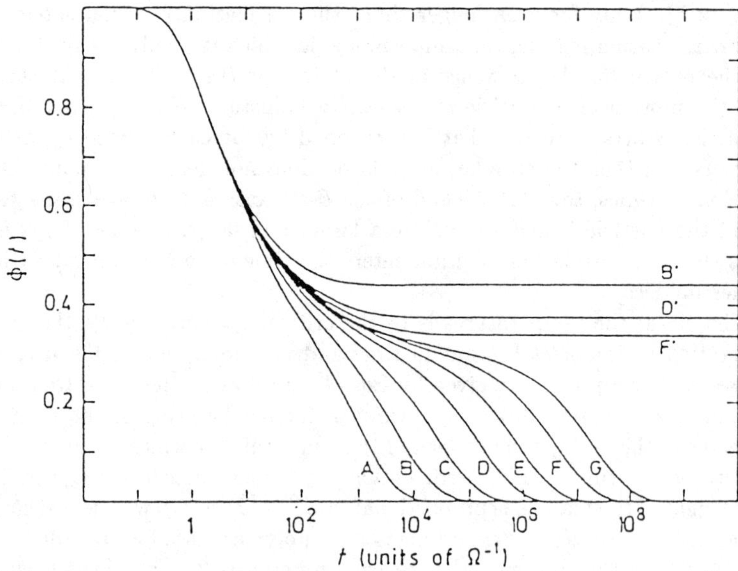

Figure 2: Time dependence of the correlation functions as computed from a schematic model without hopping processes. The different curves correspond to different values of the coupling parameters (see text for details). Reprinted with permission from reference [12]. Copyright 1988 The American Chemical Society. All rights reserved.

details of the system are very important, hardly any general predictions can be made for this time regime. This is different for the second and third relaxation processes, the aforementioned β- and α-relaxation. For these time regimes MCT makes quite a few predictions, some of which we will discuss now. Note that the predictions, as stated below, are correct only in leading order in $\sigma = (T_c - T)/T_c$. The corrections to this asymptotic behavior have recently been worked out for the case of a hard sphere system and it was found that they can be quite significant [14]. Thus for a *quantitative* comparison between experiments and MCT, these corrections should be taken into account.

For the β-relaxation regime MCT predicts that its time scale t_σ diverges upon approach to the critical temperature T_c as

$$t_\sigma \propto |T - T_c|^{1/2a} \quad , \tag{5}$$

with $0 < a < 1/2$. Note that this type of singularity is predicted to exist *above and below* T_c. This divergent time scale can be seen in Fig. 2, in that the inflection point of the correlation function in the β-relaxation regime moves to larger times when the temperature is decreased to $T_c + 0$ and that the time it takes the correlator to reach the plateau diverges when the temperature is increased to $T_c - 0$. Light scattering experiments have shown that the divergence of the time scale of the β-relaxation, as given by Eq. (5), can indeed be found in real materials [15, 16].

Furthermore the theory predicts that in the β-relaxation regime the *factorization property* holds, by which the following is meant. If we consider a real system, as opposed to a schematic model, we have correlation functions $F(q,t)$ which correspond to different values of q, see Eq. (2). The factorization property says now that in the β-relaxation regime these space-time correlation functions can be written as

$$F(q,t) = f_c(q) + h(q)\sqrt{\sigma} g_\pm(t/t_\sigma) \quad , \tag{6}$$

where the (q-dependent) constant $f_c(q)$ is the height of the plateau, and is also called *nonergodicity parameter*, the amplitude $h(q)$ is independend of temperature and time, and the $\pm$ in $g_\pm$ corresponds to $\sigma \gtrless 0$. Thus the time dependence of the correlation function enters only through the *q-independent* function $g_\pm(t)$ which is therefore, for a given system, "universal". Equation (6) means that, in the β-relaxation regime, the time correlations are completely independent from the spatial correlation, which is in stark contrast to other types of dynamical processes such as, e.g., diffusion, where the relaxation time of a mode for wave vector q is proportional to q^{-2}.

In order to check whether for a given system the factorization property holds, one can consider the space-Fourier transform of Eq. (6) which gives $F(r,t) = f_c(r) + H(r)\sqrt{\sigma} g_\pm(t)$. Making use of this last equation, it is simple to show that the ratio

$$\frac{F(r,t) - F(r,t')}{F(r',t) - F(r',t')} = \frac{H(r)}{H(r')} \tag{7}$$

is independent of time (here r and r' are arbitrary and t and t' are times in the β-relaxation regime). In Fig. 3 we show this ratio for the distinct part of the van Hove correlation function of the binary Lennard-Jones system mentioned above. Every curve corresponds to a different time t, all of which belong to the β-relaxation regime. (The value of t' is kept fixed at 3000 reduced time units and $r' = 1.05$.) We see that all the curves lie in a narrow band, which shows that the left hand side of Eq. (7) is indeed independent of time, i.e. that the factorization property holds. Thus we see that for this system the time dependence of the correlation functions are indeed given by a single function $g_-(t)$. The same results were found for the dynamics of colloidal suspensions [16].

It can be shown that the full time dependence of $g_\pm(t)$ can be computed if one number λ, the so-called *exponent parameter*, is known. This parameter can in turn be computed from the structure factor, although such a computation is rather involved and therefore λ is often used as a fit parameter in order to fit $g_\pm(t)$ to the data. The result of such a fit is shown in Fig. 4 (dotted curve) where we show the incoherent intermediate scattering function of the Lennard-Jones system discussed above, versus $t/\tau(T)$. From this figure we recognize that for this Lennard-Jones system the functional form provided by MCT is able to fit the data very well in the late part of the β-relaxation regime. However, for this system the *early* β-relaxation regime is not fitted well by the β-correlator [18]. The reason for this is likely the strong influence of the microscopic dynamics. This view is corroborated by the fact that if the dynamics is changed from a Newtonian one to a stochastic one, the observed β-relaxation is much more similar to the one predicted by the theory [19]. In addition to this, light scattering experiments have shown that in colloidal systems the *whole* β-relaxation regime can be fitted very well with the β-correlator [16, 20] thus showing that there are systems for which the β-correlator gives the correct description of the dynamics in the β-relaxation regime.

The calculation of $g_\pm(t)$ from λ is rather complicated and therefore has to be done numerically. However, for fitting experimental data, it is often more useful to have simple analytic expressions at hand, even if they are correct only in leading order in σ, and MCT provides such expressions. It can be shown that in the *early* β-relaxation regime, i.e. the time range during which the correlator is already close to the plateau, but has not reached it yet, the function $g_\pm(t)$ is a power-law, i.e.

$$g_\pm(t/t_\sigma) = (t/t_\sigma)^{-a} \quad , t/t_\sigma \ll 1, \tag{8}$$

where the exponent a is the same that appears in Eq. (5). This time dependence is often also called *critical decay.*

For times in the *late* β-relaxation regime, i.e. in the time interval where the correlator has already dropped below the plateau but is still in its vicinity, MCT predicts that $g_-(t)$ is given by a different power-law, the so-called von Schweidler law:

$$g_-(t/t_\sigma) = -B(t/t_\sigma)^b \quad , t/t_\sigma \gg 1, \tag{9}$$

where the exponent b $(0 < b \leq 1)$ is related to a via the nonlinear equation

$$\Gamma^2(1-a)/\Gamma(1-2a) = \Gamma^2(1+b)/\Gamma(1+2b) \quad , \tag{10}$$

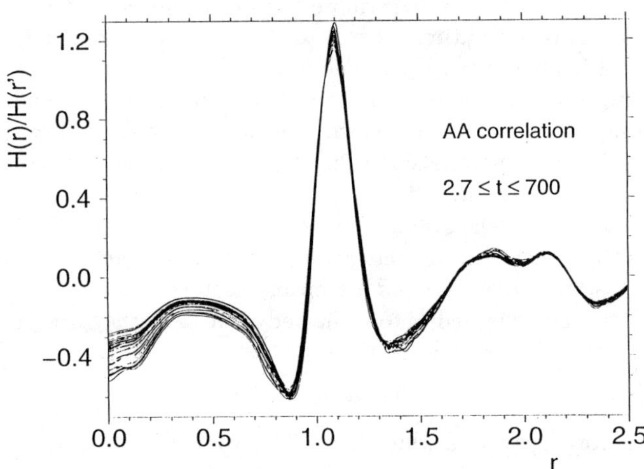

Figure 3: Ratio of critical amplitudes for different times in the β-relaxation regime for the distinct part of the van Hove correlation function in a binary Lennard-Jones system. Reprinted with permission from reference [17]. Copyright 1995 The American Physical Society. All rights reserved.

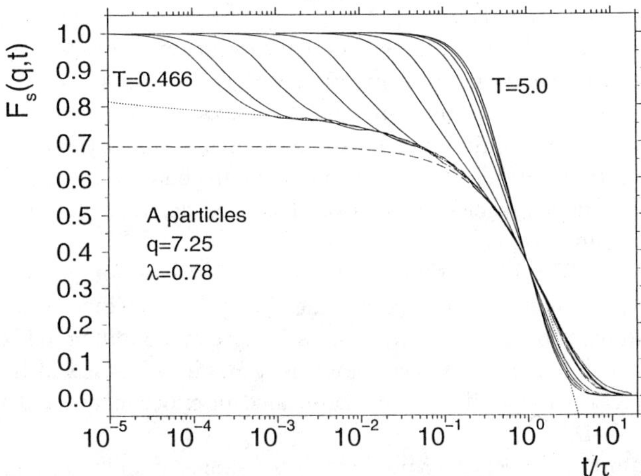

Figure 4: The incoherent intermediate scattering function $F_s(q,t)$ for a binary Lennard-Jones mixture versus rescaled time $t/\tau(T)$ for different temperatures (solid lines). The dotted curve is a fit with the functional form provided by MCT for the β-relaxation regime. The dashed curve is the result of a fit with the KWW function. Reprinted with permission from reference [18]. Copyright 1995 The American Physical Society. All rights reserved.

and $\Gamma(x)$ is the Γ-function. Furthermore it can be shown that the left (or the right) hand side of this equation is equal to the exponent parameter λ, which we have introduced before. Thus if one knows one element of the set $\{a, b, \lambda\}$, this relation and Eq. (10) can be used to compute the two other elements of this set. Light scattering experiments have shown that both power-laws, Eqs. (8) and (9), can be observed in real materials and that Eq. (10) is indeed satisfied [15, 16].

We now turn our attention to the relaxation of the correlation functions on the time scale of the α-relaxation. One of the main results of MCT concerning this regime is the so-called *time-temperature superposition principle* (TTSP), also this valid to leading order in σ. This means that the correlators for different temperatures can be collapsed onto a master curve Ψ if they are plotted versus t/τ, where τ is the α-relaxation time, i.e.

$$\Phi(t) = \Psi(t/\tau(T)) \quad . \tag{11}$$

As an example for a system in which the TTSP works very well we show in Fig. 4 the time dependence of $F_s(q,t)$, the incoherent intermediate scattering function, of the Lennard-Jones system discussed above, versus rescaled time $t/\tau(T)$. In accordance with MCT, the curves for the different temperatures fall onto a master curve, if the temperature is low enough. Furthermore the theory predicts that this master curve can be fitted well with the so-called Kohlrausch-Williams-Watts (KWW) function, $\Phi(t) = A\exp(-(t/\tau(T)^{\beta})$. The result of such a fit is included in the figure as well (dashed curve) and from it we recognize that this prediction of the theory holds true also.

Contrary to the situation in the β-relaxation regime, where the time dependence of the different correlation function was governed by the single function $g_{\pm}(t)$, MCT predicts that in the α-regime the relaxation behavior of different correlation function is not universal. This means that not only the amplitudes A in the KWW function, but also the exponent β will depend on the correlator considered. That such dependencies indeed exist has been demonstrated in experiments and computer simulations [18, 21].

Although MCT predicts that the shape of the relaxation curves will depend on the correlation function considered, the theory also predicts that all of them share a common property, namely that all the corresponding relaxation times diverge at T_c with a power-law whose exponent γ is independent of the correlator, see Eq. (4). Also this prediction was confirmed in experiments and in computer simulations [16, 18].

Furthermore the theory also predicts the existence of an interesting connection between the exponents a and b, which are important for the β-relaxation [see Eqs. (8) and (9)], and the exponent γ, which governs the time scale of the α-relaxation [see Eq. (4)], in that

$$\gamma = 1/2a + 1/2b \tag{12}$$

should hold. Thus, according to MCT, from measurements of the temperature dependence of the α-relaxation time we can learn something about the time dependence of the relaxation in the β-relaxation regime and vice versa.

Before we conclude this section we return to the extended version of MCT, i.e. that form of the theory in which the hopping processes are taken into account. In Fig. 5 we show the solution of the same schematic model which was discussed in the context of Fig. 2, but this time with the inclusion of hopping processes [12]. From this figure we recognize that the main effect of such processes is that the correlation functions decay to zero at *all* temperatures, which shows that the system is always ergodic. Thus one *might* conclude that the concept of a critical temperature does not make sense anymore, since there is no temperature at which the relaxation times diverge. This is, however, not the case. If the hopping processes are not too strong, there still will exist a temperature range in which the relaxation times will show a power-law behavior with a critical temperature T_c. However, this power-law will not extend down to T_c but deviations will be observed in the vicinity of T_c, in that the temperature dependence of τ will be weaker than a power-law. Thus, despite the presence of the hopping processes it is still possible to identify a T_c.

As a comparison between the corresponding curves in Figs. 2 and 5 shows, also the relaxation behavior of the correlation functions are not affected too much by the hopping processes, *if one is not too close to* T_c. Therefore many of the predictions that MCT makes for the relaxation behavior hold even in the presence of such processes. As an example of how important it can be to take into account the hopping processes for the interpretation of real data very close, or below, T_c, we show in Fig. 6 the results of a light scattering experiment on calcium potassium nitrate (CKN) in a temperature range which includes $T_c = 378$ K [22]. Shown is the imaginary part of the susceptibility, i.e. the time Fourier transform of the intermediate scattering function, multiplied by the frequency ω. In the left figure the data is analyzed by using the idealized version of MCT and we recognize that although the theory (smooth curves) fits the data (wiggly curves) well for temperature above T_c, significant deviations occur at lower temperatures, since there the hopping processes become important. If these are taken into account in a phenomenological way, the agreement between theory and experiment is very good for the whole temperature range (right part of the figure). It has to be noted that in the latter set of fits the theory makes use of one additional fit parameter, the strength of the hopping processes. Nevertheless, the inclusion of the hopping processes has improved the agreement between theory and experiment in such a dramatic way, that the price of one additional fit parameter seems to be well justified.

Conclusions

The goal of this article was to give a concise introduction to the mode-coupling theory of the glass transition. For this we briefly described the origin of the theory and explained some of its main predictions. Although we have presented here only a few results of the tests by which the capability of the theory to describe the dynamics of supercooled liquid were investigated, we mention that in the last few years many other such tests were performed [10]. Most of them showed that the

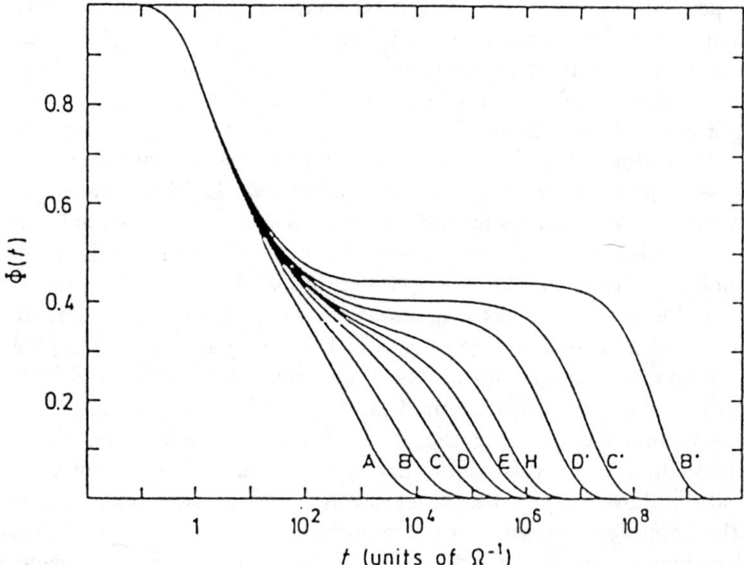

Figure 5: Time dependence of the correlation functions as computed from a schematic model with hopping processes. The different curves correspond to different values of the coupling parameters (see text for details). Reprinted with permission from reference [12].

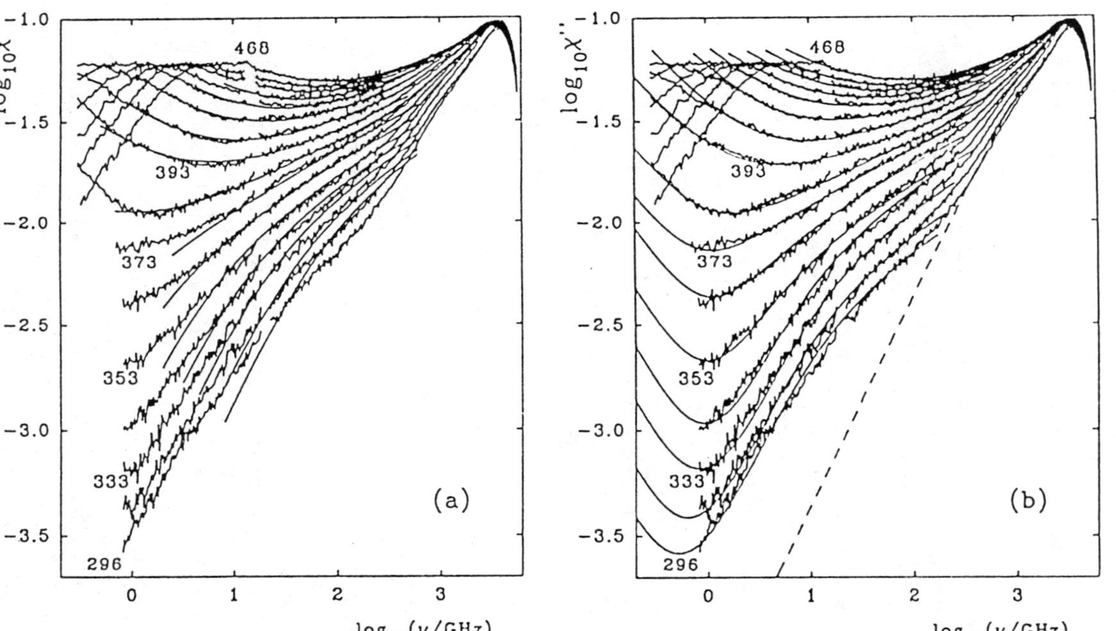

Figure 6: Susceptibility spectra of CKN and fits from MCT without hopping processes (a) and with hopping processes (b). Reprinted with permission from reference [22].

theory is *at least* able to describe *certain* aspects of this dynamics and that there exist systems for which the predictions of MCT are correct in surprisingly great detail. One unexpected result that came out of these tests is that MCT seems to work reasonably well even for systems that are very different (e.g. polymers) from the simple liquids for which the theory was originally devised. Thus one might tentatively conclude that the basic mechanism that leads to the slowing down of a liquid upon supercooling is not specific to simple liquids and can be described with the help of MCT.

It is interesting to note that most of the systems for which MCT gives a good description of the dynamics belong to the class of fragile glass formers [23], i.e. are systems that show a significant change of their activation energy if their viscosity is plotted in an Arrhenius plot. This bend occurs at a temperature that is about 10-40 K above the glass transition temperature and investigations of the dynamics of these systems have shown that the T_c of MCT is in the vicinity of this bend. In the ideal version of MCT, T_c corresponds to the temperature at which the system undergoes a structural arrest. Since in experiments it is found that at the temperature T_c the viscosity of the systems is significantly enhanced with respect to the one at the triple point but by no means large, we thus must conclude that for most real systems the hopping processes, which are neglected in the idealized theory, become important in the vicinity of T_c. Therefore for temperatures in the vicinity of T_c the extended version of the theory has to be used. However, despite the presence of these hopping processes, the signature of the sharp transition of the idealized theory is still observed and gives rise to a dynamical anomaly in the relaxation behavior of the system, e.g. the pronounced bend in the viscosity. Thus from the point of view of MCT the dynamics of supercooled liquids can be described as follows: In the temperature range where the liquid is already supercooled, but which is still above T_c, the dynamics is described well by the idealized version of MCT. In the vicinity of T_c the extended version of the theory has to be used, whereas for temperatures much below T_c it is likely that the simple way MCT takes into account the hopping processes is no longer adequate and a more sophisticated theory has to be used.

From the above one might get the impression that MCT is applicable only for fragile glass formers and not for strong ones. That this is not necessarily the case is shown by a recent calculations by Franosch *et al.* who have shown that also features in the dynamics of glycerol, a glass former that is rather strong, can be described well by simple schematic models [24]. Whether this is also true for very strong glass formers such as SiO_2 remains to be seen, however.

In this article we have shown that MCT is able to give a *qualitative* correct description of the dynamics of certain supercooled liquids in that, e.g., the relaxation of system in the β-regime shows the critical decay or the von Schweidler law predicted by the theory. To this two important comments have to be added: The first one is that the predictions of the theory that have been presented here are valid only asymptotically close to T_c. If one considers temperatures that have a finite distance from T_c, corrections to the mentioned scaling laws become important. Some of these corrections have very recently been computed for a model of hard spheres and it was found that they can be quite important in order to make a *consistent* analysis of experimental data within the framework of MCT [14]. The second comment we make is, that MCT is not only able to make *qualitative* statements on the relaxation behavior of supercooled liquids but that it is also able to predict the non-universal values of the parameters of the theory, such as T_c, the exponent parameter or the q-dependence of the nonergodicity parameter, reasonably well [16, 20, 25]. Unfortunately these last types of calculations are rather involved, since one has to take into account the full wave-vector dependence of the mode-coupling equations, and thus have been done only for a few cases. Despite these difficulties it would be very useful to have more investigations of this kind since they allow to test the range of applicability of the theory to a much larger extend that the tests of the "universal" predictions of the theory allow.

Acknowledgments: We thank W. Götze for many useful comments on this manuscript. Part of this work was supported by the DFG under SFB 262/D1.

References

[1] R. Zallen *The Physics of Amorphous Solids*, (Wiley, New York, 1983); J. Jäckle, Rep. Progr. Phys. **49**, 171 (1986); J. Zarzycki (ed.) *Materials Science and Technology, Vol. 9*, (VCH Publ., Weinheim, 1991); U. Mohanty, Adv. Chem. Phys. **89**, 89 (1995).

[2] G. Adam and J. H. Gibbs, J. Chem. Phys. **43**, 139 (1965); J. H. Gibbs and E. A. Di Marzio, J. Chem. Phys. **28**, 373 (1958).

[3] K. L. Ngai, Jour. de Phys. IV **2**, C2-61 (1992).

[4] W. Götze, p. 287 in *Liquids, Freezing and the Glass Transition* Eds.: J. P. Hansen, D. Levesque and J. Zinn-Justin, Les Houches. Session LI, 1989, (North-Holland, Amsterdam, 1991); W. Götze and L. Sjögren, Rep. Prog. Phys. **55**, 241 (1992); R. Schilling, p. 193 in *Disorder Effects on Relaxational Processes* Eds.: R. Richert and A. Blumen, (Springer, Berlin, 1994); H. Z. Cummins, G. Li, W. M. Du, and J. Hernandez, Physica A **204**, 169 (1994).

[5] K. Kawasaki, Phys. Rev. **150**, 291 (1966).

[6] L. Sjögren, Phys. Rev. A **22**, 2866 (1980); J.-P. Hansen and I. R. McDonald, *Theory of Simple Liquids* (Academic, London, 1986).

[7] U. Bengtzelius, W. Götze and A. Sjölander, J. Phys. C **17**, 5915 (1984).

[8] E. Leutheusser, Phys. Rev. A **29**, 2765 (1984).

[9] S. P. Das and G. F. Mazenko Phys. Rev **A 34**, 2265 (1986); W. Götze and L. Sjögren, Z. Phys. **B 65**, 415 (1987).

[10] Theme Issue on Relaxation Kinetics in Supercooled Liquids-Mode Coupling Theory and its Experimental Tests; Ed. S. Yip. Volume **24**, No. 6-8 (1995) of *Transport Theory and Statistical Physics.*

[11] W. Kob and H. C. Andersen, Phys. Rev. Lett. **73**, 1376 (1994).

[12] W. Götze and L. Sjögren, J. Phys. C **21**, 3407 (1988).

[13] B. Frick, B. Farago, and D. Richter, Phys. Rev. Lett. **64**, 2921 (1990).

[14] T. Franosch, M. Fuchs, W. Götze, M. R. Mayr, and A. P. Singh, (preprint 1996).

[15] G. Li, W. M. Du, X. K. Chen, H. Z. Cummins, and N. J. Tao, Phys. Rev. A **45**, 3867 (1992).

[16] W. van Megen and S. M. Underwood, Phys. Rev. E **49**, 4206 (1994).

[17] W. Kob and H. C. Andersen, Phys. Rev. E **51**, 4626 (1995).

[18] W. Kob and H. C. Andersen, Phys. Rev. E **52**, 4134 (1995).

[19] T. Gleim, W. Kob and K. Binder, (unpublished).

[20] W. Götze and L. Sjögren, Phys. Rev. A **43**, 5442 (1991).

[21] F. Mezei, W. Knaak and B. Farago, Phys. Scr. T**19**, 363 (1987).

[22] H. Z. Cummins, W. M. Du, M. Fuchs, W. Götze, S. Hildebrand, A. Latz, G. Li and N. J. Tao, Phys. Rev. E **47**, 4223 (1993).

[23] C. A. Angell, in: K. L. Ngai and G. B. Wright (eds.) *Relaxation in Complex Systems*, (US Dept. Commerce, Springfield, 1985).

[24] T. Franosch, W. Götze, M. R. Mayr, and A. P. Singh, (preprint 1996).

[25] M. Nauroth and W. Kob. Phys. Rev. E, **55**, 657 (1997).

Chapter 4

Basic Physics of the Coupling Model: Direct Experimental Evidences

K. L. Ngai and R. W. Rendell

Naval Research Laboratory, Code 6807, 4555 Overlook Avenue, SW, Washington, DC 20375–5320

The fundamental results of the Coupling Model, *i.e.* crossover from independent relaxation at short-times to slowed down cooperative relaxation at a temperature independent time t_c, are represented by three equations. They are shown to be in accord with experimental data in molecular glass formers, glass-forming, glassy or crystalline ionic conductors, and concentrated colloidal suspensions. Emphasis is given in this work to the colloidal suspensions which have the advantage of the absence of vibrational contributions and the experimentally measured intermediate scattering function is attributed entirely to diffusion dynamics. Direct evidences of such crossover are demonstrated. These, together with the many successful applications of the predictions to several fields and many materials have led us to believe that the Coupling Model has captured the basic physics of relaxation in materials with many-body interactions. We mention in passing that the theoretical basis of the Coupling Model is closely related to chaos in Hamiltonian dynamical systems.

The coupling model (CM) (1-3) is a general approach to dynamics of constrained or interacting systems, that has been shown to be applicable in depth to many problems of relaxation in different materials (4-6). Interaction between relaxing units implies cooperativity between them and vice versa. Thus, the effect of the many-body interactions on relaxation can be rephrased as cooperativity in the context of the CM. Several approaches to this problem have been proposed. Recent versions of the coupling theory are based on classical mechanics for systems that exhibit chaos caused by the anharmonic (nonlinear) nature of the interactions between the basic molecular units (2,3,7). For example, (*i*) intermolecular interaction between monomer units in polymers and small molecules in a glass-forming van der Waals liquid modeled by the Lennard-Jones potential; (*ii*) the entanglement interaction between polymer chains; (*iii*) Coulomb interaction between ions in glass-forming electrolytes including the much studied $0.4Ca(NO_3)\cdot 0.6KNO_3$ (CKN); and (*iv*) hard-sphere like interaction between colloidal particles (8-10) are all highly anharmonic and they give rise to chaotic dynamics (6). Although a rigorous theory based on first principles is still at large, there are theoretical results supporting the basic elements of the CM (1-3). The common misconception that the CM is entirely phenomenological is not accurate. Future development in theory may best be in the hands of experts in the dynamics of nonlinear Hamiltonian systems (7) and not necessary by us who have no formal training in this field. Whatever the future course of development of the theory, we are quite confident that the results of the CM are essentially correct. This confidence is gained by: (*a*) the results being borne out by simple models (2,3) and the fact that manifestation of the effect of chaos is usually general, so that behavior found in simple systems carries over to more complicated ones; (*b*) the many successes of the CM in its applications to a variety of challenging problems in several fields (4-6) which represent the bulk of the effort of our group of coworkers, and (*c*) the direct verification of the fundamental physics of the CM by recent experiments. The latter include quasielastic neutron scattering experiments in several polymers including polyvinylchloride (11), polyisoprene and polybutadiene (12); very high frequency dielectric measurements for glassy and non-glassy ionic conductors (13,14) and glass-forming molten salts (14,15). D.c. conductivity measurements of glassy and non-glassy ionic conductors up to high temperatures where the conductivity relaxation time is of the order of a picosecond or less (16-20); and molecular dynamics simulations of small molecule liquids (21). The fundamental results of the coupling theory, which have remained unchanged since its inception fifteen years ago (1), are restated here as follows. There exists a *temperature insensitive* cross-over time, t_c, before which ($t<t_c$) the basic units relax independently with correlation function

$$\phi(t) = \exp(-\frac{t}{\tau_o}) \qquad (1)$$

and afterwards ($t>t_c$) with a slowed-down nonexponential correlation function. A particularly convenient function which is compatible with both computer simulations and experimental data for coupled systems is

$$\phi(t) = \exp[-(\frac{t}{\tau^*})^{1-n}] \qquad (2)$$

where n is the coupling parameter whose value lies within the range $0\leq n<1$ and depends on the intermolecular interaction. What distinguish the CM from other models that may also have the simple exponential decay at short times and later a slowed-down decay is that in the other models the crossover time can be shifted by external variables such as temperature. On the other hand, t_c of the CM is determined by the interactions and the onset of chaos and is insensitive to changes of external variables. For motion of monomers in polymers (11,12), small molecules in van der Waals liquids (21) and diffusion of ions in ionic glasses (13-20), such a cross-over has been observed by various experimental techniques to occur at $t_c \cong O(1\text{ps})$. For motion of entangled polymer chains in melts which involves the longer length scale of the entanglement distance and hence weaker interaction, $t_c \cong O(10^{-9}\text{s})$. In semidilute solutions (22-29) interaction is further weakened and $t_c \cong O(5\times10^{-5}\text{s})$. Large colloidal particles with radius$\cong O(10^2\text{nm})$ has a broad crossover with an onset at $t_c \cong 2\times10^{-3}$s, as will be discussed in more detail later.

Continuity of the two pieces of the correlation function at $t=t_c$ leads to the important relation

$$\tau^* = [t_c^{-n}\tau_o]^{\frac{1}{1-n}} \tag{3}$$

which links the effective (i.e. after cooperative dynamical constraints between the relaxing molecular units have been taken into account) relaxation time, τ^*, to their independent (i.e. without taking into account of the cooperative dynamical constraints) relaxation time, τ_o. It is the fact that t_c of the CM is insensitive to external variables such as temperature that makes equation 3 nontrivial. In other models having deceptively similar crossover but the dependence of their t_c on external variables being the same as τ_o, equation 3 formally still holds but it has degenerated to become a trivial relation. It is useful to note that when considering macroscopic relaxation phenomena of molecular systems occurring at long times (i.e. when τ^*/t_c>>1), the condition, τ_o/t_c>>1, holds by virtue of equation 3. As a result, the linear exponential $\exp(-t/\tau_o)$ has decayed by an insignificant amount at $t=t_c$, and the correlation function $\phi(t)$ is practically given at all times by the stretched exponential piece. Nevertheless, relation (3) still holds and we have shown that it spawns many important consequences that can explain many intriguing properties of different interacting systems (4-6,23,28). For example, the various thermorheological complexities found in the viscoelastic spectrum of amorphous polymers (4,5), and the miscellaneous properties of ionic diffusion and conductivity in glass-forming ionic conductors (4,6) can be explained. The ability of the CM to address the fundamental short-time (microscopic) dynamics and to solve many practical long time (macroscopic) relaxation problems that confront most experimentalists all the time distinguish it from all other models. Free volume theories (30) and configurational entropy theory (31), have no specific predictions on short time dynamics and, when applied to polymer viscoelasticity, they cannot explain the thermorheological complexities (4,5). These deficiencies do not mean that these theories are wrong, rather that they are incomplete. The effects of interaction between the relaxing units have not been taken into account at a sufficiently fundamental level, as they have in the CM, to capture the cooperative dynamics. However, we hasten to point out that when applying to the problem of the temperature dependence of the α-relaxation time of supercooled liquids, the CM is not complete either. The temperature dependence of τ_o is not given within the CM in the current formulation, and it needs to be imported from elsewhere. As temperature T changes, other thermodynamic variables such as density and entropy change and these changes determine in turn the T-dependence of τ_o through some theory such as those mentioned. The Mode Coupling Theory (MCT) (32) has addressed the short-time dynamics of density fluctuations of liquids, but so far has not demonstrated its use in solving any practical problem in the long-time regime and its application to solve problems in polymer viscoelasticity, ionics and etc., as the CM has been able to do. We understand that often a reader is interested in only in one problem (e.g. the glass transition) and maybe even in one aspect of the problem (e.g. existence of a fast relaxation at ps time scale, or the non-

Arrhenius T-dependence of the α-relaxation time). This decision of considering a limited subset of all available facts will make the task easier but it risks the danger of not being able to explain or even contradicting experimental facts that were excluded from consideration. These comparisons with other theories are used to bring out what the CM can do and also what it cannot do.

The Basic Physics of the CM (Direct Experimental Evidences)

The basic physics of the CM is embodied by the three equations (1)-(3). Even from a different theoretical standpoint, people will likely agree on equation 1; that it is reasonable to expect, at sufficiently short times, each unit relaxes independently. Equation 2 is the simplest representation of a non-linear-exponential process. We do not believe that it necessarily represents *exactly* all or even any one of the relaxation processes that we have applied the CM to. If we choose to use a different non-exponential function in place of equation 2 which is also compatible with the data at long times, continuity of this function with equation 1 would again result in an equation analogous to equation 3 relating the independent relaxation time τ_o to some other cooperative relaxation time τ^*. Equation 3 is important for application of the CM and it provides the key to explanations of many nontrivial problems (4-6, 23,28). The many applications of the CM to different materials all originate as consequences of these three very simple equations. The last claim may be hard to believe, particularly for someone who has no previous knowledge of the many applications of the CM. We invite anyone to look into these achievements, although we realize that it is difficult for those who do not have expertise in the diverse subjects covered to make any judgment. Nevertheless, we hope that more people will realize how pervasive are the applications of these apparently innocuous three equations. Although these equations have been shown to be at least a consequence of simple Hamiltonian systems that exhibit chaos (1-3), further discussion of the former in this work will be confined to reexamining their validity from the experimental point of view. These three equations are amenable to direct experimental tests that has the experimental time window spanning across the crossover time t_c.

Molecular Systems. The first experimental evidence comes from quasielastic neutron scattering experiments in the polymers polyvinylchloride (PVC) by Colmenero *et al.* (11), and later in other polymers, polyisoprene and polybutadiene by Zorn et al.(12). These authors followed the procedure of Kiebel *et al.* (33) and assumed that the intermediate scattering function $I(Q,t)$ is the product of a vibrational contribution $I_{vib}(Q,t)$ and a relaxational contribution $I_{rel}(Q,t)$. The former was obtained by scaling a spectrum measured at low T, where there is negligible relaxation to the temperature of interest by the ratio of Bose-Einstein and Debye-Waller factors. $I_{rel}(Q,t)$ was then obtained by dividing $I(Q,t)$ by the scaled $I_{vib}(Q,t)$. Using this procedure, it was found by Colmenero *et al.* and Zorn *et al.* that $I_{rel}(Q,t)$ exhibits a crossover from equation 1 to equation 2 at $t_c \approx 2$ ps in these polymers. In polyvinylchloride, the τ_o found has all the signatures of independent motion which include its Q^{-2} dependence of normal diffusion and its activation energy being close to the conformation energy barrier of rotation of monomers in a single chain. Equation 3 holds and relates these 'normal' dependencies of τ_o to the 'anomalous' dependencies of τ^*. Similar success has been found from analysis of molecular dynamics simulation data of *ortho*-terphenyl (21).

However, the results have been questioned because the procedure is based on the assumption of harmonic behavior of the vibrational modes and there is uncertainty of the results at high temperatures if anharmonic effects are important. However Colmenero (34) has experimental facts such as the comparison between coherent and incoherent scattering to support the harmonic phonon hypothesis for his data. In our view, the task of removing the vibrational contribution in order to isolate out $I_{rel}(Q,t)$ is never going to be easy. The uncertainty of the validity of the harmonic approximation should not overshadow the possibility that the fast process seen in the experimental data is consistent with the independent relaxation given by equation 1 for $t<t_c$, which we have also called the α_{fast}-process (15,35). The cooperative relaxation given by equation 2 for $t>t_c$ is naturally called the α_{slow}-process. Even in the event that the vibrations are anharmonic and the Debye-Waller factor at high temperatures can no longer be obtained by extrapolation of its values deduced from $I(Q,t)$ measured at low temperatures under the harmonic phonon hypothesis, all is not lost in showing that the fast process seen in these 'fragile' glass-formers (36) can be identified with the α_{fast}-process of the CM and that there is a crossover at some t_c. Roland *et al.*(37) have demonstrated that varying the Debye-Waller factor at high temperatures from the harmonic phonon values to simulate the effect of anharmonicity to calculate $I_{vib}(Q,t)$ and then obtaining $I_{rel}(Q,t)$ by taking the ratio of $I(Q,t)$ to $I_{vib}(Q,t)$ does not drastically change the physical picture. The resulting $I_{rel}(Q,t)$ is still consistent with a crossover from the α_{fast}-process to the α_{slow}-process at a slightly different t_c. The parameters τ_o, n and τ^* are different from the corresponding values determined with the harmonic phonon hypothesis, but not by very much to suggest breakdown of the CM interpretation. Admittedly, without knowing exactly the anharmonicity of the vibrations there is some uncertainty in the determinations of the parameters τ_o, n and τ^*. However, the two main points to make are: (1) failure of the harmonic phonon approximation in the procedure to isolate $I_{rel}(Q,t)$ does *not* imply failure of the CM to describe short-time relaxation, and (2) there is evidence that the CM interpretation of the short-time dynamics is robust under any circumstances of the presence or absence of anharmonicity effects. Kartini *et al.* (38) have found from inelastic neutron scattering studies of CKN that the Debye-Waller factor deviates from the harmonic-phonon-like behavior at $T>T_g$=335 K. CKN is definitely anharmonic and it is moot to check the CM using the harmonic phonon approximation.

Since the criticism of the harmonic approximation comes mainly from supporters of the Mode Coupling Theory (MCT) (39), it is fair to examine the impact of the vibrational contribution on the interpretation of neutron and light scattering data by MCT. In the earlier comparison of MCT with incoherent neutron scattering data by Kiebel *et al.* and coherent neutron scattering data by Bartsch *et al.* (40), the harmonic phonon hypothesis was used in the procedure to isolate out $I_{rel}(Q,t)$. Nevertheless, good agreements of $I_{rel}(Q,t)$ with the MCT predictions were found (33,40). If the harmonic phonon hypothesis were not valid, then these good fits to illegitimately deduced relaxation data by MCT are fortuitous and suggests that the good fit to the susceptibility minimum data is neither necessary nor sufficient to conclude that MCT correctly captures the physics of the short-time dynamics. Most comparisons of experimental data with MCT were made without removing the vibrational contribution from the data. The fast β-process of the MCT fit was laid near the $I(Q,t)$ data which increases with decreasing t or the susceptibility $\chi''(\omega)$ data which increases with increasing ω. Without knowing the vibrational contribution in this time or frequency region, it is hard to evaluate how good the fit really is. Recent inelastic coherent neutron scattering data (41) indicates the importance of the vibrational contribution in the region of the fast process in polybutadiene. Finally, if the comparison of the fast dynamics of the CM with experiment were carried out in the same manner as that involves MCT (i.e. leaving the vibrational contribution being uncertain), then good fits to the data by the CM would be already guaranteed (36).

Ionic Glass Formers CKN is a good example of this class of ion-containing materials which have the structural relaxation time or the shear relaxation time $<\tau_s>$ being nearly the same as the ionic conductivity relaxation time $<\tau_\sigma>$ at least at high temperatures where these relaxation times become of the order of picoseconds. This condition ensures that one can study the fast dynamics of structural relaxation by making ionic conductivity relaxation measurements. If the CM holds and there is no contribution to the a.c. conductivity $\sigma(\omega)$ other than the diffusing ions, then it predicts that $\sigma(\omega)$ consists of three stages (41- 44). At high frequencies, $\omega>(t_c)^{-1}$, where ions execute independent motions with correlation function given by equation 1, $\sigma(\omega)$ is a frequency independent number which is denoted by σ_o. At lower frequencies, $\omega<(t_c)^{-1}$, when cooperative motion of the ions prevails and the correlation function is now given by equation 2, on decreasing ω, initially $\sigma(\omega)$ has the ω^{-n}-dependence and at about an $\omega\approx(\tau^*)^{-1}$ it levels off to the frequency-independent d.c. conductivity σ_{dc} level. This prediction from the CM has been made several years ago and reiterated in a recent paper (15) in which good agreement of the CM with conductivity relaxation data of CKN (14,15) were made. Measurement of CKN by Cramer *et al.* (14,15) were made at frequencies above the 10^{11} Hz (corresponding to about 2 ps) through the THz and into the far infrared region, where the vibrational contribution is evident as absorption peaks. It was established phenomenologically in *glasses* (where there is no fast relaxational process because $T<T_g$) (45) that the vibrational contribution to $\sigma(\omega)$ extends down to low frequencies with an ω^2-dependence. A similar frequency dependence of $\sigma(\omega)$ is found in CKN melt at high temperatures. These facts suggest the removal of the vibrational contribution from $\sigma(\omega)$ at lower frequencies by subtracting off the ω^2-dependent contribution extrapolated from the data at high frequencies. The difference at high temperatures is solely coming from the conductivity relaxation contribution from ion diffusion, $\sigma_{ion}(\omega)$, and as shown in Ref.15 it exhibits the crossover at about 10^{11} Hz which corresponds to $t_c\approx1$ ps (see Figure 1). The relation between τ^* and τ_o given by equation 3 and the corresponding relation between σ_{dc} and σ_o have been verified including the relation between their different temperature dependencies (42-45). Details can be found in Ref.15. Lunkenheimer *et al.*(46) have measured $\sigma(\omega)$ of CKN up to only $10^{10.45}$ Hz. Their data are in agreement with that of Cramer *et al.*(14,15). However, without the higher frequency data, it is not possible for Lunkenheimer *et al.* to remove the vibrational contribution. There is a minimum in the dielectric loss $\varepsilon''(\omega)$ data of Lunkenheimer *et al.*, calculated by the expression, $\sigma(\omega)/e_o\omega$, and has been fit to the MCT predictions. Nearly the same minimum in $\varepsilon''(\omega)$ is evident from the data of Cramer *et al.* from Figure 2. From this figure it is clear that much of the information on relaxation at higher frequencies above $10^{10.45}$ Hz and the contribution from vibrations available in Cramer *et al.* is missing in Lunkenheimer *et al.* The latter are fully aware of this and they have pointed out the possibility that the minimum in $\varepsilon''(\omega)$ is caused by the presence of the

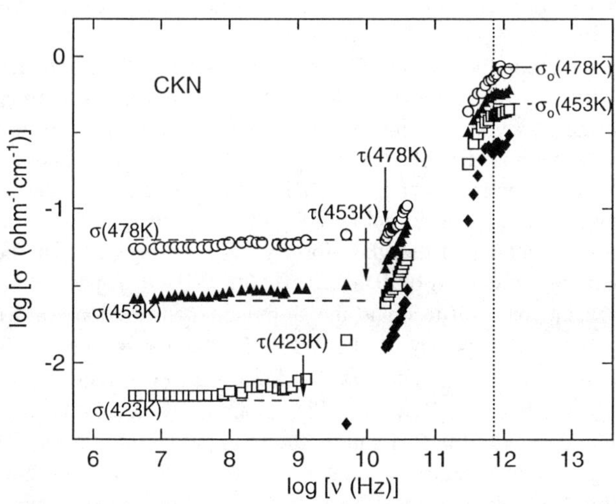

Figure 1. Log-log plot of the frequency dependence of the contribution by ionic motions to the a.c. conductivity in CKN. Circles (478 K); triangles (453 K); squares (423 K); filled diamonds (383 K). The vertical dotted line indicates the crossover frequency.

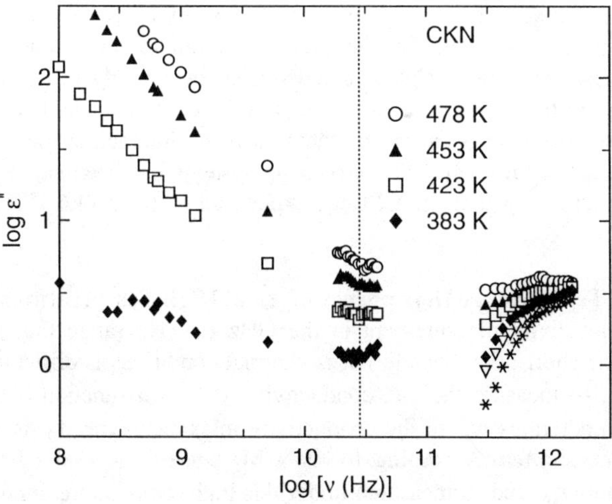

Figure 2. Frequency dependence of the dielectric loss in CKN at high frequencies and several temperatures. The vertical dotted line indicates the highest frequency reached in the work of Reference 47. Open circles (478 K); filled triangles (453 K); open squares (423 K); filled diamonds (383 K); inverted triangles (363 K); and stars (333 K).

vibrational contribution at higher frequencies and not related to the fast β process of MCT (32).

Fast glassy ionic conductors (FGIC), unlike CKN which is a poor ionic conductor, can have conductivity relaxation time becoming short by raising temperature but still staying within the glassy state. The short-time dynamics of ions can be probed by the a.c. conductivity method described in the previous paragraph. This is an example of relaxation of an interacting system which has little to do with the glass transition problem because T here is always below T_g and yet the phenomenology are very similar to the short-time dynamics of the glass-former CKN in the melt. The best example is the work of Cramer *et al.* (13) on $0.44LiBr\text{-}0.56Li_2O\text{-}B_2O_3$. Again, as in CKN, at high temperatures when ionic diffusion is the dominant contribution to $\sigma(\omega)$, there is a crossover of $\sigma(\omega)$ at a frequency of about 10^{11} Hz from a frequency independent σ_o to the ω^{-n}-dependence and at about an $\omega \approx (\tau^*)^{-1}$ it levels off to the frequency-independent d.c. conductivity σ_{dc} (see figure in Reference 13). Quantitative details of correspondence of the experimental data with the CM can be found in paragraph (*k*) and Table I of Ref.6 and will not be reiterated here. To dramatize this further, we can include crystalline ionic conductors Na β-alumina and $RbAgIO_4$, whose short-time conductivity relaxation behavior (42,48,49) is no different from the glassy ionic conductors and CKN. There is also a minimum at high frequencies in the $\varepsilon''(\omega)$ of these solid state materials and possibly an MCT-fit may even work. It does not seem likely that a fluid theory like MCT is applicable to ionic motions in the solid state, glassy or crystalline. In these cases, the minimum in the $\varepsilon''(\omega)$ is certainly due to vibrational absorption, causing the rise of $\varepsilon''(\omega)$ at high frequencies and hence the minimum. In view of the connection that has been made between the conductivity relaxation of CKN to MCT (47), it seems obligatory for workers in the field of glass transition in supercooled liquids to pay homage to these remarkably similar behaviors in the entirely different systems. The common behavior found in the glass-forming CKN melt, the fast ionic conductors in the glassy or crystalline state is caused by them all being interacting ionic systems. If the physics of relaxation in many-body interacting systems has been captured by the CM, then it is not surprising that the CM can explain all of them (4,6,15,17,42-45) at the same time.

Crossover of Temperature Dependence of σ_{dc} at High Temperatures. High frequency conductivity measurement in the GHz to THz range that is necessary to characterize the short-time dynamics at isothermal conditions is difficult to perform. It is much easier to measure the d.c. conductivity, σ_{dc}, as a function of temperature to reach high temperatures where the conductivity relaxation time, τ_σ, is of the order of picoseconds or shorter. According to the CM, since t_c is also of the order of a picosecond, both σ_{dc} and τ_σ measured under this high temperature condition pertain to the independent motion of the ions with correlation function given by equation 1 and τ_σ can be identified with $\tau_o \equiv \tau_\infty \exp(E_a/RT)$. Since $\sigma_{dc} = e_o\varepsilon_\infty/\tau_\sigma$ (50), the temperature dependence of τ_σ or τ_o is mirrored by that of the measured σ_{dc}. Therefore, the CM expects that by increasing T, the T-dependence of σ_{dc} will eventually cross over to reveal the T-dependence of τ_o. The latter, calculable from σ_{dc}, should have readily interpretable physical meaning because τ_o is unencumbered by complicating effects including cooperativity. For example, its apparent activation energy can be identified with a realistic energy barrier of independent relaxation, and the reciprocal of its prefactor is the attempt angular frequency which should correspond to some peak frequency of the infrared or Raman spectrum. These properties associated with the expected crossover of σ_{dc} at high temperatures have been found in many ionic conductors, including glass-formers such as CKN, fast glassy ionic conductors and crystalline ionic conductors (17-20,42,48).

For glass-formers at temperatures below T_g under isostructural conditions, τ_σ is long compared with t_c, the ion diffusion correlation function is given by equation 2, $\tau_\sigma=[\Gamma(1/\beta)/\beta]\tau^*$, where $\beta\equiv 1-n$, and τ^* is related to the independent relaxation time $\tau_o^{(g)}$ in the glassy state by equation 3. In the glassy state, σ_{dc} and hence τ_σ have Arrhenius T-dependence. Writing τ_σ and $\tau_o^{(g)}$ as $\tau_\sigma\equiv\tau_\infty^*\exp(E_a^*/RT)$ and $\tau_o^{(g)}\equiv\tau_\infty^{(g)}\exp(E_a^{(g)}/RT)$ respectively, equation 3 enables the prefactor $\tau_\infty^{(g)}$ and the activation enthalpy $E_a^{(g)}$ of independent diffusion in the glassy state to be deduced from the experimental quantities, i.e. $\tau_\infty^{(g)}=(t_c)^n(\tau_\infty^*)^{1-n}$ and $E_a^{(g)}=(1-n)\ E_a^*$. From a large data base of glass-forming ionic conductors, we (20) found: (1) the reciprocal of $\tau_\infty^{(g)}$ is in rough agreement with τ_∞ deduced from the high temperature σ_{dc} data and corresponds well to the peak angular frequency in the measured vibrational spectrum; (2) $E_a^{(g)}$ is about the same as, or slightly larger than (but by no more than 20%) the value of E_a deduced from the high temperature σ_{dc} data which have the τ_σ's shorter than t_c as described above. These good correspondences between $\tau_\infty^{(g)}$ and $E_a^{(g)}$ with their counterparts at high temperature are expected by the CM because both sets of parameters are for independent diffusion of the ion, and the difference between them is caused only by the changes of thermodynamic variables which include T and the density. These changes should not have a large effect on the attempt frequency and the activation enthalpy of the independent motion of the ion.

The crossover of temperature dependence of σ_{dc} at high temperatures in glassy ionic conductors is general and occurs in non-glassy ionic conductors as well. An example of the latter is shown in Figure 3 for yttria stabilized zirconia containing 9.5 mol % yttria, an oxygen ion conductor (19). The activation energy at high temperature or σ_{dc}, E_a,is clearly smaller than that, E_a^*, at low temperatures or σ_{dc}. Again, when the conductivity relaxation time is long the frequency dispersion is well described by equation 2 with n=0.57. The relation $E_a=(1-n)\ E_a^*$ that follows from equation 3 has again been verified (19).

Entangled Polymer Chains. This subject is not as closely related to the dynamics of supercooled liquids as the others given in this work. We mention it briefly here because the length scale of entanglement interaction (i.e. the entanglement distance which is of the order of a few nanometers and the radius of gyration of the Gaussian chain which depends on the chain length and is even larger) between the chains in polymer melts is much larger than that between small molecular units and between ions. The strength of interaction between entangled chains is correspondingly weaker and we expect that the crossover time of Rouse dynamics (corresponding to independent motion of the chains) to slowed-down dynamics would be much longer than 1 or 2 ps found for local segmental motion in polymers and for motions in small molecular liquids

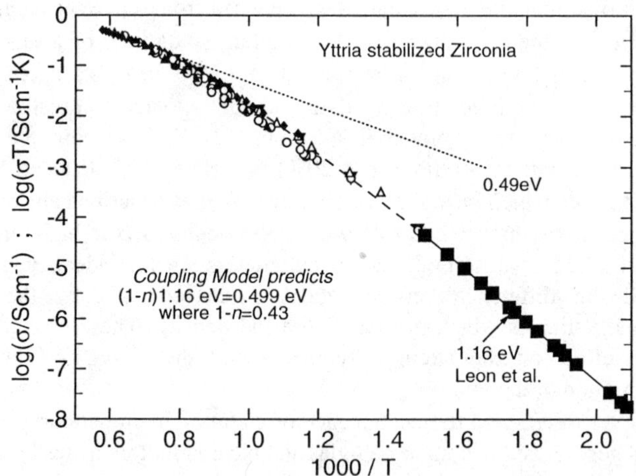

Figure 3. D.c. conductivity data of yttria stabilized zirconia (O) and (Δ). The dashed line indicates the high temperature activation energy of 0.50 eV.

and ionic conductors. Experimentally, it is found that Rouse dynamics ceases and is slowed down after a time of the order of nanoseconds. The length scale of another many-body interacting system that we shall discuss in more detail in the next subsection, concentrated colloidal particles, as measured by their radius is even larger, of the order of a hundred nanometers. The interaction strength of these objects with larger length scale is expected to be weaker and the crossover time t_c becomes much longer. There are experimental evidences in semidilute entangled polymer solutions, associating polymer solutions and interpenetrating polymer cluster solutions (22-29), that the correlation function measured is initially a linear exponential function of time and crosses over to a stretched exponential at some time t_c which is of the order of μs, 10 μs or even larger depending on the system (23-29). The mean squared displacement of these diffusing polymer systems has the same time dependence as ionic conductors (4,45) and the concentrated colloidal particles to be discussed next. These isomorphic time dependencies of diffusion in different interacting systems are in accord with the CM (45).

Concentrated Colloidal Suspensions. The Brownian motions of colloidal particles (8,10) and polymer-micronetwork colloidal particles (9) suspended at high concentrations in a liquid constitute another problem of relaxations in a many-body interacting system that bear some resemblances to the motions of atoms in a liquid. The particles are nearly monodisperse with a mean radius of about 10^2 nm. In contrast to molecular systems, the diffusion of the large colloidal particles occurs at macroscopic times longer than a μs and there is no vibrational contribution to complicate the analysis of the diffusional dynamics. This advantage of colloidal particles has enabled a direct observation of the crossover in dynamics proposed by the CM. Previous dynamic light scattering measurements and analyses (8,9) have concluded that all the predictions of the idealized version of MCT were verified. However, recently, Segrè and Pusey (SP) (10) have found new results of Brownian motions of colloidal suspensions in their equilibrium and metastable fluid states also by dynamic light scattering (DLS). The technique enables them to obtain the normalized intermediate scattering function $f(Q,t) = F(Q,t)/F(Q,0)$, where $F(Q,t) = N^{-1}\sum_{i=1}^{N}\sum_{j=1}^{N}\langle \exp\{i\vec{Q}\cdot[\vec{r}_i(0) - \vec{r}_j(t)]\}\rangle$. Here $\vec{Q}$ is the scattering vector, N is the number of particles, and $\vec{r}_i(t)$ is the position of particle i at time t. $F(Q,0)$ is the static structure factor usually denoted by $S(Q)$. Information on dynamics are contained in $f(Q,t)$.

SP found at short times the decay of $f(Q,t)$ is linear in t, $f(Q,t)=1-D_S(Q)Q^2t+ ...$, which defines the short-time, Q-dependent diffusion coefficient, $D_S(Q)$. Equivalently stated, $f(Q,t)$ decays exponentially, i.e. $f(Q,t)=\exp(-Q^2D_S(Q)t)$, at short times. As pointed out by SP, the theory for $D_S(Q)$ is well understood (10) and the result is $D_S(Q)=D_oH(Q)/S(Q)$, where $H(Q)$ originates from hydrodynamic interactions between the particles and D_o is the free particle diffusion constant.

At longer times $f(Q,t)$ departs strongly from the single exponential decay (10). This crossover to a dispersive time dependence was seen in earlier works (8,9) but the short-time diffusion part was not taken into account in the fit to MCT. We shall see later the short-time diffusion is the independent Brownian motion of the CM, and the observed crossover is related to t_c.

Eventually, at sufficiently long times, the decay of $f(Q,t)$ returns to another exponential form (10), $f(Q,\text{large } t) \propto \exp(-Q^2 D_L(Q)t) \equiv \exp(-Q^2 \langle r^2(t)\rangle/6)$, where $\langle r^2(t)\rangle = 6D_L(Q)t$ and $D_L(Q)$ is appropriately called the long-time diffusion coefficient. The return to another exponential form was apparently missed in the analysis of data in previous works (8,9) which fit the long-time decay of $f(Q,t)$ to the Kohlrausch function according to MCT.

This evolution of $f(Q,t)$ from the short- to long-time regimes can be represented by a time-dependent diffusion coefficient $D(Q,t)$ defined by

$$D(Q,t) \equiv -(1/Q^2)d\ln f(Q,t)/dt. \qquad (4)$$

At short-times, $D(Q,t)$ is equal to $D_S(Q)$, but starting at some characteristic time it shows a decrease with time continuously over an intermediate time regime. Finally, at some long time it levels off to the terminal value of $D_L(Q)$. The ratio $D_L(Q)/D_S(Q)$ is about 0.23 at volume fraction $\phi=0.465$ and decreases with increasing ϕ to approximately 10^{-3} at $\phi=0.570$. The two crossovers can be seen also through the time dependence of $-\ln f(Q,t)$ as it changes from being proportional to t at short times to a slower increase with time at intermediate times, and later returns back to being proportional to t again.

SP discovered a simple Q-scaling property of the measured $f(Q,t)$ and $D(Q,t)$ for $QR \geq 2.5$, where R is the mean radius of the particles. This scaling property is embodied by either equations, $[\ln f(Q,t)/Q^2 D_S(Q)] = -\chi(t)$ or $[D(Q,t)/D_S(Q)] = \chi'(t)$, where $\chi(t)$ and $\chi'(t) = d\chi(t)/dt$ are Q-independent functions of time. SP pointed out that at large Q ($QR>>3.5$), for all times, the coherent scattering function $f(Q,t)$ tends to the self (incoherent) intermediate scattering function $F^{self}(Q,t)$ which has the form

$$F^{self}(Q,t) = \exp[-Q^2 \langle \Delta r^2(t) \rangle / 6] \qquad (5)$$

where $\Delta \vec{r}(t) \equiv \vec{r}(t) - \vec{r}(0)$ is the displacement and $\langle \Delta r^2(t) \rangle$ the mean-squared displacement of a single particle. Requiring $f(Q,t)/Q^2 D_S(Q)$ reduce to equation 5 in the high Q-limit, $f(Q,t)$ can be written

$$f(Q,t) = \exp(-\frac{D_S(Q)}{D_S^{self}} Q^2 \langle \Delta r^2(t) \rangle / 6) \qquad (6)$$

where $D_S^{self} = \lim_{Q\to\infty} D_S(Q)$ is the short-time self diffusion coefficient and

$$\langle r^2(t) \rangle = 6D_S^{self} t \qquad \text{for short } t \qquad (7)$$

Thus equation 6 indicates that for QR sufficiently large, the coherent scattering function is determined by the diffusion of a single particle, scaled by the ratio of collective to self short-time diffusion coefficients. Identifying the observed long-time dependence of $f(Q,t)$ with equation 6, the long-time self diffusion coefficient D_L^{self} defined by

$$\langle r^2(t) \rangle = 6D_L^{self} t \qquad \text{for long } t \qquad (8)$$

are related to other quantities by

$$D_L^{self} / D_S^{self} = D_L(Q) / D_S(Q) \qquad (9)$$

Since the ratio on the left-hand-side of equation 9 has no Q-dependence, consequently $D_L(Q)$ and $D_S(Q)$ have the same dependence on Q as observed experimentally.

SP pointed out that this scaling has not been predicted theoretically by MCT. In this paper we show that the scaling property is consistent with the coupling model (CM) results for diffusion of particles with many-body interactions. More importantly, we show the time dependencies of the scaled functions $\chi(t)$ and $\chi'(t)$, as well as their rapid changes with increasing ϕ, can be derived from an extant prediction of the CM on the mean squared displacement, $\langle r^2(t) \rangle$, for diffusion of units that are interacting with each other. This extant prediction also follows from the three equations 1-3. From now on the term 'diffusion' refers to motions that can be indefinitely long-range in a translational sense, and should be distinguished from other local relaxation processes of different nature such as density and reorientational fluctuations. The coherent as well as the incoherent intermediate scattering functions for diffusion and for these other relaxations are different. This difference is evident from results of mean squared displacement obtained previously for diffusion by the CM (45,51-54).

In applying the CM to diffusion in the past, we have used the formalism of continuous time random walk (CTRW) of Montroll and Weiss (55). CTRW is a generalization of random walks through the introduction of a probablity density function, $\Psi(t)$, for waiting times between successive steps in the walk. Taking c_o as the jump distance, D_o in equation 5 is rewritten as $D_o \equiv c_o^2/6\tau_o$. According to the CM, since the correlation function is given by equation 1, diffusion is normal for $t<t_c$. Correspondingly, in CTRW the waiting time distribution is

$$\Psi(t) = \tau_o^{-1} \exp(-t/\tau_o) \qquad \text{for } t<t_c \qquad (10)$$

which recaptures the mean-square displacement

$$\langle \Delta r^2(t) \rangle = c_o^2 (t/\tau_o) \equiv 6 D_S^{self} t \qquad \text{for } t<t_c \text{ (regime I)} \quad (11)$$

For $t>t_c$, the crossover of the correlation function to the stretched exponential in equation 3 is mirrored in CTRW by the cross-over of $\Psi(t)$ from equation 10 for $t<t_c$ to

$$\Psi(t) = -\frac{d}{dt} \exp(-(t/\tau^*)^{1-n} \qquad \text{for } t>t_c \qquad (12)$$

where again τ^* is related to τ_o and n by equation 3. The entire $\Psi(t)$ consisting of equations 10 and 12 has to be renormalized in order for it to be a probability distribution function.

The mean-square displacement corresponding to equation 12 has been calculated before from the CTRW formalism analytically in limiting time regimes (51) and numerically for all times (52). The numerical result obtained for all times $t>t_c$ agree with the analytical results which consists of two stages:

$$\langle \Delta r^2(t)\rangle = [\frac{c_o^2(1-n)\Gamma(1-n)}{\Gamma(2-n)}](t/\tau^*)^{1-n},\ t_c<t<<\Gamma(\frac{2-n}{1-n})\tau^* \equiv \tilde{\tau} \text{ (regime II)} \quad (13)$$

and

$$\langle \Delta r^2(t)\rangle = [c_o^2/\Gamma(\frac{2-n}{1-n})](t/\tau^*) = 6D_L^{self}t,\ t>\Gamma(\frac{2-n}{1-n})\tau^* \equiv \tilde{\tau} \text{ (regime III)}. \quad (14)$$

Defining the generalized diffusion coefficient by $D^{self}(t)= (1/6)d\langle\Delta r^2(t)\rangle/dt$, we can easily verify from equations 4 and 6 that

$$D^{self}(t)/D_S^{self} = D(Q,t)/D_S(Q) \qquad (15)$$

which recaptures equation 6 when t becomes long.

Combining equations 11, 13 and 14, these extant general results of the CM indicate that dynamics of diffusion in interacting many-body systems are comprised of *three* distinct relaxation regimes. Similarly, the generalized diffusion coefficients $D^{self}(t)$ and $D(Q,t)$ have correspondingly three different time regimes. These general behaviors of $\langle\Delta r^2(t)\rangle$ have been observed in molecular dynamics and Monte Carlo simulations of monodisperse entangled polymer melts (56-58), Monte Carlo simulations of a disordered Coulomb lattice gas model (18,59) and in ionic diffusion and electrical conductivity measurements of glassy as well as crystalline fast ion conductors (4,6,15,42-45). For ionic conductors when cross correlation can be neglected, the a.c. conductivity $\sigma(\omega)$ is related to $\langle\Delta r^2(t)\rangle$ via the relation (59): $\sigma(\omega) \propto -\omega^2 \lim_{\varepsilon\to 0}\int_0^\infty \langle r^2(t)\rangle e^{i\omega t-\varepsilon t}dt$ and the three time regimes of $\langle\Delta r^2(t)\rangle$ given by equations 11,13 and 14 correspond to the three frequency regimes of $\sigma(\omega)=\sigma_o$ for $\omega\geq(t_c)^{-1}$, $\sigma(\omega)\propto\omega^n$ for $(\tilde{\tau})^{-1}<<\omega<<(t_c)^{-1}$, and $\sigma(\omega)=\sigma_{dc}$ for $(\tilde{\tau})^{-1}>>\omega$ that have been discussed in a previous subsection.

It is important to point out that the cross-over time t_c, the coupling parameter n and hence $\langle\Delta r^2(t)\rangle$ of the CM given by the expressions from equations 11,13 and 14 depend only on the nature of many-body interactions and not on other variables such as the scattering vector Q. Thus, the Q-scaling property of the measured $f(Q,t)$ and $D(Q,t)$ found by SP follows naturally as a consequence from equation 6.

Let us apply this general result for $\langle\Delta r^2(t)\rangle$ to concentrated colloidal suspensions and calculate $f(Q,t)$ according to equation 6. SP have found their experimental intermediate scattering function exhibits also three distinct relaxation regimes. We identify the short time behavior as that given by equation 11 of the CM results in the time regime I of $t<t_c$. The cross-over time t_c can be estimated as the time when the initial exponential decay of $f(Q,t)$ or the $(-t)$-dependence of $\ln f(Q,t)$ ceases. At times longer than t_c, the effect expected by the CM is $\langle\Delta r^2(t)\rangle$ having a sublinear power law increase according to equation 13 in the time regime II, leading to a stretched exponential like decay of $f(Q,t)$ and a power law decrease, t^{-n}, of $D(Q,t)$ defined by equation 4. Thus, the coupling parameter n can be roughly estimated from these time dependencies. Finally, in regime III $\langle\Delta r^2(t)\rangle$ returns to normal diffusion according to equation 14, albeit with a long-time diffusion coefficient, D_L^{self}, which is smaller than the short-time counterpart D_S^{self}, and $f(Q,t)$ decays exponentially again. The onset time of these dependencies give an estimate of $\tilde{\tau}$. When the condition

$\tau_o > t_c$ holds, we can see from equation 3 that τ^* and also $\tilde{\tau}$ are longer than τ_o. The ratios $\tau_o/\tilde{\tau}$ and $D_L^{self} / D_S^{self} = D_L(Q) / D_S(Q)$, increase rapidly with increasing n. An increase in n can be realized by an enhancement of the interaction between the particles through an increase of their volume fraction ϕ.

The results given so far have been obtained assuming a sharp crossover at t_c. However, the cross-over at t_c is not expected to be sharp, due to the following reasons. Firstly, results of simple models demonstrates that the cross-over is broadened over a finite time range (2,3). Secondly, the particles are made of PMMA cores stabilized by thin layers of poly-12-hydroxystearic acid and the latter may further broaden the cross-over. We have found that the essential features of the $<\Delta r^2(t)>$ data can be described using the CM with a sharp cross-over. However, the features of the data for the derivative $d<\Delta r^2(t)>/dt$ are rounded in the vicinity of the cross-over and thus deviate from the model over a short time interval Δt_c if a sharp crossover is used. In order to improve the fit a smoothing of the cross-over need to be introduced in the present calculation. However, this will not be attempted here.

Figure 4 shows the colloidal suspension data of SP (10) with $\phi = 0.465$ and the CM fit for the normalized time-dependent diffusion coefficient $D(Q,t)/D_S(Q)$, which is proportional to the derivative $d[<\Delta r^2(t)>/6D_o]/dt$, with $n= 0.35$, $\tau_o = 0.033$ s , and $t_c = 3.8\times10^{-3}$ s. The other quantities τ^* and $\tilde{\tau}$ are automatically determined by equations 3 and 13. By inspection of equation 6, we can identify $<\Delta r^2(t)>/R^2$ with the scaled function $(-6D_S^{self}/R^2)\ln f(Q,t) / Q^2D_S(Q)$ and this quantity is shown in Figure 5 comparing the data and the CM fit. The prefactor `is known from experiment. We have also plotted $\ln f(Q,t) / Q^2D_S(Q)$ against time in Figure 6. These results reproduce the essential features of the colloidal suspension data with $\phi = 0.465$ quite well.

Careful scrutiny of fig.4 reveals that the dashed curve deviates from the data as the crossover from Regime II to Regime III is approached. There is a possibility that this deviation is due to the neglect of the mitigation of the dynamic constraints (or cage effects) when approaching the long term diffusion of Regime III from Regime II. The cage effects which firmly take hold in the earlier part of time regime II have to be eventually dissolved somewhat before the terminal diffusion in Regime III can be established. The existence of constraint mitigation is reaffirmed by the fact that what is experimentally studied is the decay of the static structure factor $F(Q,0)$ which reflects the presence of the cage in the first place. Previously, we have encountered an example of such constraints mitigation in the diffusion of polymer chains which are entangled together (22). The characteristic features of the coupling model (CM) relaxation function, equations 1-3 with a constant n are for time-independent or permanent constraints (cage). In many of our previous applications of the CM which do not involve long term diffusion or flow, we have found that a single value of n (determined experimentally by the relation between τ^* and τ_0 in equation 3) could accurately describe the entire relaxation for $t > t_c$. In these systems, the constraints imposed on the relaxing unit remain permanent and thus the coupling parameter remains unchanged. This is not the case for the concentrated colloidal particles studied here. The diffusive motions of the colloidal particles become impeded and slowed down by its neighbors which form a cage surrounding each particle. This cage is expected to be well-defined until timescales are reached which are on the order of the mobility of the cage particles, which is approximately $t \cong \tau^*$. For $t > \tau^*$, the cage will begin disintegrating as these particle diffuse and the constraints of the cage on the

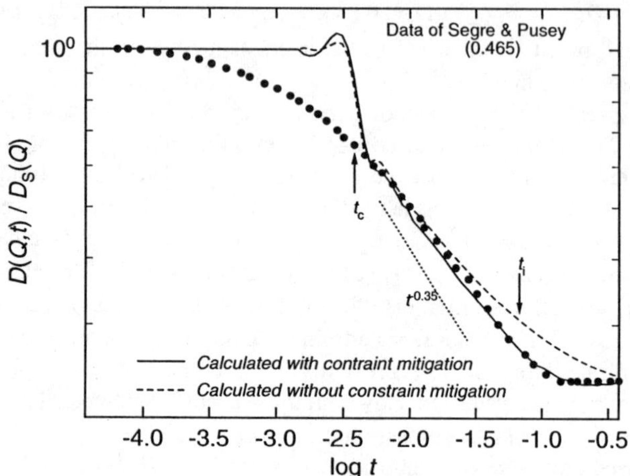

Figure 4. Normalized time-dependent diffusion coefficient for colloidal suspensions with volume fraction $\phi = 0.465$. Symbols are data replotted from Figure 2a of Ref.10. Solid (dashed) line is CM fit with (without) constraint mitigation. The parameters used are given in the text.

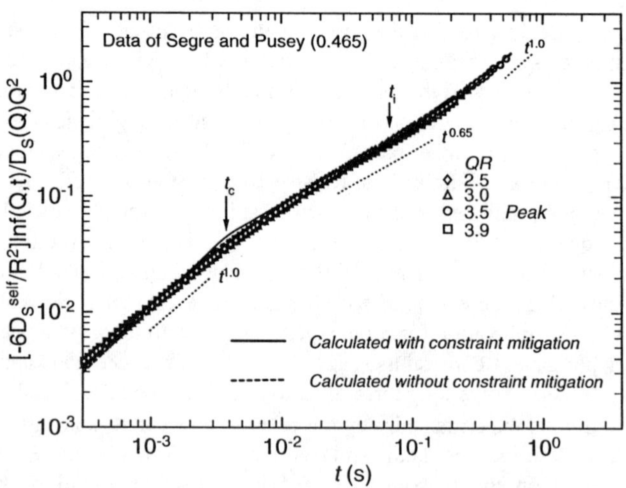

Figure 5. Scaled scattering function $(-6D_S^{self}/R^2)\ln f(Q,t) / Q^2 D_S(Q)$ for colloidal suspensions with volume fraction $\phi = 0.465$. Symbols are data replotted from Figure 3b of Ref.10. Solid (dashed) line is CM fit with (without) constraint mitigation. The parameters used are given in the text.

enclosed colloidal particle will be softened or mitigated. For situations in which such constraint mitigation occurs, we would expect a constant value of n for $t < \tau^*$, followed by decreasing values of n for $t > \tau^*$. This can be incorporated into the CM by introducing a time dependence for n and using this in a relaxation rate formulation of the CM. Note that the CM relations, eqs.1-3, are solutions of the following time-dependent rate equation :

$$d\phi / dt = -W(t)\phi(t) \quad (16)$$

$$W(t) = \tau_0^{-1}, t < t_c, \quad (17)$$

and

$$W(t) = \tau_0^{-1}(1-n)(t/t_c)^{-n}, t > t_c \quad (18)$$

This has the solution :

$$\phi_{cm}(t) = \exp(-t/\tau_0), \quad t < t_c \quad (19)$$

and

$$\phi_{cm}(t) = \phi_0 \exp(-\int_{t_i}^{t} (1-n)\tau_0^{-1}(\tilde{t}/t_c)^{-n})d\tilde{t}, \quad t > t_c \quad (20)$$

where the subscript 'cm' is used to indicate the inclusion of constraint mitigation. We now parameterize $n(t)$ to reflect constraint mitigation for $t > t_i \cong \tau^*$:

$$n(t) = n_f + (n_i - n_f)\exp(-(t-t_i)/\tau_n), \quad t > t_i \quad (21)$$

where τ_n is expected to be of order τ^*. The ϕ_{cm} of CM with constraint mitigation is described by eq.19 for $t<t_c$, by $\phi_{cm} = \exp(-(t/\tau^*)^{1-n})$ for $t_c,<t<t_i$ and eqs.20 and 21 with $\phi_o = \exp(-(t_i/\tau^*)^{1-n})$ for $t > t_i$. Without constraint mitigation $n(t) = n_i=n_f$, $t_i \rightarrow\infty$, and ϕ_{cm} reduces to the usual CM equations 1-2.

From $\phi_{cm}(t)$ we can calculate $\langle\Delta r^2(t)\rangle$, $\ln f(Q,t)/Q^2D_S(Q)$ and $D(Q,t)/D_S(Q)$ by the same procedure discussed before and, wherever applicable, replacing $\phi(t)$ by $\phi_{cm}(t)$ in the equations. To fit the calculated quantities to the experimental data, the parameters t_c, τ_o and n_f have to be consistent with that required by the experimental facts and they can be considered to be predetermined. The two new parameters t_i and n_i are adjusted to improve the fit to $D(Q,t)/D_S(Q)$ in fig.4 for ϕ=0.465. The new fit with constraint mitigation and t_c=3.8×10^{-3}s, τ_o=0.077s, n_i=0.35, n_f=0.27 and t_i=0.078s is shown as the solid curve in figs.4 and 5. Significant improvement in fitting $D(Q,t)/D_S(Q)$ is evident from fig.4. The position of t_i is indicated in the figure and its location is near the crossover from Regime II to Regime III. At this time, $\sqrt{\langle r^2(t)\rangle}$ ≈0.58R. Although the results for ϕ=0.465 with or without constraint mitigation for $\langle\Delta r^2(t)\rangle$ and $\ln f(Q,t)/Q^2D_S(Q)$ are hardly distinguishable (figs.5 and 6), they are quite different for ϕ=0.57 as we shall discuss immediately.

Next, we include constraint mitigation and calculate the corresponding results for ϕ=0.57 by using $\phi_{cm}(t)$. A typical good fit to the $D(Q,t)/D_S(Q)$ data are shown by the solid curve in Figure 7 for the following choice of parameter values: t_c=8.7×10^{-3}s, τ_o=0.027s, n_i=0.85, n_f=0.75 and t_i=3.28s. The dotted line has the $t^{-0.85}$-dependence which approximates

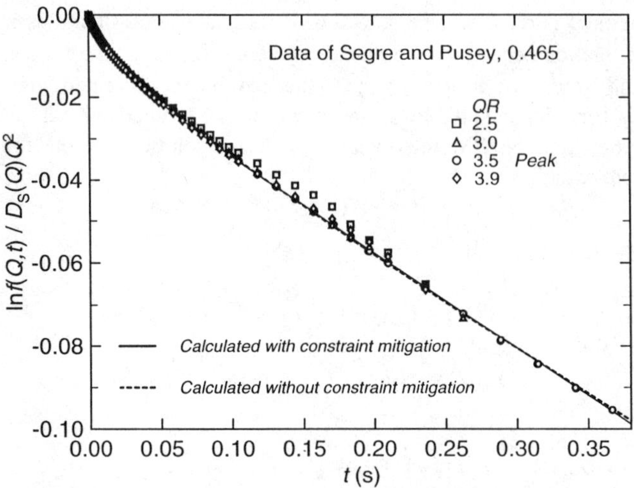

Figure 6. Scaled scattering function $\ln[f(Q,t)/Q^2 D_S(Q)]$ for colloidal suspensions with volume fraction $\phi = 0.465$. Symbols are data replotted from Figure 3b of Ref.10. Solid (dashed) line is CM fit with (without) constraint mitigation. The parameters used are given in the text.

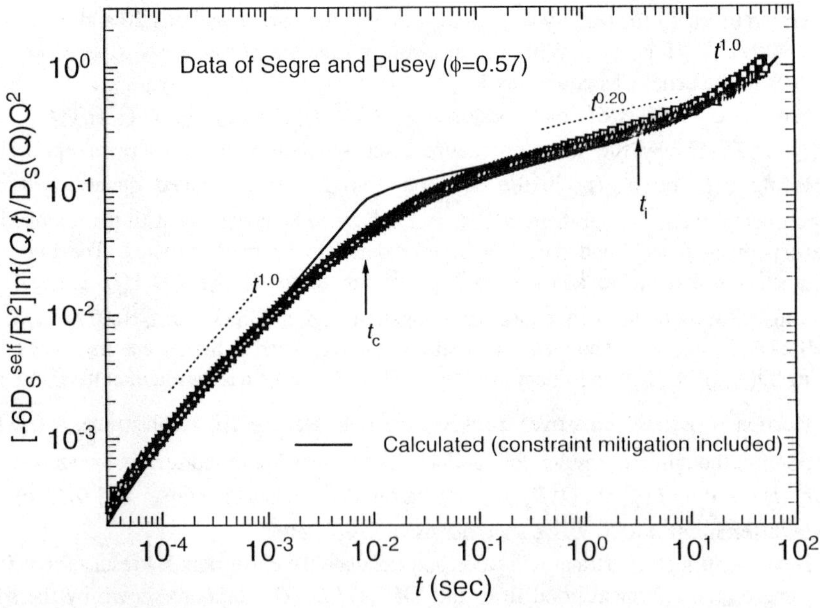

Figure 7. Scaled scattering function $(-6D_S^{self}/R^2)\ln f(Q,t) / Q^2 D_S(Q)$ for colloidal suspensions with volume fraction $\phi = 0.57$. Symbols are data from Ref.10. Solid line is CM fit with constraint mitigation. The parameters used are given in the text.

the time dependence of the data in Regime II. The values of t_c, τ_o and n_i are within the predetermined ranges and the results presented are therefore acceptable. The location of t_i shown in fig.7 is consistent with the onset of constraint mitigation when the terminal diffusion regime III is drawing near. The ratio $D_L(Q)/D_S(Q)\approx 10^{-3}$ is correctly reproduced. The calculated results fit well also the time dependencies of the $\langle \Delta r^2(t) \rangle$ and $\ln[f(Q,t)/Q^2 D_S(Q)]$ for all times except in a neighborhood of the crossover time t_c. Due to space limitations, these plots will not be shown here. This discrepancy occurs also for ϕ=0.465 and is attributed to the broader crossover in the colloidal particles which cannot be reproduced by the sharp crossover at specific value of t_c assumed in the CM for the sake of simplicity. Except for this blemish, the calculations with the inclusion of constraint mitigation are in good agreement with the experimental data. This blemish can be remedied by introducing a range of crossover times. The values of $n(t)$ for ϕ=0.57 are significantly higher than those for ϕ=0.465 as expected for increased interaction strength at higher volume fraction, and the corresponding τ^* is more than two orders of magnitude larger.

Comparing the new data of Segre and Pusey (10) with the previous works (8,9) that fit data to MCT predictions, several deficiencies of the latter become apparent. First, the α-process of MCT which has the form of a Kohlrausch function, $f(Q,t)=f_c\exp[-(t/\tau_e)^\beta]$, supposedly is responsible for the long time decay of the intermediate scattering function. (8,9,60). However, a Kohlrausch function will remain at long times as a stretched exponential function with no return to a linear exponential function to give a time independent long-time diffusion coefficient, $D_L(Q)$, as found by SP. Second, the short-time exponential relaxation seen in all works is left out of consideration in the MCT fits. There is no effort in removing the exponential relaxation process from the raw data. The MCT fit is just laid near the experimental curve and their departure from each other at shorter times is left to the imagination of the reader. This is to be contrasted with the CM fit in which the short-time exponential relaxation, $\exp[-D_S Q^2 t]$, is not only taken into account (equation 1) but also plays some role of determining the subsequent time dependence of $f(Q,t)$ in the longer time regimes, and enables $D_L(Q)$ to be calculated from $D_S(Q)$. Although it must be pointed out that in the case of high volume fraction, ϕ=0.57, in the time region of $5\times10^{-2}<t<5\times10^{-1}$s the CM fit is not as good as the MCT fit (8) to the data due to the sharp cross-over assumed.

The behaviors of $<\Delta r^2(t)>$ of concentrated colloidal particles described both experimentally and theoretically in previous sections turn out to be general and have been observed in other materials (see Tables 4.4a, 4.4b and 4.5 in Ref.4). These include ions in glassy and nonglassy ionic conductors (4,13-18,59), entangled polymer chains in monodisperse melts (4,53-58) and in semidilute solutions (28). The materials and the species that diffuse in them are totally different from the colloidal hard spheres. From the CM point of view, the general behavior is no surprise because these different systems share the common characteristic that many-body interactions between the diffusing species are important in determining the dynamics of the diffusion. The many-body interactions in these systems, modeled by the Lennard-Jones potential for polymers, Coulomb interaction for ionic conductors and hard sphere interactions for colloidal particles, are all nonlinear. Hence, the general mechanism in the CM of slowing down the relaxation rate by nonlinear Hamiltonian dynamics (chaos) should be operative in all such systems, leading to the general behaviors of $<\Delta r^2(t)>$ as found.

Conclusion

We have reasons (1-3) to believe that the basic physics of relaxation in any system with many-body interactions is related to chaos of nonintegrable Hamiltonian systems (7). Relaxation processes in many complex materials involves many-body interactions, however, most theoretical treatments of them, including the formally exact memory equation for the autocorrelation function of density fluctuation, have not gone back to the basic physics of chaos. Simple models of relaxation based on chaos have led us since 1979 (1-3) to the same results, which are represented by the three equations, 1-3. These three coupled equations may look deceptively simple, but they have many successful and nontrivial applications to various fields (4-6,15-17,19,20-23,28,42-45). Even if we pretend that these three equations had no theoretical basis, their remarkable ability to explain many experimental facts should be worth noticing. These applications in the macroscopic time regime, though important, could still be considered as *indirect* evidences of these three equations. In this work we focus on the *direct* experimental evidences for the validity of the three equations, drawn from more fundamental observations made in molecular glass-formers, ionic conductors, and colloidal hard spheres. Special attention is paid to concentrated colloidal suspensions because they have no contribution from vibrations, making the crossover of the diffusional contribution clear. The basic physics of the crossover from independent motion, equation 1, to cooperative motion, equation 2, at a characteristic time t_c and the quantitative relation between the two corresponding relaxation times, equation 3, are shown to be general for relaxations in materials with many-body interactions between the relaxing units.

Acknowledgement

This work is supported by ONR. We thank Phil. Segrè for sending us the published data of concentrated colloidal suspensions.

Literature Cited

(1) Ngai K.L., *Comment Solid State Phys.* **1979**, *9* 127; Ngai K.L. and White C.T., *Phys.Rev. B* **1979** *20* 2475.

(2) Ngai K.L. and Tsang K.Y., *Macromol. Chem. Macromolec.Symp.***1995,***90*,95; Ngai K.L.; Peng S.L.;Tsang K.Y.*Physica A* **1993**, *191*,523; Ngai K.L. and Rendell R.W., *J. Non-Cryst.Solids* **1991**,*131-133,* 233; Tsang K.Y.;Ngai K.L.,*Phys.Rev.E,* **1996**,*54*, R3067; Rendell R.W. *Phys.Rev.E* **1993,***48,*R17; Rendell R.W. *J.Appl.Phys.* **1994,***75,*7626. Tsang K.Y.;Ngai K.L.,*Phys.Rev.E*, submitted **1997**.

(3) We like to include the recent work, Tsironis G. P. and. Aubry S, **1996**, *Phys. Rev. Lett.* *77*, 5225, which shows a similar crossover from exponential relaxation at short times to a slower relaxation at longer times. See also Flach S. and Mutschke G. *Phys. Rev. E* **1994** *49*, 5018.

(4) Ngai K.L., in "Relaxational Processes in Disordered Systems", Richert R. and Blumen A. Eds.,(Springer-Verlag, Berlin, **1993**) p.89.

(5) For applications to polymer viscoelasticity, see Ngai K.L. and Plazek D.J., *Rubber Chem. Tech. Rubber Review* **1995,***68,* 376.

(6) Ngai, K.L. *J.Non-Cryst.Solids* **1996**, *203*, 232.

(7) MacKay R.S. and Meiss J.D., "Hamiltonian Dynamical Systems" (Adam Hilger, Bristol, **1987**).
(8) van Megan, W.; Underwood, S.M. *Phys.Rev.E,* **1994**, *49*, 4206.
(9) Bartsch, E.; Antonietti, M.; Schupp, W.; Sillescu, H. *J.Chem.Phys.* **1992**, *97*, 3950.
(10) Segre, P.N.; Pusey, P.N. *Phys.Rev.Letters*, **1996,** *77*, 771.
(11) Colmenero J., Arbe A. and Alegria A., *Phys.Rev.Lett.* **1993**,*71*,2603; *J. Non-Cryst.Solids* **1994,***172-174*,126; Ngai K.L., Colmenero J., Arbe A., and Alegria A., *Macromolecules* **1992,**25, 6727.
(12) Zorn R., Arbe A., Colmenero J., Frick B., Richter D., and Buchenau U., *Phys. Rev. E* **1995,** *52,* 781.
(13) Cramer C., Funke K. and Saatkamp T., *Philos. Mag. B* **1995**, *71*, 701.
(14) Cramer C., Funke K., Buscher M., and Happe A., Philos. Mag. *B* **1995** *71*, 713.
(15) Ngai K.L., Cramer C., Saatkamp T., and Funke K., Non-Equilibrium Phenomena in Supercooled Fluids, Glasses and Amorphous Materials, World Scientific, Singapore, M. Giorano,D. Leporini, M. P. Tosi, Eds. **1996**, pp.3-24.
(16) J. Kincs and S.W. Martin, *Phys.Rev.Lett.* **1995,***76*,70.
(17) K.L. Ngai and A.K. Rizos, *Phys.Rev.Lett.* **1996,** *76*, 1296.
(18) P. Maass, M. Meyer, A. Bunde and W. Dieterich, *Phys.Rev.Lett.* **1996**,*77*, 1528.
(19) K.L. Ngai, **1996** to be submitted for publication.
(20) Ngai, K.L.; Greaves, G.N.; Moynihan, C.T. **1996**, under preparation.
(21) Lewis L.J. and Wahnström G., *Phys. Rev. E* **1994,** *50*, 3865; Roland C.M., Ngai K.L. and Lewis L.J., *J. Chem. Phys.* **1995**, 103, 4632.
(22) Ngai, K.L.; Rendell, Plazek, D.J., **1997** *J.Phys.(Paris) Colloq.C*, in press.
(23) Ngai, K.L. *Advan.Colloid Interf.Sci.***1996**, *64*, pp.1-43.
(24) Adam, M.; Delasanti, M.; Munch, J.P.; Durand, D. *Phys.Rev.Lett.***1988**, *61*, 706.
(25) Nyström, B.; Walderhaug, H.; Hansen, F.N. *J.Phys.Chem.***1993**, *97*, 7743.
(26) Ren, S.Z.; Shi, W.F.; Zhang, W.B.; Sorensen, *Phys.Rev.A***, 1992**,*45*, 2416.
(27) Phillies, G.D.J.; Richardson, C.; Quinlan, C.A.; Ren, S.Z. *Macromolecules* **1993**,*26*,6849.
(28) Ngai, K.L.; Phillies, G.D.J. *J.Chem.Phys.* **1996**, *105* 8385.
(29) Ngai, K.L. in NATO ASI Series C, **1992**, *369*, pp.221-232.
(30) For basic free volume theory, see Ferry, D.J. *Viscoelastic Properties of Polymers*, J.Wiley, New York, N.Y. 1980.
(31) Adam, G.; Gibbs, J.H. *J.Chem.Phys.***1965**,*43*,139.
(32) Götze, W.; Söjgren, L. *Rep.Prog.Phys.* **1992**, *55*, 241.
(33) Kiebel, M.; Bartsch, E.; Debus, O.; Fujara,F.; Petry, W.; Sillescu, H. *Phys.Rev.B*, **1992**, *45*,10310.
(34) Colmenero, J. private communication **1996,** and *Phys.Rev.Lett.* January 1997.
(35) Roland, C.M.; Ngai, K.L. *J.Chem.Phys.* **1995,** *103* 1152; **1996,** *104* 8171.
(36) Ngai, K.L.; Greaves, G.N. in *Diffusion in Amorphous Materials*, Jain H.; Gupta, D. Eds.; TMS, Warrendale, PA; pp.17-41.
(37) Roland, C.M.; Ngai, K.L.; *J.Chem.Phys.* **1996,** *104* 2967; *Phys.Rev.E*, **1996**,*54*, 6969
(38) Kartini, E. *et al.* **1996,** *54*, 6292.
(39) Wuttke, J.; Petry, W.; Coddens, G.; Fujara, F. *Phys.Rev.E*, **1995**,*52*, 4026.

(40) Bartsch, E.; Fujara, F.; Legrand, J.F.; Petry, W.; Sillescu, H.; Wuttke, J. *Phys.Rev.E* **1995**, *52*, 738.
(41) Buchenau, U. *et al. Phys.Rev.Letters*, **1996**, *77*, 4035.
(42) K.L. Ngai and U. Strom, *Phys.Rev.B* **1988**, *38*, 10350.
(43) K.L. Ngai, *J.Phys.(Paris) Colloq. C2* **1992**, *2,* 61.
(44) K.L. Ngai and O. Kanert, *Solid State Ionics* **1992**, *53-55*, 936.
(45) See Tables 4.4a and 4.4b in Reference 4, and Table 3 in Rajagopal, A.K.; Ngai, K.L.; Teitler, S. *J.Non-Cryst.Solids* **1991**,*131-133*, 282.
(46)U. Strom, J.R. Hendrickson, R.J. Wagner and P.C. Taylor, *Solid State Commun.***15** (1974) 1871; U. Strom and P.C. Taylor, *Phys.Rev*. B **1977,** *16,* 5512.
(47) Lunkenheimer, P. *et al. Phys.Rev.Letters* **1996**, *77*
(48) K. Funke, *Prog.Solid St.Chem.* **1993,** *22*, 111.
(49) C. León, L. Lucia and J. Santamaria, *Phys.Rev.B*, in press.
(50) C.A. Angell, *Chem.Rev.***1990**, *90,*523; C.A. Angell, *Annu.Rev.Phys.Chem.* **1992,** *172,* 1.
(51) Ngai, K.L.; Liu, F.S., *Phys.Rev.B*, **1980**, *24*, 1049.
(52) Rendell, R.W.; Ngai, K.L. *J.Non-Cryst.Solids* **1991**,*131-133*, 667.
(53) Ngai K.L.; Skolnick, J. *Macromolecules* **1991**, *24*,1561.
(54) Ngai K.L. *Macromolecules* **1992,** *25,* 2184.
(55) Montroll, E.W.; Weiss, G.H., *J.Math.Phys.* **1965**, *6*, 167.
(56) Skolnick J., Kolinski A.,*Adv. Chem. Phys.* **1990,** *78*,223
(57) Pakula, T.and Geyler S., *Macromolecules* **1987,** *20,* 2909; **1988,** *21,* 1665.
(58) Kremer K.and Grest G., *J.Chem.Phys.* **1990,** *92,* 5057.
(59) Maass, P.; Petersen, J.; Bunde, A.; Dieterich, W.; Roman, H.E. *Phys.Rev.Lett.* **1991**, *66*, 52.
(60) Fuchs, M.; Hofacker, I.; Latz, A. *Phys.Rev.A.***1992**, *45*, 898.

Chapter 5

Frustration-Limited Domain Theory of Supercooled Liquids and the Glass Transition

Gilles Tarjus[1], Daniel Kivelson[2], and Steven Kivelson[3]

[1]Laboratoire de Physique Theorique des Liquides, Université Pierre et Marie Curie, 4 Place Jussieu, 75252 Paris Cedex 05, France
Departments of [2]Chemistry and Biochemistry, and [3]Physics, University of California, Los Angeles, CA 90095

We present a novel, and what we believe to be a reasonably successful, way of looking at supercooled liquids. It attributes the salient properties of these latter to the formation of supermolecular domains that, because of inherent structural frustration, grow only at a modest rate when the temperature is decreased. The theoretical approach is based on the postulated existence of a narrowly avoided critical point at a temperature T* and makes use of scaling about but below this T*. Application to the description of the structural (α) relaxations and to the explanation of various apparently anomalous properties of supercooled liquids is discussed.

Although below the melting point (T_m) the crystal is presumably the stable state, liquids can remain uncrystallized over times long compared to the experimentally accessible times and thus exist in a metastable supercooled state. Besides their increasing sluggishness with decreasing temperature (that leads to the dynamic arrest on experimental time scale known as the "glass transition"), supercooled liquids exhibit a rich phenomenology that departs in many ways from that of ordinary liquids above the melting point (*1,2*). There is no obviously unique way of selecting the salient features in the whole body of thermodynamic and relaxational properties of supercooled liquids. The emphasis one places on specific aspects is usually guided by an underlying view of what is to be explained about supercooled liquids and glass formation. This having been said, we take as particularly significant in the phenomenology of glass-forming liquids the following points (we shall not be concerned here with nonequilibrated systems, i.e. glasses):

1) *Strong temperature dependences.* The distinctive property of supercooled liquids is of course the stupendous increase of the shear viscosity and the structural (α) relaxation times with decreasing temperature T when approaching the glass transition temperature T_g. These transport coefficients can increase by 15 orders of magnitude over a T-range of perhaps 150 K (*1*). Not as spectacular (the entropy difference between the liquid and the crystal often exhibits a factor of 3 to 5 decrease

from T_m to T_g), but still intriguing, is the marked decrease of the entropy of the liquid as T is lowered towards T_g (*1*), a result that leads to the Kauzmann paradox (*3*) (the vanishing of the difference between the entropy obtained by extrapolation of the liquid curve and the entropy of the crystal at a temperature $T_K < T_g$).

2) *A high degree of universality.* Strong, superArrhenius variation with T of the α-relaxation times is observed for virtually all glass-formers, with the exception of a minority of so-called "strong" liquids; for a given substance, this behavior is seen for diverse probes, such as dielectric susceptibility, light and neutron scattering, shear modulus, specific heat, viscosity. This T-dependence has been described with reasonable success by empirical formulas such as the Vogel-Fulcher-Tamman expression which make use of a small number of species-dependent, but T-independent parameters (*1*). Also generic to glass-forming systems is the nonexponential character of the relaxation functions. As shown by Nagel and coworkers (*4,5*), the frequency-dependent susceptibility for α-relaxations associated with a variety of measurements on many different liquids (including the power-law high-frequency behavior of the α-relaxation also known as the von Schweidler relaxation) can be placed with good accuracy on a universal master-curve. All data at all temperatures, from close to T_m down to T_g, can be collapsed on this curve by using three T-dependent parameters.

3) *Apparently anomalous behavior.* Supercooled liquids display "anomalous" behavior relative to ordinary liquids above T_m. In addition to anomalies already mentioned (e.g., seemingly diverging α-relaxation times as well as the Kauzmann paradox and increasingly nonexponential relaxations with decreasing T), one can cite the bifurcation into α-relaxations and various faster β-relaxations with T-dependence significantly weaker and often Arrhenius-like (*6,7*), as well as the "decoupling" between translational and rotational diffusion (*8,9*). This latter effect may seem rather puzzling : whereas the rotational diffusion coefficient is quite well related to the viscosity by the Stokes-Einstein-Debye formula throughout the whole liquid and supercooled liquid range, strong positive deviations in the Stokes-Einstein relation between translational diffusion coefficient and shear viscosity occur when approaching T_g (*8,9*). This breakdown of the Stokes-Einstein relation goes well beyond the small deviations that are often detected in ordinary liquids, and it has been observed for several glass-formers.

We interpret all these features (strong T-dependences, large degree of universality, "anomalous" behavior) as a call for a nonmolecular or supermolecular, scaling theoretical approach to supercooled liquids. This idea is not new, and several attempts have already been made (*2*). While there is evidence supporting this approach, there are also severe constraints. The supermolecular character is supported by a number of recent dynamical experiments on supercooled liquids which provide evidence for the presence of long-lived heterogeneities. The heterogeneous nature of relaxational processes has been shown by 4-dimensional NMR (*10,11*), fluorescence bleaching (*12*), and dielectric hole-burning (*13*) experiments. However, the size of these heterogeneities have been estimated to be about 5 to 10 molecular diameters close to T_g (*12,14*), that is fairly modest. Many theories have postulated the existence of a continuous transition to an "ideal glass" that would take place at a temperature T_o below T_g if one were able to keep the liquid equilibrated (*2*); this is the only temperature at which one might expect a clear-cut divergence. Standard scaling about the critical point at T_0 would lead to power-law behavior with small exponents for structural and dynamical properties. However, power-law fits to the α-relaxation times and the viscosity can only be obtained with very large values of the exponents of the order of 12 (*15*). This *per se* does not forbid the existence of a low-T critical point, but it requires at least peculiar critical behavior that has been so far found only in systems with imposed quenched disorder (*16,17*) (which is not the case of glass-forming liquids). Other than the speculated divergence of the relaxation times (and related dynamic quantities) near T_0 below T_g, no diverging, or even rapidly growing, structural correlation lengths have been detected in supercooled liquids (2,*18,19*), which reduces the potential of critical theories for these systems.

We present in this article a new theoretical approach which is partly based on scaling and which we shall refer to as the frustration-limited domain (FLD) theory (*20*). However, rather than using scaling about but above a low-T critical point at a T_0 below T_g that represents a point of ultimate congestion for the liquid, as in the above mentioned theories, we consider scaling about but below a high-T "narrowly avoided" critical point at T*. We interpret T*, which is close to and usually slightly above T_m, as a crossover between "molecular" behavior characteristic of ordinary liquids and "collective" behavior that can be described in terms of supermolecular domains.

Physical picture underlying the FLD theory (*20*).

We envisage the molecules in a liquid above the melting point as organized in a locally preferred structure. For simple systems, this structure may be readily identified : for example, hexagons for disks on a plane, icosahedra for spheres in 3-dimensional space. We postulate that such a locally preferred structure, obtained by minimizing some appropriately defined local free energy, exists for all liquids. As the liquid is cooled, the local structure becomes more preferred and more extended. Were it possible, the system would undergo a continuous phase transition to an ideal crystal characterized by a macroscopic, periodic extension of the local structure. However, this does not happen because of inherent "structural (or geometric) frustration".

The mechanism of geometric frustration has been studied in detail for some simple systems. For disks on a plane, the hexagonal short-range order that dominates in the liquid can be easily extended to form a hexagonal close-packed (triangular) lattice and, if one is willing to overlook the subtleties associated with 2-dimensional ordering, the freezing can be described as continuous (*21*). As a consequence, there is no supercooled state and no glass formation for such systems. Quite different is the situation for identical spherical particles in 3-dimensional Euclidean space. As first shown by Frank (*22*), the local cluster which is most stable is an icosahedron. Icosahedral short-range order may be prominent in the liquid phase, but there is no icosahedral crystal of identical spheres since the fivefold rotational symmetry characteristic of icosahedra is not consistent with translational periodicity (*23*). Instead, the liquid freezes in a hcp or fcc crystal via a first-order phase transition involving a change in the local structure. The concept of geometric frustration has been put forward to describe this inability of a locally preferred structure to tile space periodically. For spheres, it has been realized that frustration could be removed in a curved space, namely on the surface of a 4-dimensional hypersphere, and this has led to the development of quite sophisticated theories of metallic glasses (*24*).

For good glass-formers, in which the molecules are not spherical, no such detailed description is available, and we shall use the concept of frustration without specifying the precise mechanism by which it operates for each system. Frustration leads to a build-up of strain that resists the extension of the locally preferred structure and completely forbids the formation of the ideal crystal. When the temperature is decreased, the free energy loss due to the ordering process competes with the free energy increase due to the strain. At some point (not necessarily sharply defined), a balance is reached and the system breaks up into ordered domains whose size and further growth with decreasing temperature are limited by the frustration. Hence the name, theory of frustration-limited domains.

The above picture is summarized in the schematic frustration-temperature diagram in Figure 1 (note that the part of phase space associated with the real crystal is excluded from the picture). Temperature is the physical control parameter. Frustration tells us how much strain the extension of the locally preferred structure produces. Zero frustration corresponds to the reference state in which the hypothetical liquid happily freezes to an ideal crystal via a continuous phase transition at a temperature T*. A real liquid is described by a given, nonzero value of the frustration. The critical point at T* is *isolated*, by which we mean that no other critical points are present in its neighborhood. The solid curve indicates a line of possibly first-order transitions to a defect-ordered phase, a phase in which the defects that must accompany the spatial extension of the locally preferred structure arrange themselves to form a periodic structure. This line intersects the zero-frustration axis at a T that is distinctly below T*. For nonzero frustration, the critical point at T* is *avoided* and the shaded area marks the crossover region between high-T liquids that have only short-range order and liquids characterized by supermolecular frustration-limited domains. We expect a scaling approach to be useful in a region of the diagram that lies at T's *below* T*, provided the liquid under consideration has weak enough frustration. For such a liquid we have tentatively placed the melting point T_m (i.e., the transition to the real crystal) slightly below T* although such crystallization is not incorporated into the theory. The corresponding T_g may be either above or below the temperature T_D of the transition to the defect-ordered phase (*25*).

Theoretical model.

Motivated by the physical picture just described, we have introduced a model that is intended to provide a *coarse-grained* description of supercooled liquids. It is a spin model in which a spin **S** represents an order variable associated not with an individual molecule but with the locally preferred liquid structure and takes on a set of discrete values which correspond to the possible "orientations" of the local structure. The simplest choice would be an Ising variable, but a clock model in which each spin can have more than two orientations might be more appropriate. The hamiltonian is the sum of two terms (*20*) :

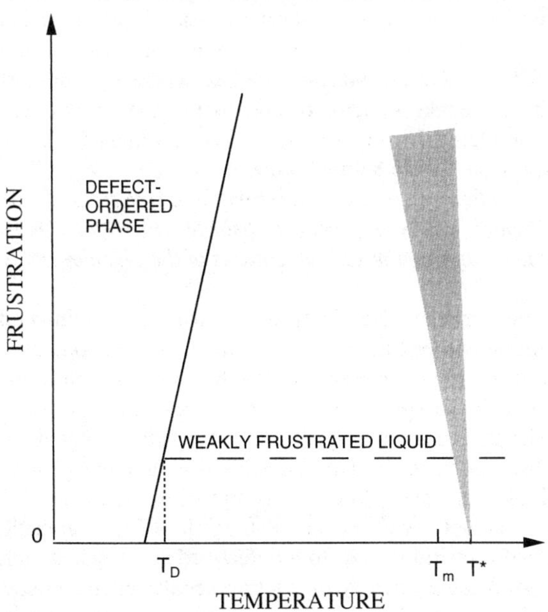

Figure 1. Schematic frustration-temperature diagram for glass-forming liquids. Details are given in the text.

$$H = - J \sum_{<ij>} \mathbf{S}_i \cdot \mathbf{S}_j + (Q/2) \sum_{i \neq j} \frac{\mathbf{S}_i \cdot \mathbf{S}_j}{|\mathbf{r}_i - \mathbf{r}_j|}, \qquad (1)$$

where J>0, Q>0, and $\mathbf{r}_i$ denotes the position of site i. The first term describes the reference system. Short-range ferromagnetic interactions between the spins lead to a continuous phase change at a temperature T* of order J. For simplicity, we consider nearest-neighbor interactions denoted by the symbol <ij>, but such short-range details are irrelevant because we only want to study collective, mesoscopic effects. Uniform frustration is described by the second term which introduces long-range competing (anti-ferromagnetic) interactions between the spins. Due to its Coulombic form, it leads to a superextensive growth of strain that opposes the extension of the locally preferred structure (i.e., ferromagnetic order) and macroscopic ferromagnetic ordering is forbidden for any nonzero value of Q. Since we are interested in a narrowly avoided critical point at T*, we always consider $Q \ll Ja_o$, where a_o is the lattice spacing that, although ill-defined, we take as being of the order of the intermolecular distance. The model can be completed by introducing Glauber dynamics or other short-range dynamics in order to study relaxational properties.

Our claim, so far only partially substantiated, is that this model, which does not include any artificially imposed quenched disorder, provides a minimum theoretical framework for understanding the salient features of the phenomenology of supercooled liquids.

Despite the apparent simplicity of the frustrated spin model described by equation 1, obtaining numerical, not to mention analytical, solutions still represents a formidable task. The only such problem that has been solved is the mean spherical model (*26*), in which the spins are taken to be real numbers subject to the global constraint that the thermally averaged norm of a spin be equal to 1. This is not a physically sensible model, but its solution exhibits a number of interesting features. Its phase diagram looks very much like that illustrated in Figure 1 with frustration being measured by the ratio of coupling constants, Q/Ja_o (*26*) : the critical point at T* is isolated, the line of transitions (T_D) to a defect-ordered phase which is characterized by some sort of modulated order (with supermolecular characteristic length) is found for 3-dimensional systems to intersect the zero-frustration (Q=0) axis at a temperature significantly below T*; for $Q \neq 0$, the critical point at T* is avoided, with no anomaly in the heat capacity, and, although the correlation length of the reference system remains infinite below T* (a peculiarity of the spherical models), there is a second supermolecular characteristic length that increases as T is lowered. These features remain true when 1/n corrections to the spherical model are considered (*27*). The presence of phases with modulated order has also been shown in zero-T analysis of 2-d (*28*) and 3-d (*29*) Ising versions of the Hamiltonian in equation 1. In addition to these exact results, Monte Carlo simulations of various spin models with long-range frustration have confirmed the existence of low-T transitions to defect-ordered phases with modulated order as well as the appearance at higher T, but below T*, of mesoscale domains (*29,30,31*). Despite incomplete information on the relaxational properties, these results are promising since they corroborate the scenario presented in the preceding section.

Phenomenological scaling approach

In the absence of exact solutions for realistic models, we have developed a phenomenological scaling description for the system described by equation 1. It is based on the postulate of an isolated, narrowly avoided critical point at T*, a property supported by the calculations discussed above. The critical behavior of the reference system, i.e., the first term in the Hamiltonian of equation 1, is characterized by a correlation length

$$\xi \propto |1-T/T^*|^{-\nu}, \qquad (2)$$

where, as for usual 3-d systems without quenched disorder, we take ν to be close to 2/3.

By considering the size at which the adiabatic approximation in the perturbation theory of finite-size systems breaks down, we have shown that a second supermolecular length R appears at temperatures sufficiently below T* in the presence of weak frustration ($Q \ll J\, a_o$) (*20*). This length scales according to

$$R \approx a_o^2 (Q/J\, a_o)^{-1/2}\, \xi^{-1} \propto Q^{-1/2}(1-T/T^*)^{\nu}. \qquad (3)$$

We interpret R(T) as the mean distance at temperature T over which a nonzero ferromagnetic order parameter can be specified and we thus take R as the mean linear size of the frustration-limited domains. Note that since this applies below T*, the correlation length ξ *decreases* whereas the mean domain size R *increases* when T is lowered.

Before proceeding further in the development of the scaling approach, one has first to address the basic question : why would frustration-limited domains give rise to a dramatic slowing down of the structural relaxation? The answer comes from standard arguments on finite-size systems below their critical temperature. For illustration, consider a simple Ising ferromagnet of volume L^3, where L is much larger than the lattice spacing. Below the critical point, the system has two low free-energy states characterized by nonzero order parameters of the same magnitude, but different orientation (up or down). Relaxation of the order parameter, i.e., the passage from one state to another, is possible via an activated process which proceeds by creation across the system of a domain wall that separates a region of mostly up spins from a region of mostly down spins. The corresponding activation free energy is proportional to $\sigma(T)L^2$, where $\sigma(T)$ is the surface tension which scales like ξ^{-2} for L large enough compared to the correlation length, ξ. Hence, the characteristic restructuring time $\tau(L,T)$, obtained from

$$T \ln[\tau(L,T)] \propto (L/\xi)^2, \qquad (4)$$

becomes longer when L increases and of course diverges in the limit $L \rightarrow +\infty$, thereby signaling a thermodynamic phase transition. Similar reasoning applies to the relaxation of the order parameter in a frustration-limited domain of size L, provided one takes into account the reduction in activation free energy that comes with the creation of a wall of defects through a domain (see below). Note that the time discussed above is associated with collective motions leading to full structural relaxation within a domain. Individual spins keep flipping on a microscopic time scale, so that partial mobility is always possible in a domain.

At temperatures below T* where the condition $a_o << \xi << R$ is fulfilled, one can represent the free-energy density of a frustration-limited domain of volume L^3 as

$$\psi = \sigma/L - \phi + sL^2, \tag{5}$$

which follows from simple considerations about nucleation in the presence of strain (*32,33*); $\sigma \propto \xi^{-2}$ is a domain surface tension (and the first term represents the domain surface free-energy density), s is a strain coefficient (and the third term represents the frustration-induced strain free-energy density), and $\phi \propto \xi^{-3}$ is the free-energy density of an unfrustrated system that has undergone a continuous transition. The coefficient ϕ is negative for T > T* and positive for T < T*. This free-energy density is discussed elsewhere (*32*). Although ψ is the free-energy density of a single domain in a liquid, we take it to be representative of the free-energy density of a domain in a system of domains; it follows, by minimization, that the mean domain size R is related to the surface tension and the strain coefficient according to

$$R \approx (\sigma/2s)^{1/3}. \tag{6}$$

The above equation indicates that the surface and the strain terms for a typical domain of size R have the same leading T-dependence. As a result, the characteristic activation free energy for restructuring of frustration-limited domains is given by (*20*)

$$\Delta E(T) \propto R(T)^2/\xi(T)^2 \propto Q^{-1}(1-T/T^*)^{8/3}, \qquad T < T^*, \tag{7}$$

where we have set $\nu=2/3$. The slow (α) relaxation dynamics thus have a characteristic time scale

$$\tau_\alpha(T) = \tau_o(T)\exp[\Delta E(T)/T], \tag{8}$$

where $\tau_o(T)$ is a characteristic time for molecular relaxation in the absence of domains. Thus, frustration leads to spontaneously forming domains whose restructuring takes place on an exponentially long and strongly T-dependent timescale, even though the mean domain size as given by equation 3 grows only modestly with decreasing T below T*.

Taking for the molecular contribution τ_o the simplest, well-behaved T-dependence, i.e., an Arrhenius dependence, $\tau_o(T) = \tau_\infty \exp(E_\infty/T)$, one can use equations 6 and 7 to fit α-relaxation time data on glass-forming liquids (analogous expressions apply to the viscosity). We have shown that with only four species-dependent, but T-independent parameters, experimental data on all kinds of glass-formers and for a large range of T can be collapsed with very good accuracy on a single mastercurve (*20,34*),

$$T \ln[\tau_\alpha/\tau_\infty] = E(T) = E_\infty, \qquad T > T^*,$$
$$= E_\infty + BT^*(1-T/T^*)^{8/3}, \qquad T < T^*, \qquad (9)$$

where τ_∞ (or η_∞ for viscosity data), E_∞, $B \propto 1/Q$, and T^* are the adjustable parameters. Note that when the exponent 4ν is also treated as an adjustable parameter, the best fit to the data is for 4ν close to 8/3 and reasonable fits are obtained with 4ν between 7/3 and 3. From these fits, we have estimated the mean domain size to be typically between 5 and 10 a_o at T_g. This estimate is however tentative since it involves several unknown numbers of order one. More details and discussions are given in references 20 and 34.

α-relaxation functions

Having described above how frustration-limited domains give rise to α-relaxation times with superArrhenius T-dependences, one can also easily understand how these domains induce nonexponential behavior in the α-relaxation functions. The liquid below T* is broken up into domains whose size distribution, although fairly peaked around the mean value R(T), is polydisperse. Since the activation free energy for the restructuring of a domain is size-dependent, as discussed in the preceding section, one finds a continuous distribution of structural relaxation times. This argument can be made more precise. In most experiments on dynamics in supercooled liquids, the measured quantities involve local variables (e.g., molecular dipoles in dielectric susceptibility measurements) which are expected to be correlated and to complete their relaxation within a single domain. Exceptions will be discussed in the next section. The slow part of the relaxation comes, via dynamic coupling to the local order variable, from the relaxation of the mesoscopic order parameter of the domain (dynamic coupling can be envisaged as coupling through a memory function) (*35*).

With the assumption that the molecular relaxation and the relaxation within a domain are exponential, a simple picture of independently relaxing domains yields then the following expression for the normalized α-relaxation function :

$$f_\alpha(t) = \int_0^\infty dL L^2 \rho(L) \exp\{(-t/\tau_\infty)\exp[-E(L)/T]\}, \qquad (10)$$

where τ_∞ is a T-independent, L-independent microscopic time, $\rho(L)$ is the distribution of domain sizes, and E(L) is the activation free energy for a domain of size L; as before, E(L) is taken as the sum of a "molecular" contribution (E_∞) and a "collective" (or "cooperative") contribution that appears only below T*.

The scaling approach presented above can be pursued to derive expressions for $\rho(L,T)$ and E(L,T). As detailed elsewhere (*35*), one obtains on the basis of equations 5 and 6 a scaling expression for the size distribution function, $\rho(L,T)=\rho(q,\gamma(T))$, where $q=L/R$, $\gamma(T)=T^{-1}\sigma R^2$, and $\rho(q,\gamma)$ is given by

$$\rho(q,\gamma) \propto \exp\{-\gamma[\kappa q^2-(3/2)q^3+(1/2)q^5]\}, \qquad T < T^*. \tag{11}$$

κ is a constant of order 1 that we have inserted to account in an *ad hoc* fashion for minor interdomain or shape effects. A scaling formula can also be derived for the activation free energy, $E(L,T)=E(q,\gamma(T))$ with $E(q,\gamma)$ given by

$$E(q,\gamma)/T = E_\infty/T + b\gamma[q^2-mq^5], \qquad T < T^*, \tag{12}$$

where b and m are numbers of order 1. To understand equation 12, note that E_∞/T is the molecular contribution, $b\gamma q^2$ is the free energy for wall creation appearing in equation 4, and $-b\gamma mq^5$ is the reduction in free energy due to the release of strain that accompanies wall creation. A quantitative comparison with experimental relaxation data thus requires 3 species-dependent, but T-independent parameters, κ, b, m, in addition to the 4 T-independent, species-dependent parameters already used to fit the characteristic α-relaxation times (namely, τ_∞, E_∞, and the 2 parameters that specify the T-dependence of γ).

The above expressions, equations 10 to 12, can be used to reproduce the main characteristics of the α-relaxation. As a preliminary investigation, we have used them to fit the frequency-dependent dielectric susceptibility $X''(\omega)$ of the fragile liquid Salol measured by Dixon et al (*4*). As shown elsewhere (*35*), one obtains a good agreement between theory and experiment over 12 decades of frequencies with the following set of parameters : $\kappa=0.86$, $b=1.46$, $m=0.22$. The three different frequency regimes, namely Debye-like behavior at frequencies significantly below the maximum of the α-peak, stretched exponential or Cole-Davidson behavior around the maximum, and von Schweidler (power law) at higher frequencies, are all described by the model. The predicted susceptibility curves for Salol at different temperatures can also be transformed to be compared to the masterplot of Nagel and coworkers (*4,5*). As illustrated in Figure 2, the scaled curves at different temperatures obtained by this procedure can be superimposed with good accuracy, which also provides support for the FLD theoretical framework.

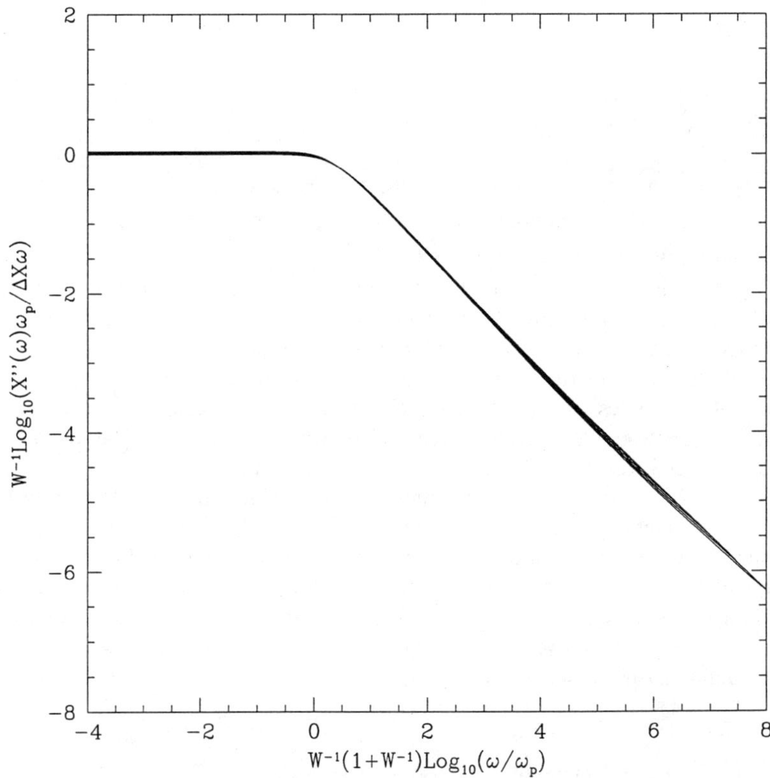

Figure 2. Masterplot of Nagel and coworkers (*4,5*) for frequency-dependent susceptibility data X"(ω). ω_p is the frequency at which the maximum of the α-peak occurs in the suceptibility X"(ω), the shape factor W is such that 1.14W is the full width at half-height of the X" vs $\log_{10}\omega$ spectrum, and ΔX is a normalization factor. The theoretical curves are obtained from the expressions and the values of the parameters that are given in the text and correspond to liquid Salol; 6 curves are incorporated, corresponding to W=1.56,1.61,1.65,1.67,1.70,1.74.

Decoupling between translational and rotational diffusion

The preceding discussion of α-relaxations applies when the probed motions are local enough to be completed within a single domain. Then, the α-relaxation time is simply given by (*35*)

$$\tau_\alpha = \int_0^\infty dt\, f_\alpha(t) = \tau_\infty \int_0^\infty dL L^2\, \rho(L) \exp[E(L)/T], \qquad (13)$$

where we have used equation 10. Note that, as required by consistency, the above equation leads back to equations 8 and 9 when the scaling forms for ρ(L) and E(L) are introduced, provided γ(T) is not too small (*36*).

If the relaxation is not completed within a single domain, as for long wavelength concentration fluctuations (*8,9*), an averaging over domains different than that in equations 10 and 13 must be considered (an intermediate case occurs when the reorientational motions of large solute molecules is studied (*14*); then, the size of the solute molecule is less than, but comparable to the mean domain size). For instance, translational motion over a scattering wavelength much larger than the mean domain size, as in forced Rayleigh (*8*) and holographic fluorescence recovery (*9*) experiments, involves passage through many different domains, which give rise to a "motionally narrowed" relaxation function that is *exponential* (*37*). In each domain, translational diffusion, just like rotational diffusion or viscous flow, is controlled by the relaxation of the mesoscopic order parameter and thus involves an activation free energy that is domain size dependent. A simple model of random walk in a random environment formed by independently relaxing domains then leads to the following expression for the translational diffusion constant (*37*) :

$$D_T = \langle D(\mathbf{r})\rangle - (1/6) \int_0^\infty dt \langle \nabla D(\mathbf{r}) . \nabla D(\mathbf{r}(t))\rangle, \qquad (14)$$

where D(**r**) is a spatially varying diffusion coefficient. In the limiting case where the domains have sharp boundaries, the above expression reduces to

$$D_T \approx \langle D\rangle \approx D_\infty \int_0^\infty dL L^2\, \rho(L) \exp[-E(L)/T], \qquad (15)$$

where D_∞ is a T-independent microscopic diffusion constant. Above T*, $E(L)=E_\infty$ and equations 13 and 15 give $D_T\tau_\alpha$ = constant, as predicted by the Stokes-Einstein-Debye formulas. When domains are present, i.e. at temperatures below T*, the averaging in equation 15 is quite different than that in equation 13 : it is more sensitive to the small domains (with, therefore, smaller activation energies) which, in our picture, are also responsible for the von Schweidler relaxation. As a consequence, translational diffusion is "enhanced" when compared to rotational diffusion or viscosity. This induces a breakdown of the Stokes-Einstein relation at temperatures sufficiently below T* (*37*), as observed experimentally. In fact, the increase of $D_T\tau_\alpha$ predicted by equations 13 and 15 when T approaches T_g is much larger than that observed, presumably because the requirement of sharp domain boundaries needed to derive equation 15 from equation 14 is not met (*35*).

Conclusion.

In summary, the FLD theory envisages the high-T liquid as a molecular fluid which may not be understood in detail, but which can be described approximately by exponential relaxation functions with relaxation times that exhibit Arrhenius-like behavior. Below a crossover temperature T* that we associate with an avoided critical point, the liquid forms supermolecular frustration-limited domains with a distribution of sizes, each domain being specified by a nonzero mesoscopic order parameter. As a result, the α-relaxation becomes heterogeneous and strongly nonexponential with relaxation times that show strong superArrhenius behavior. The FLD theory is based on a well-defined statistical-mechanical model for which we have proposed a phenomenological scaling description. As discussed in this article, this allows us to reproduce the main features of the α-relaxations in glass-forming liquids, including temperature and frequency dependences, at a quantitative level with a limited number of species-dependent, but T-independent adjustable parameters.

Although not yet worked out in all details, the theory also provides a consistent framework to explain most of the apparent anomalies observed in supercooled liquids. We have already discussed the decoupling between translational and rotational diffusion. It is tempting to interpret the bifurcation between α- and β-relaxations as a consequence of the existence of two *supermolecular* characteristic lengths below T* : whereas α-relaxations occur on a lengthscale corresponding to the mean domain size R, additional faster relaxations presumably take place on the smaller lengthscale associated with the correlation length ξ of the reference system. Although we have not yet studied these shorter-range processes, it seems natural to take them as the fast β-relaxations. Likewise, the rapid decrease of the entropy difference between the supercooled liquid and the crystal can be understood as resulting from formation and growth of frustration-limited domains. Elsewhere (*35*), we show that the scaling description, when extrapolated to low temperatures, leads to an entropy crisis as described by Kauzmann, but that this crisis is avoided in a more exact description because of the breakdown far from T* of the scaling formulas and the interposition of a transition to a defect-ordered phase. Such a transition has been described in the above sections, and it may indeed have been observed experimentally (*25,38*).

Clearly, additional work is needed in order to obtain a complete (at least numerical) solution of a realistic frustrated spin model and thus to put the scaling approach on firmer ground. A fully satisfying description of supercooled liquids and glass formation would also require a molecular interpretation of the order variable associated with the locally preferred liquid structure. However, even at the present stage, the physical picture and the comparison to experiment drawn from the FLD theory are quite encouraging.

We wish to thank the C.N.R.S., the N.S.F., the Research Corporation, and NATO for their support. We are indebted to Pascal Viot for many useful discussions.

References

1. C. A. Angell, J. Non-Cryst. Solids, **131-133**, 13 (1991).
2. For a recent review, see e.g. M. D. Ediger, C. A. Angell, and S. R. Nagel, J. Phys. Chem. (1996).
3. W. Kauzmann, Chem. Rev. **43**, 219 (1948).
4. P.K. Dixon, L. Wu, S.R. Nagel, B.D. Williams and J.P. Carini, Phys. Rev. Lett. **65**, 1108 (1990).
5. L. Wu, P. K. Dixon, S. R. Nagel, B. D. Williams, and J. P. Carini, J. Non-Cryst. Solids, **131-133**, 32 (1991). S. R. Nagel, in *"Phase Transitions and Relaxation in Systems with Competing Energy Scales"*, Ed. T. Riste and D. Sherrington (Kluwer Academic, Netherlands, 1993), p. 259.
6. G. P. Johari and M. Goldstein, J. Chem. Phys. **53**, 2372 (1970). G. P. Johari, J. Chem. Phys. **58**, 1766 (1973); Ann NY Acad. Sci. **279**, 117 (1976).
7. E. Rössler, Phys. Rev. Lett. **65**, 1595 (1990). E. Rössler, U. Warschewske, P. Eiermann, A. P. Sokolov, D. Quitmann, J. Non-Cryst. Solids, **172-174**, 113 (1994).
8. F. Fujara, B. Geil, H. Sillescu and G. Fleischer, Z. Phys. B-Condensed Matter **88**, 195 (1992). M. Lohfink and H. Sillescu, in AIP Conf. Proc. No. **256**, Ed. K. Kawasaki, M. Tokuyama, and T. Kawakatsu (American Institute of Physics, New York, 1992), p. 30. I. Chang, F. Fujara, G. Heuberger, T. Mangel, and H. Sillescu, J. Non-Cryst. Solids **172-174**, 248 (1994).
9. M. T. Cicerone and M. D. Ediger, J. Chem. Phys. **104**, 7210 (1996).
10. K. Schmidt-Rohr and H.W. Spiess, Phys. Rev. Lett. **66**, 3020 (1991). A. Heuer, M. Wilhelm, H. Zimmermann, and H. W. Spiess, Phys. Rev. Lett.**75**, 2851 (1995).
11. G. Dietzemann, R. Böhmer, G. Hinze, and H. Sillescu, Abstract #**133**, Div. of Phys. Chem., 212 ACS National Meeting (1996). R. Bohmer. G. Hinze, G. Diezemann, B. Geil, and H. Sillescu, Europhys. Lett. **36**, 55 (1996).
12. M. T. Cicerone and M. D. Ediger, J. Chem. Phys. **103**, 5684 (1995).
13. B. Schiener, R. Böhmer, A Löidl, and R. V. Chamberlin, Science **274**, 752 (1996).
14. M. T. Cicerone, F. R. Blackburn, and M. D. Ediger, J. Chem. Phys. **102**, 471 (1995).
15. S. S. N. Murthy, J. Phys. Chem. **93**, 3347 (1989).

16. T. R. Kirkpatrick, D. Thirumalai, and P. G. Wolynes, Phys. Rev. A **40**, 1045 (1989).
17. J. P. Sethna, J. D. Shore, and M. Huang, Phys. Rev. B **44**, 4943 (1991).
18. R. M. Ernst, S. R. Nagel, and G. S. Grest, Phys. Rev. B **43**, 8070 (1991).
19. R. L. Leheny, N. Menon, S. R. Nagel, D. L. Price, K. Suzuya, and P. Thiyagarjan, J. Chem. Phys. **105**, 7783 (1996).
20. D. Kivelson, S. A. Kivelson, X.-L. Zhao, Z. Nussinov, and G. Tarjus, Physica A **219**, 27 (1995).
21. K. J. Standburg, Rev. Mod. Phys. **60**, 161 (1988).
22. F.C. Frank, Proc. Royal Soc. London **215A**, 43 (1952).
23. D. R. Nelson and F. Spaepen, Solid State Phys. **42**, 1 (1989).
24. M. Kleman and J. F. Sadoc, J. Phys. Lett. (Paris) **40**, L569 (1979). J. P. Sethna, Phys. Rev. Lett. **51**, 2198 (1983). S. Sachdev and D. R. Nelson, Phys. Rev. Lett. **53**, 1947 (1984); Phys. Rev. B **32**, 1480 (1985).
25. D. Kivelson, M. Lee, J.-C. Pereda, K. Luu, H. Sakai, A. Ha, I. Cohen, and G. Tarjus, these Proceedings.
26. L. Chayes, V. J. Emery, S.A. Kivelson, Z. Nussinov, and G. Tarjus, Physica A **225**, 129 (1996).
27. Z. Nussinov, private communication.
28. U. Low, V. J. Emery, K. Fabricus, and S. A. Kivelson, Phys. Rev. Lett. **72**, 1918 (1994).
29. P. Viot, private communication.
30. D. Wu, D. Chandler, and B. Smit, J. Phys. Chem. **96**, 4077 (1992).
31. I. Booth, A. B. MacIsaac, J. P. Whitehead, and K. De'Bell, Phys. Rev. Lett. **75**, 950 (1995).
32. S. A. Kivelson, X-L Zhao, D. Kivelson, C. M. Knobler and T. Fischer, J. Chem. Phys. **101**, 2391 (1994). D. Kivelson, S. A. Kivelson, X-L Zhao, C. M. Knobler and T. Fischer, in "*Lectures on Thermodynamics and Statistical Mechanics*" Ed. M. Costas, R. Rodrigues, and A. L. Benavides (World Scientific, London, 1994), p. 32.
33. D. Kivelson, G. Tarjus, and S. A. Kivelson, Prog. Theoret. Physics (Japan), to appear.
34. D. Kivelson, G. Tarjus, X-L Zhao, and S. A. Kivelson, Phys. Rev. E **53**, 751 (1996).
35. G. Tarjus and D. Kivelson, in preparation.
36. Experimentally, there may be different ways to determine the characteristic α-time for a given motion; see for instance the discussion of retardation vs relaxation times in I. Chang and H. Sillescu, preprint (1996). This may require slight modifications of equation 13 and may induce differences in the T-dependences of the various times. However, these differences seem to remain typically less than an order of magnitude around T_g and will not be considered here.
37. G. Tarjus and D. Kivelson, J. Chem. Phys. **103**, 3071 (1995).
38. I. Cohen, A. Ha, X-L. Zhao, M. Lee, T. Fischer, M. J. Strouse, and D. Kivelson, J. Phys. Chem. **100**, 8518 (1996).

Chapter 6

Soft Modes in Glass-Forming Liquids: The Role of Local Stress

U. Zürcher and T. Keyes

Department of Chemistry, Boston University, Boston, MA 02215

Liquids and glasses have localized low-frequency vibrational modes associated with disorder. These modes represent relaxational motion in double wells, and quasi-harmonic motions in single wells. Single- and double-well potentials are described by the soft-potential model, which is an extension of the two-level-system model for glasses. We use soft modes to derive the unstable frequency spectrum of instantaneous normal modes in liquids. In agreement with recent molecular-dynamics simulations, we find different frequency and temperature dependence of the spectrum for liquids in the normal and supercooled phase. We relate this crossover behavior to the presence of shear stress in the liquid. Assuming that the diffusion of particles requires hopping over potential barriers, we find exponential temperature dependence of the shear viscosity. Arrhenius and Zwanzig-Bässler behavior follows for liquids in the normal and supercooled phase, respectively. We discuss properties of the energy landscape in glass-forming liquids. Possible applications to protein dynamics are mentioned.

Introduction and Physical Motivation

Dynamic and thermal properties of systems in condensed phases are governed by low-lying collective excitations. Perhaps the best known example of such modes are the phonons in crystal lattices, where the Hamiltonian is expanded to second order in the deviations of the particles' coordinates from their (mechanical) equilibrium values. Phonons describe oscillatory motions throughout the entire system and have a Debye-like spectrum of long-wavelength modes, i.e., $g(\omega) \sim \omega^{d-1}$ for $\omega \to 0$, where d is the (spatial) dimensionality. The oscillatory nature of these excitations is a consequence of the rigidity of the crystal *(1)*. In contrast, systems in the liquid state are fluid and have no elastic resistance to shear stress. Despite this qualitative difference between solids and liquids, Maxwell proposed that atoms in a liquid vibrate around equilibrium positions *(2)*. This apparent contradiction was later reconciled by Frenkel who pointed out that equilibrium positions in a liquid have only temporary character *(3)*. More recently, Zwanzig has shown that a relation between self-diffusion and the viscosity of liquids (Stokes-Einstein relation) emerges from this approach to liquid dynamics *(4)*.

Upon cooling from the melt, the many-body system is typically not in the unique crystalline state but, is rather in one of the many local minima of the potential energy representing supercooled or glassy states *(5)*. In these states, long-range order of the positions of particles is destroyed and the specific heat has contributions from low-frequency modes that can be identified with two-level-systems (TLS) *(6)*. Stillinger and Weber studied rigid aperiodic structures in glasses and showed how group of atoms create local bistability and how they move from one equilibrium position to another along a collective coordinate *(7)*. More recently. Heuer and Silbey have developed a qualitative method for finding TLSs in computer simulations, and they proposed a universal theory of low-temperature properties of structural glasses *(8)*. Cotterill and Madsen pointed out that the potential profile along eigendirections of the dynamic matrix can be used to further characterize the modes *(9)*.

In the supercooled and normal liquid phase, the eigenmodes of the dynamic matrix can be identified with "instantaneous normal modes" *(10)*, *(11)*. In molecular dynamics simulations, a random configuration of atomic positions, $\vec{R}_0$, is chosen from the trajectory, and the potential energy $\Phi(\vec{R})$ is expanded in a Taylor series,

$$\Phi(\vec{R}) = \Phi(\vec{R}_0) - \vec{F} \cdot \left(\vec{R} - \vec{R}_0\right) + \frac{1}{2}\left(\vec{R} - \vec{R}_0\right) \cdot \mathbf{K} \cdot \left(\vec{R} - \vec{R}_0\right) + ..., \qquad (1)$$

where $\vec{F} = -\nabla\Phi(\vec{R}_0)$ is the force and $\mathbf{K} = \nabla\nabla\Phi(\vec{R}_0)$ is the dynamic matrix. Diagonalizing the dynamic matrix yields a set of $3N$ eigenfrequencies ω_i and eigenvectors $\vec{e}_i$ $(i = 1, 2, ..., 3N)$ for each configuration $\vec{R}_0$. The INM density of states is then given by

$$g(\omega) = \left\langle \frac{1}{3N} \sum_{i=1}^{3N} \delta(\omega - \omega_i) \right\rangle . \qquad (2)$$

The INM spectrum contains both real ($\omega^2 > 0$) and imaginary frequency modes ($\omega^2 < 0$), and has bimodal character due to the linear behavior around $\omega = 0$, $g(\omega) \sim |\omega|$ (using the standard convention that the imaginary lobe is plotted along the negative frequency axis). In liquids and glasses, this linear behavior is found for intermediate frequencies $0.2\text{THz} < \omega < 1.0\text{THz}$, $g(\omega) \sim \omega^{\delta-1}$, where $\delta \simeq 2$ is the spectral dimensionality *(12)*. The "fracton" model of amorphous material relates the spectral dimensionality to the scaling behavior of the conductance with length, $\zeta(L) \sim L^{\beta}$, where $\beta \propto [1 - 2/\delta]$ *(13)*. Since $\beta < 0$ for

$\delta < 2$, the linear behavior of the DOS is consistent with localization of low-frequency vibrational modes. It was shown in Refs. *(14)* and *(15)* that similar to the situation in glasses, the potential energy surface along eigendirections of low-frequency modes are either single- or double-well potentials.

The *soft-potential model* describes both the tunneling and the soft vibrational modes in a glass *(16)*. The soft-potential model adds two parameters to the standard tunneling model. One parameter is the frequency of the lowest maximum in the vibrational density of states and is directly accessible to experiments. The second parameter is the product of the effective mass and the square of the atomic displacements, at which point the anharmonic part of the potential is dominant. This second parameter thus describes vibrational localization and anharmonicity in the glass. The number of particles participating in a localized mode has been estimated to be 10-100.

The Green-Kubo formalism relates transport quantities to (auto-) correlation functions of the response of the system to external forces. The total stress in the system $\sigma^{\alpha\beta}$, where α and β are the Cartesian coordinates, is the response of the system to an infinitesimal strain field $\epsilon^{\alpha\beta}$. The shear viscosity η is given as the time integral of the correlation function *(17)*,

$$\eta = \frac{1}{VT}\int_0^\infty \left\langle \sigma^{\alpha\beta}(0)\sigma^{\alpha\beta}(t)\right\rangle dt, \quad (\alpha \neq \beta), \tag{3}$$

where $\langle\,\rangle$ indicates the thermal average, $\sigma^{\alpha\beta}(t)$ is the total shear stress at time t, V is the volume, and T is the temperature (in units such that $k_B = 1$). Local shear stresses in a monatomic liquid have been investigated in Ref. *(18)*. The authors find that the increase in viscosity leads to the development of spatial correlations of local shear stress. In this view, the percolation of correlated regions then leads to the glass transition. Long-range stress fields are interrupted by thermal motions of the particles in the liquid and then completely disappear at some temperature above the glass temperature, $T_c > T_g$. Below T_c, the supercooled liquid supports long-range stress fields.

Physical properties of glass-forming liquids are not easily derived from theories that are based on spatial characteristics of particles. Ziman pointed out that quantities such as the radial distribution function cannot quantitatively describe rare rearrangements which involve a few atoms interacting through interatomic forces *(19)*. In random, close packed structures, unstable local configurations are produced under the influence of shear stress, and the fluidity is associated with the motion along such local coordinates.

Goldstein *(20)* and later Stillinger *(5)* have argued that topographic considerations offer an alternative view of viscous and dynamic properties of glass-forming liquids. In the normal and moderately supercooled regime, the susceptibility spectrum has a single absorption peak. In the supercooled phase, this peak splits into a pair of maxima which correspond to slow α- and fast β-processes *(21)*. Fast processes have Arrhenius temperature dependence and are operative at the glass temperature. Slow processes have non-Arrhenius temperature dependence and are associated with structural changes in the liquid that are frozen-out at T_g.

That is, β-processes correspond to transitions between local minima ("basins"), that are in turn organized in deeper potential energy wells ("craters"). At high temperatures, particles explore regions of the configuration space with uniformly rough topography, while at lower temperatures, they surmount larger and wider potential energy barriers.

In recent papers *(22)* - *(24)*, we have followed up on the suggestion by Buchenau *(25)* and identified low-frequency modes in liquids with soft localized modes. Here, we mainly summarize our results, and emphasize where the shear stress enters in determining the liquid properties. Our findings support the view that the dynamics in liquids becomes more solid-like at some temperature above the glass temperature, $T_c > T_g$. In particular, we relate our results to recent ideas on the origin of the glass transition *(26)*.

Soft Modes

The soft-potential model describes the potential energy surface along a local coordinate,

$$V_0(x) = W\left[-D_2x^2 + x^4\right], \tag{4}$$

where x is a reduced coordinate, $[x] = 1$, and the restoring force constant is dimensionless as well, $[D_2] = 1$. It follows that the square of the frequency, $\omega^2 = d^2V/dx^2$, has the dimension of an energy, $[\omega^2] = [W]$. From Eq. (4), we readily find the density of states at zero temperature, $g(\omega) = 2\omega p_2(\omega^2/4W)$, where $p_2(D_2)$ is probability distribution of the restoring force. It follows that a linear frequency dependence of the DOS implies a constant distribution,

$$p_2(D_2) = p_2^0 = \text{const}, \;\; 0 < D_2 < \frac{\Omega^2}{4W}. \tag{5}$$

In Ref. *(27)* configurational modes have been shown to explain the additional specific heat of supercooled liquids. These configurational modes are identified as local stress and are described by a linear term in the soft potential,

$$V(x) = W\left[D_1x - D_2x^2 + x^4\right]. \tag{6}$$

Assuming a Boltzmann distribution of the coordinate x, $\exp(-V(x)/T)$ (in units such that $k_B = 1$), the mean value $\langle x\rangle$ depends on the external field, $\langle x\rangle = \langle x\rangle(D_1)$. We do not include the external field in our definition of the internal energy of the system. Thus, the change in $\delta V = WD_1\langle x\rangle$ follows $d\delta V = WD_1 d\langle x\rangle$. Since, $d\langle x\rangle = (\partial\langle x\rangle/\partial D_1)dD_1$, the internal energy follows by integration,

$$\delta V = W\int D_1\frac{\partial\langle x\rangle}{\partial D_1}dD_1. \tag{7}$$

We find $\partial \langle x \rangle / \partial D_1 = -(W/T)[\langle x^2 \rangle_0 - \langle x \rangle_0^2]$, where the averages are evaluated with $D_1 = 0$ and $D_2 = 0$. Since $\langle x^2 \rangle_0 = 0.338(T/W)^{1/2}$ and $\langle x \rangle_0 = 0$, Eq. (7) gives the internal energy associated with a small stress in the liquid,

$$\delta V = -0.169 \frac{W^{3/2}}{T^{1/2}} D_1^2. \tag{8}$$

The energy required to generate a small stress follows $\delta W = -\delta V$. Buchenau and coworkers argue *(28)* that thermal stress is frozen in the liquid at the glass temperature T_g and propose that the Boltzmann factor $\exp(-\delta W/T_g)$ gives the distribution of the linear coefficient D_1,

$$p_1(D_1) = 0.231 \left(\frac{T_g}{W}\right)^{3/4} \exp\left(-0.169 \left(\frac{W}{T_g}\right)^{3/2} D_1^2\right). \tag{9}$$

That is, D_1 is a Gaussian random variable with variance $\langle D_1^2 \rangle = 3.96(T_g/W)^{3/2}$. For temperatures $T >> T_g$, the liquid is capable of rearranging its atomic configurations such that thermal stress vanishes. In the limit $T_g \to 0$, thermal stress vanishes at any non-zero temperature,

$$p_1(D_1) = \delta(D_1), \quad T_g \to 0. \tag{10}$$

Imaginary Modes Density of States

The density of states is defined as

$$g(\omega) = 2\omega\, G(\omega^2), \tag{11}$$

where $G(\omega^2)$ is the density of the square of frequencies,

$$G(\omega^2) = \left\langle \delta\left(\frac{d^2V}{dx^2} - \omega^2\right)\right\rangle. \tag{12}$$

Here, the coordinate x is weighed by the Boltzmann factor $\exp(-V(x)/T)$, and the average is taken with respect to both x and the parameters of the soft potential D_1 and D_2. At non-zero temperatures, the spectrum contains both stable ($\omega^2 > 0$) as well as unstable modes ($\omega^2 < 0$). The fraction of unstable modes increases with increasing temperatures.

A temperature scale enters via the variance of the local stress, $\langle D_1^2 \rangle = 3.96\,(T_g/W)^{3/2}$. In this paper, we only consider temperatures above the glass temperature $T > T_g$. Introducing scaled coefficients and coordinates,

$$D_1 = \left(\frac{T_g}{W}\right)^{3/4} \widetilde{D}_1, \tag{13}$$

$$D_2 = \left(\frac{T_g}{W}\right)^{1/2} \widetilde{D}_2, \tag{14}$$

$$x = \left(\frac{T_g}{W}\right)^{1/4} \widetilde{x}, \tag{15}$$

the ratio $V(x)/T$ entering the Boltzmann factor then follows

$$\frac{V(x)}{T} = \frac{T_g}{T}\left[\widetilde{D}_1\widetilde{x} - \widetilde{D}_2\widetilde{x}^2 + \widetilde{x}^4\right], \tag{16}$$

so that the temperature enters the density of states via the combination T_g/T only. In particular, the Boltzmann factor is independent of the energy scale W. In Ref. *(22)* we have taken the limit $W \to 0$ and have then considered the limits $T \to 0$ and $T_g \to 0$, such that the ratio T/T_g is constant. We have shown that in this limit, the density of states is uniquely defined. Furthermore, we have calculated the density of states for a constant temperature T and then studied two limiting cases by varying the glass temperature T_g. For $T/T_g = \mathcal{O}(1)$ and $T/T_g \to \infty$, the low- and high-temperature limits follow, respectively. In the latter case $T_g = 0$, and the local stress vanishes, $p(D_1) = \delta(D_1)$. That is, the soft-potential model describes the dependence of liquid properties on stress fields that has been found by Chen et. al. in a molecular dynamics simulation *(18)*.

Unstable modes have imaginary frequencies, $\omega = i\nu$. From Eq. (12), the density of states follows $G_u(\nu^2) = \langle\delta(d^2V/dx^2 + \nu^2)\rangle$. Unstable modes originate from both single- and double-well potentials *(15)*. We find the potential energy associated with soft modes and thermal stress by setting $D_1 = 0$ and $D_2 = 0$, respectively, $V_{soft} \sim D_2^2$ and $V_{stress} \sim D_1^{4/3}$. We put $V_{soft} = V_{stress}$ and find an inequality characterizing double wells, $D_2^3 > 3.375 D_1^2$. We replace D_1^2 by its average $\langle D_1^2\rangle = 3.96\,(T_g/W)^{3/2}$. Since $\nu^2 = -d^2V/dx^2 = W[-2D_2 + 12x^2]$, the coefficient is bound from below, $D_2 > \nu^2/2W$. We conclude that for large frequencies, $\nu > \nu_c$, the density of states is dominated by contributions from double wells, while for small frequencies, $\nu < \nu_c$, it is dominated by contributions from single wells. This frequency cutoff ν_c depends on the product of the glass temperature T_g and the energy scale W,

$$\nu_c = 3W\left\langle D_1^2\right\rangle^{1/3} = 2.18\,(T_g W)^{1/4}. \tag{17}$$

In the limit $T_g \to 0$, the thermal stress vanishes and the cutoff is arbitrarily small, $\nu_c \to 0$, so that all unstable modes originate from double-well potentials.

We have calculated the unstable density of states *(22)*, *(29)*,

$$G_u(\nu^2) = G_u(0,T)\,G_u'(\nu^2,T), \tag{18}$$

where

$$G_u(0) = \exp\left(-\frac{8}{59}\frac{E_\beta}{T} - \frac{1}{5}\frac{E_\alpha E_\beta}{T^2}\right). \tag{19}$$

Here, we have introduced an energy associated with thermal stress,

$$E_\alpha = 16\frac{T_g^{3/2}}{\Omega^2}, \tag{20}$$

and an energy of soft vibrational modes,

$$E_\beta = \frac{\Omega^4}{64}. \tag{21}$$

In the high-temperature limit, thermal stress is negligibly small and the soft potential describes symmetric double wells,

$$G'_u(\nu, T) = \exp\left(-E_0\left(\frac{\nu}{\nu_0 T}\right)^2\right), \quad T >> T_g. \tag{22}$$

Recalling that thermal stress is represented by the linear term in the soft-potential model, contributions from highly asymmetric double-well potentials dominate at low temperatures,

$$G'_u(\nu, T) = \exp\left(-E_0\left(\frac{\nu}{\nu_0 T}\right)^4\right), \quad T \gtrsim T_g. \tag{23}$$

Here, the constant E_0 is the ratio of the energy scales E_α and E_β,

$$E_0 = \frac{2E_\beta}{5E_\alpha}, \tag{24}$$

and the frequency scale ν_0 is given by

$$\nu_0 = \frac{1}{\sqrt{18}}\frac{\Omega^2}{T_g^{3/4}}. \tag{25}$$

In molecular dynamics simulations, the above frequency and temperature dependence has been found by Keyes for liquids in the supercooled phase, $-\log G'_u(\nu, T) \propto \nu^4/T^2$ *(30)* and by Vijayadamodar and Nitzan for liquids in the normal phase, $-\log G'_u(\nu, T) \propto \nu^2/T$ *(31)*.

At intermediate temperatures, the unstable density of states interpolates smoothly between the two limiting forms. We find that a linear superposition gives an excellent fit without introducing an additional parameter *(24)*, *(32)*,

$$G'_u(\nu, T) = R(T)\exp\left(-E_0\left(\frac{\nu}{\nu_0 T}\right)^4\right) + [1 - R(T)]\exp\left(-E_0\left(\frac{\nu}{\nu_0 T}\right)^2\right). \tag{26}$$

Here, the fraction of configuration space with large stress, $R(T)$, has the temperature dependence of a thermally activated process,

$$R(T) = \tanh\left(\frac{E_0}{T}\right). \tag{27}$$

Eqs. (26) and (27) suggests spatial inhomogeneity as the shear stress does not grow uniformly across the entire system. Rather, at temperatures well above T_g, stress is large in small regions of the liquid. These regions grow as the temperature decreases, perhaps via some percolation process as Chen et. al. have suggested *(18)*. Near the glass temperature, most particles in the liquid are in regions with large shear stress.

Viscous Properties: Self-Diffusion Constant

Viscous properties describe dynamical properties of liquids on *long* timescales. On the other hand, instantaneous normal modes are defined via the *short* time expansion of the equations of motion of particles in the liquid. The topographic view of liquids *(5)* suggests a connection between long- and short-time liquid properties. The self-diffusion coefficient (or the inverse of the shear viscosity) is approximately equal to the ratio of the time spent in a valley τ_v to the time spent in crossing a barrier, τ_b, $D \simeq T^{1/2}\tau_b/\tau_v$. Since unstable modes are the signature of barrier crossings, it follows that any strong temperature dependence of D follows from the relation $D \sim f_u(T)$, where $f_u(T)$ is the fraction of unstable modes at temperature T. As we have seen, unstable modes originate from both single- and double-well potentials. Because hopping over barriers leads to particle transport, unstable modes from double wells should only contribute to diffusion. Since the frequency ν_c separates contributions from single and double wells, we have the estimate

$$D \sim G_u(\nu_c). \tag{28}$$

Inserting Eq. (22) into Eq. (28), we find Arrhenius behavior for high temperatures,

$$D \sim \exp\left(-\frac{1}{36}\frac{\Omega^2 T_g^{1/2}}{T}\right), \quad T >> T_g, \tag{29}$$

while for low temperatures above the glass temperature, Zwanzig-Bässler *(33)* behavior follows from Eq. (23),

$$D \sim \exp\left(-2\frac{T_g^{5/2}}{\Omega^2 T^2}\right), \quad T \gtrsim T_g. \tag{30}$$

Because the self-diffusion constant and the shear viscosity are related to each other via the Stokes-Einstein relation, $D\eta/T = \text{const}$, Eqs. (29) and (30) describe an increase of the viscosity by several orders of magnitude as the temperature descreases towards the glass temperature. Angell proposed a classification of glass formation based on this behavior *(34)*. Strong liquids show Arrhenius behavior, $\eta \propto \exp(\epsilon/T)$, over the entire supercooled temperature range, while for fragile liquids, the exponential temperature dependence cannot be described by a single, temperature-independent activation energy, i.e., $\epsilon = \epsilon(T)$. Typical examples of strong liquids are open network systems such as SiO_2 and GeO_2, while fragile liquids such as *o*-terphenyl (OTP) have non-directional Coulomb and van der Waals interactions. In Ref. *(18)*, long-range correlation of stress has been investigated only for a system with non-directional interactions. From our results, we expect that a similar investigation for systems with partially directional interactions would show a much weaker long-range correlation of shear stress.

Properties of the Energy Landscape

In the topographic view of glass formation *(5)* the distribution of barrier heights is of particular interest. For each pair of parameters (D_1, D_2) and frequency ν, the soft potential defines a unique barrier,

$$\Delta V = \Delta V(D_1, D_2; \nu). \tag{31}$$

The distribution of barrier heights is then defined as

$$\mathcal{P}(E; \nu) = \langle \delta\, (E - \Delta V) \rangle_{D_1, D_2} \,. \tag{32}$$

An exact calculation gives the barrier height distribution associated with Arrhenius and Zwanzig-Bässler temperature dependence *(24)*. We have for high temperatures,

$$\mathcal{P}(E; \nu) = \delta \left(E - \frac{4}{27} E_\beta - \frac{1}{18} \sqrt{E_\beta}\, \nu^2 \right), \quad T >> T_g, \tag{33}$$

and for low temperatures

$$\mathcal{P}(E; \nu) = \frac{1}{\sqrt{\pi E_\alpha \nu^4 / 32}} \exp \left(- \frac{32}{E_\alpha \nu^4} \left[E - \frac{4}{27} E_\beta \right]^2 \right), \quad T \gtrsim T_g. \tag{34}$$

We thus conclude that in the absence of thermal stress, the energy landscape is uniformly rough such that each unstable mode defines a unique barrier, $E_\nu \propto \nu^2$. We identify E_β with the characteristic energy of local "basins" *(5)*. In the presence of stress, the system is in deep "craters" of the potential energy landscape. The barrier associated with these craters is random, and we find a Gaussian barrier height distribution for each unstable mode, $\langle E^2 \rangle_\nu \propto \nu^4$. We identify E_α with the characteristic energy of craters.

Discussion and Outlook

Recent experimental and theoretical investigations have shown that solid-like aspects of liquid dynamics are important for understanding properties of liquids near the glass transition *(26)*. In dielectric measurements, the structural relaxation time τ indicates qualitative changes at some temperature T_c above the glass temperature, $T_c > T_g$ *(35)*. Above T_c, structural relaxation in the normal liquid is characterized by a single activation energy E_A, $\tau \propto \exp(-E_A/T)$. Below T_c, the relaxation is characterized by spatial and dynamical heterogeneities, and the temperature dependence of τ no longer follows Arrhenius behavior. Spatial inhomogeneities imply a "decoupling" between rotational and translational diffusion *(36)*, *(37)*.

Qualitative changes in the dynamics at some temperature above T_g has been predicted from mode coupling theory (MCT) *(38)*. MCT is a hydrodynamic theory of nonlinear interactions of density fluctuations in which glass formation is an ideal kinetic transition. At T_c, different relaxation mechanisms decouple and the Johari-Goldstein β-process appears below T_c *(21)*. Fast dynamic processes are evident as excitations in the frequency range $\omega \sim 10$ to 100 GHz *(39)*. The spectrum has contributions from quasi-harmonic modes (single-well potentials) and relaxational modes (double-well potentials). The contributions from quasi-harmonic modes are referred to as the "boson peak."

Fast dynamic processes are observed in measurements of the mean square displacement (Debye-Waller factor). A transition-like behavior is observed in the temperature dependence of the slope at a temperature above T_g *(34)*. Buchenau and Zorn reported neutron time-of-flight measurements of atomic displacements for frequencies above 10 GHz in glassy, liquid and crystalline selenium *(40)*. They found a linear relation between the logarithm of the viscosity and the inverse of the enhancement of the mean square displacement. Since the viscosity is inversely proportional to the diffusion constant, this result is in agreement with our prediction from the soft-potential model *(23)*,

$$D \sim \exp\left(-\frac{1}{3}\frac{\langle x^2\rangle_{static}}{\langle x^2\rangle_\beta}\right), \quad T >> T_g. \tag{35}$$

Here, $\langle x^2\rangle_\beta$ is the displacement of fast processes, $\langle x^2\rangle_\beta = 4T/\Omega^2$, and $\langle x^2\rangle_{static}$ is the displacement for zero restoring force constant $D_2 = 0$, $\langle x^2\rangle_{static} = 0.338T_g^{1/2}$.

Instantaneous normal modes (INM) are an ideal tool for investigating solid-like properties of liquid dynamics. The INM spectrum consists of both stable and unstable frequencies. The real frequencies describes oscillations around rigid aperiodic structures, while the unstable frequencies are a signature of the fluidity of the liquid. Structural rearrangements of 20-100 particles give rise to low-frequency vibrational modes, which we have identified with soft modes. The potential energy profiles of soft modes are double wells. These double wells become single wells under the influence of sufficiently large stress. Single- and double-well potentials in liquids are described by the soft-potential model, which is an extension of the two-level-system (TLS) model for glasses. The TLS model assumes that atoms (or a group of atoms) reside in either of only two minima of the local potential surface.

The complex energy landscape of glasses is represented by an ensemble of TLSs characterized by a distribution of TLS parameters. The many-body system visits all configurations accessible at each temperature, and thus properties of its energy landscape depend on temperature. In fact, our results from the soft-potential model suggest qualitative changes in the energy landscape at some temperature above the glass temperature. At high temperatures, $T >> T_g$, the landscape consists of local "basins" characterized by a uniform barrier height distribution. For $T \gtrsim T_g$, the system visits deep "craters" of the potential energy, and each unstable frequency defines a Gaussian distribution of barrier heights, $\langle E^2\rangle_\nu \propto \nu^4$. We have shown furthermore that this change implies a crossover in the temperature dependence of the shear viscosity (or the self-diffusion constant) from Arrhenius-, $\log\eta \propto 1/T$, to Zwanzig-Bässler-behavior, $\log\eta \propto 1/T^2$.

Some ideas presented in this paper may prove useful in understanding protein dynamics. In fact, structural relaxation of proteins and glass-forming liquids have similar properties. "Glass-like" transitions in globular proteins have been observed in neutron scattering *(41)* and molecular dynamics simulations *(42)*. It is found that the line shape of the inelastic-scattering function follows the scaling predictions of mode coupling theory. Alternatively, the glass-like behavior observed in the Mössbauer spectrum of the iron atom in myoglobin has been interpreted as softening of low-frequency vibrational modes *(43)*. The complexity of the energy landscape in proteins is evident from hierarchical structure of local minima along conformational coordinates. These minima correspond to the *conformational substates* which were first used to interpret the nonexponential time dependence of CO rebinding on myogobin after photodissociation *(44)*.

Recently, picosecond vibrational echo experiments of myoglobin-CO have been reported that examine the influence of protein dynamics on the CO ligand bound to the active site of the protein at physiologically relevant temperatures (60 to 300K) *(45)*. The vibrational echo of the CO-stretch decays exponentially in time which implies a Lorentzian lineshape with the width $\Gamma = (\pi T_2)^{-1} = (\pi T_2^*)^{-1} + (2\pi T_1)^{-1}$, where T_2 is the homogeneous dephasing, T_1 is the vibrational lifetime, and T_2^* is the pure dephasing. The pure dephasing has power-law temperature dependence below the glass temperature of the solvent ($T_g \simeq 185\,\mathrm{K}$ for glycerol/water) and then becomes exponentially activated, $1/T_2^* = aT^\alpha + b\exp(-\Delta E/T)$ with $\alpha = 1.3 \pm 0.1$. This power-law behavior is a "universal" feature of glasses and has been explained using the TLS model for glasses *(46)*. In particular, the exponent $\alpha \simeq 1.3$ implies flat distributions for both the tunneling parameters and the energy splittings of TLSs. The authors of Ref. *(45)* propose protein two-level-systems to explain the temperature dependence of pure dephasing below T_g. Above T_g, barriers of conformational changes are lowered and the protein becomes more flexible. In future work, the soft-potential model can be used to describe vibrational dephasing from both tunneling and thermally activated processes. Such an investigation will be of particular interest as it will allow us to study in detail the combined effect of dynamic and spatial heterogeneity on protein dynamics at elevated temperatures.

Acknowledgments. This work has been supported by NSF grant CHE9415216.

Literature Cited

(1) Anderson P. W. *Basic Notions of Condensed Matter Physics*; Benjamin: Menlo Park, CA, 1984.
(2) Maxwell, J. C. *Phil. Mag.* **1867**, 165, 49.
(3) Frenkel, J. *Kinetic Theory of Liquids*; Dover: New York, NY, 1955.
(4) Zwanzig, R. *J. Chem. Phys.* **1983**, 97, 4507.
(5) Stillinger, F. H. *Science* **1995**, 267, 1935.
(6) Anderson, P. W.; Halperin, B. I.; Varma, C. M. *Philos. Mag.* **1972**, 25, 1. Phillips, W. A. *J. Low Temp.* **1972**, 7, 351.

(7) Stillinger, F. H.; Weber, T. A. *Science* **1984**, 225, 983.
(8) Heuer, A.; Silbey, R. J. *Phys. Rev. Lett.* **1993**, 70, 3911. *Phys. Rev. B* **1993**, 48, 9411.
(9) Cotterill, R. M. J.; Madsen, J. U. *Phys. Rev. B* **1986**, 33, 262.
(10) Keyes, T. *J. Phys. Chem.* (submitted, 1996); and references therein.
(11) Stratt, R. M. *Acc. Chem. Res.* **1995**, 28, 201; and references therein.
(12) Phillips, W. A.; Buchenau, U.; Nücker, N.; Dianoux, A.-J.; Petry, W. *Phys. Rev. Lett.* **1989**, 63, 2381.
(13) Alexander, S.; Orbach, R. *J. Phys. (Paris) Lett.* **1982**, 43, L-625.
(14) Rosenberg, R. O.; Thirumalai, D.; Mountain, R. D. *J. Phys. Cond. Mat.* **1989**, 1, 2109.
(15) Bembenek, S. D.; Laird, B. B. *Phys. Rev. Lett.* **1995**, 74, 936. *J. Chem. Phys.* **1996**, 104, 5199.
(16) Buchenau, U.; Galperin, Yu. M.; Gurevich, V. L.; Schober, H. R. *Phys. Rev. B* **1991**, 43, 5039.
(17) Hansen, J. P.; McDonald, I. R. *Theory of Simple Liquids*; Academic: New York, NY, 1986.
(18) Chen, S.-P.; Egami, T.; Vitek, V. *Phys. Rev. B* **1988**, 37, 2440.
(19) Ziman, J. M. *Models of Disorder*; Cambridge: New York, NY, 1979.
(20) Goldstein, M. *J. Chem. Phys.* **1969**, 51, 3728.
(21) Johari, G. P.; Goldstein, M. *J. Chem. Phys.* **1970**, 53, 2372.
(22) Zürcher, U.; Keyes, T.; Laird, B. B. *J. Phys. Cond. Mat.* (submitted, 1996).
(23) Zürcher, U.; Keyes, T. *Phys. Rev. E* (accepted, 1997).
(24) Zürcher, U.; Keyes, T. manuscript in preparation (1997).
(25) Buchenau, U. *Philos. Mag.* **1992**, 65, 303.
(26) Sokolov, A. P. *Science* **1996**, 273, 1675.
(27) Ferrari, L.; Phillips, W. A.; Russo, G. *Europhys. Lett.* **1987**, 3, 611.
(28) Gil, L.; Ramos, M. A.; Bringer, A.; Buchenau, U. *Phys. Rev. Lett.* **1993**, 70, 182.
(29) In fact, the RHS of Eq. (18) must be multiplied with a temperature-independent factor that describes the topology of the landscape near barriers; see Ref. *(30)* below.
(30) Keyes, T. *J. Chem. Phys.* **1992**, 101, 5081.
(31) Vijayadamodar G. V.; Nitzan, A. *J. Chem. Phys.* **1995**, 103, 2169.
(32) An alternative fit of the unstable INM spectrum has been proposed in Keyes, T.; Vijayadamodar, G. V.; Zürcher, U. *J. Chem. Phys.* (accepted, 1996).
(33) Zwanzig, R. *Proc. Natl. Acad. Sci. USA* **1988**, 85, 2029. Bässler, H. *Phys. Rev. Lett.* **1987**, 58, 767. An earlier derivation is is given by De Gennes, P. G. *J. Stat. Phys.* **1975**, 12, 463.
(34) Angell, C. A. *Science* **1995**, 267, 1924.
(35) Stickel, F.; Fischer, E. W.; Richert, R. *J. Chem. Phys.* **1995**, 102, 6251. *ibid* **1996**, 104, 2043.
(36) Cicerone M. T.; Ediger, M. D. *J. Chem. Phys.* **1995**, 103, 5684. *ibid* **1996**, 104, 7210.

(37) Tarjus G.; Kivelson, D. *J. Chem. Phys.* **1995**, 103, 3071.
(38) Götze W.; Sjögren, L. *Rep. Prog. Phys.* **1992**, 55, 241.
(39) Jäckle, J. in Phillips, W. A., ed., *Amorphous Solids*; Springer-Verlag: New York, NY, 1981.
(40) Buchenau, U.; Zorn, R. *Europhys. Lett.* **1992**, 18, 523.
41) Doster, W.; Cusack, S.; Petry, W.; *Nature* **1989**, 337, 754. *Phys. Rev. Lett.* **1990**, 65, 1080.
(42) Smith, J.; Kuczera, K.; Karplus, M. *Proc. Natl. Acad. Sci. USA* **1990**, 87, 1601.
43) Bialek W.; Goldstein, R. F. *Biophys. J.* **1985**, 48, 1027.
(44) For a review, see, e.g., Frauenfelder, H.; Sligar, S. G.; Wolynes, P. G. *Science* **1991**, 254, 1598.
(45) Rella, C. W.; Rector, K. D.; Kwok, A.; Hill, J. R.; Schwettman, H. A.; Dlott, D. D.; Fayer, M. D. *J. Phys. Chem.* **1996**, 100, 15620.
(46) Bai Y. S.; Fayer, M. D. *Phys. Rev. B* **1989**, 39, 11066.

Chapter 7

Entropic Approach to Relaxation Behavior in Glass-Forming Liquids

Udayan Mohanty

Eugene F. Merkert Chemistry Center, Boston College, Chestnut Hill, MA 02167

Adam-Gibbs picture of a "cooperative rearranging region" in glass-forming liquids is generalized. The extension leads to differential equations which are suggestive of the differential equations emerging in Wilson's theory of critical phenomena. A relation between the size of a "cooperative rearranging region" and the entropy emerges from solving these differential equations. The size of the region does not diverge near the glass-transition temperature. The line of metastable states of the supercooled liquid leads to a "twice unstable" fixed point. This elusive fixed point is identified with the Kauzmann temperature. Polymorphism between fragile and strong liquids is accounted for in terms of a "discontinuity" critical "end point". The "discontinuity" fixed point leads to a non-critical phase via a first-order phase transition. The relaxation time of supercooled liquids is expressed in terms of the topography of the potential energy hypersurface in configuration space via a non-equilibrium generalization of the Adam-Gibbs model.

There has been renewed experimental and theoretical activity in trying to unravel and understand glass-forming liquids and the amorphous states of matter (*1,2*). Theoretical techniques to describe the supercooled and the glassy states are hampered by the fact that these states are far away from equilibrium (*2,3*).

A theoretical framework is proposed that extends the seminal Adam-Gibbs picture (*4*) of a "cooperative rearranging region". The technique leads to differential equations that appear in Wilson's renormalization group (*5,6*) approach to critical phenomena. The polymorphism of the glassy state as well as the Kauzmann temperature is controlled by fixed points of these differential equations. A non-equilibrium generalization of the Adam-Gibbs model is also described in terms of the inherent structures of the supercooled state (*2,3,7*). The notion of a cross-over temperature T_x below which motions are governed by "entropic" barriers emerges from the formulation.

Adam-Gibbs Model. In elucidating the features of supercooled liquids, Adam-Gibbs (AG) introduced the idea of a "cooperatively rearranging region" or domains (*2,4*). It is defined to be part of the system, i.e. a "subsystem" that on "fluctuations in enthalpy" is capable of rearranging itself in a cooperative manner unconstrained of its surroundings (*2,4*). In this model, the probability per unit time for "cooperative rearrangements" is given by (*2,4*)

$$W\left(z^*, T\right) = A\exp\left(\frac{z^*\Delta\mu}{k_B T}\right). \tag{1}$$

Here, k_B is the Boltzmann's constant, $\Delta\mu$ is the molar enthalpy, T is the absolute temperature, A is a constant which is weakly temperature dependent, and z^* is the "minimum" number of molecules in a domain that leads to "cooperative rearrangements".

In the Adam-Gibbs picture, the "minimum" size z^* is intimately related to the molar configurational entropy $S_c(T)$ of the undercooled melt (*2,4*)

$$z^*(T) = \frac{s^* N_A}{S_c(T)}, \tag{2}$$

where N_A is the Avogadro's number. The quantity s^* appearing in Eq. (2) is the configuration entropy of "minimum" size domains (*2,4*). These are the smallest domains that allow cooperative relaxation. Since the specific heat difference ΔC_p between the equilibrium melt and the glassy state at T_g is approximately constant, the configuration entropy $S_c(T)$ is obtained via (*2,4*)

$$S_c(T) = \int_{T_2}^{T} \frac{\Delta C_P(T)}{T} dT. \tag{3}$$

Here, T_2 is a reference temperature usually identified with the Kauzmann temperature. By definition, the configuration entropy vanishes at the Kauzmann temperature T_2 .

A few comments are in order. First, Eq. (2) makes *no* assumption about the existence or the nature of the Kauzmann temperature. Second, according to Eq. (3), configurational entropy varies with temperature as $\Delta C_p \ln\left(\frac{T}{T_2}\right)$ (*2,4*). Third, the change of Gibbs-free energy $\Delta G = z^* \Delta\mu$ (*2,4*). On combining Eqs. (2) and (3) one observes that ΔG is singular as $T \to T_2$.

New Picture of "Cooperative Rearranging Region". Adam-Gibbs considered an isothermal-isobaric ensemble of "cooperatively rearranging regions". Any given member of the ensemble has the same number of molecules (*2,4*). The partition function of the supercooled liquid is constructed so that the phase space associated with crystal-like packing configurations has been excluded. The partition function $\Delta(T,P,z)$ of the ensemble is (*2,4*)

$$\Delta(z,P,T) = \sum_{E,V} \omega(E,V,z)\exp(-\beta E - \beta PV), \qquad (4)$$

where $\beta = \frac{1}{k_B T}$ and $\omega(E,V,z)$ is the density of states at energy E and V. Since the Gibbs-free energy is defined as $\beta G(z,P,T) = -\ln\Delta(z,T,P)$, the ratio of the subsystems that allows transitions, to the total number of subsystems in the ensemble, is proportional to $\exp(\beta\Delta G)$ (*2,4*).

Each term in the sum over E and V in Eq. (4) is analytic in T,P, and z. But, near the temperature T_2, AG find a "singular" behavior of the partition function (*2,4*). Various approximations to the partition function then lead to the expression for transition probability $W(z^*,T)$ for "cooperative rearrangements" as given by Eq. (1). However, $W(z^*,T)$ is *not* an analytic function as $T \to T_2$.

To deduce such "singular" behavior in thermodynamic functions, such as the free-energy, by explicitly evaluating the partition function is not a simple task for several reasons. First, evaluation of the "singular" part of the free-energy requires precision that is hard to accomplish by conventional techniques and approximations (*2,5,9*). Second, "singular" feature of the partition function can appear only in the so-called thermodynamic limit (*2,5,9*).

It is suggested that generalization of the Adam-Gibbs picture in terms of differential equations of the renormalization group may provide insights into the low temperature relaxation dynamics of glass-forming liquids (*2,9,10*). Since it is easier for singularities to arise by solving differential equations, the hope is that these differential equations would lead to non-analytic behavior (if any) in the free-energy.

The total volume, L^3, of the system is partitioned into cubic "blocks" or "cells", each of length L (*9,2*). Each block has z_L number of molecules. As the temperature of the system is lowered, the configurations available to a subsystem for "cooperative rearrangements" decrease (*9,2*). What this suggests is the relevance as well as the importance of accounting for correlations between the "blocks" (*9,2*). Consequently, one introduces the "correlation length" ξ to denote the size over which the blocks are correlated.

As the temperature is lowered, the "blocks" grow in size compared with typical molecular scale length a_0. Long wavelength fluctuations govern the relaxation behavior of supercooled liquids in the AG model. To put it differently, the short wavelength fluctuations are integrated out form the partition function under renormalization group transformations (*5,6*). The exact nature of the renormalization group transformation is not necessary, except to note that the partition function is the same before and after such a transformation (*5,6*). Consequently, the Gibbs free-energy of a "block" and the Gibbs free-energy per unit volume of the subsystem obey the relation (*2,9,5*)

$$g(z,\alpha) = L^{-3} g(z_L,\alpha_L). \tag{5}$$

Further, the correlation length $\xi(z_L,\alpha_L)$ of a "block" is linked to the correlation length $\xi(z,\alpha)$ of the subsystem (*2,9,5*)

$$\xi(z,\alpha) = L\xi(z_L,\alpha_L) \ . \tag{6}$$

In the framework of renormalization group, the α's include the "relevant" as well as the "irrelevant" variables (*5*). Example of a "relevant" variable is the number of flex bonds introduced by Gibbs-DiMarzio in their theory of amorphous polymers (*2,8*).

Differentiating Eq. (5) with respect to temperature leads to the relation (*9,2*)

$$s(z,\alpha) = L^{-3} s(z_L,\alpha_L) + \left(\frac{3}{L^4}\right)\left(\frac{dL}{dT}\right) g(z_L,\alpha_L), \tag{7}$$

where $s(z_L, \alpha_L)$ is the entropy of a "block". The ratio of the volumes $\mathbf{L}^3/L^3$ is a measure of the number of "blocks". If one takes a mole of the substance, then there are also $\frac{N_A}{z_L}$ "blocks". This observation together with Eq. (7) provides the link between the size of a "block" and the entropy of the subsystem (*9,2*)

$$z_L = \frac{N_A s(z_L, \alpha_L)}{S(z, \alpha)} + \left\{\frac{3N_A}{S(z, \alpha)}\right\}\left(\frac{1}{L}\left\{\frac{dL}{dT}\right\}\right) g(z_L, \alpha_L). \tag{8}$$

The Adam-Gibbs relation (2) emerges if (a) the second term in Eq. (8) is disregarded, (b) one approximates z_L by its "minimum" value z^*, and (c) the total entropy which includes vibrational anharmonicity is approximated by the configurational entropy.

If z_L and α_L are solutions of Eqs. (5) and (6), then we can differentiate both sides of the equations with respect to L. The variations of the "effective couplings" z_L and α_L are (*2,5,9*)

$$\frac{dz_L}{dL} = \Gamma(z_L, \alpha_L), \tag{9}$$

$$\frac{d\alpha_L}{dL} = \Omega(z_L, \alpha_L). \tag{10}$$

Eqs. (9)-(10) define the functions $\Gamma(z_L, \alpha_L)$ and $\Omega(z_L, \alpha_L)$; they are L dependent entirely via z_L and α_L. The differential equations for z_L and α_L are the desired generalization of the AG model. They are reminiscent of Wilson's renormalization group equations in critical phenomena (*5,6*).

Correlation Length. Let τ_i denote the relaxation time for a local "rearranging region" in the volume V. Various subregions of size z may span the volume. It is fruitful to considers how $\ln \tau_i(T)$ deviates from $\ln \tau(T)$. The temperature dependence of the relaxation time is assumed to be given by Eq. (1); however, no assumption relating $z^*(T)$ to the configurational entropy of the system is necessary (*2,11*). By making use of fluctuations of the size $z^*(T)$ in the volume V (*2,11*)

$$\frac{\left\langle(\Delta z^*)^2\right\rangle}{V} = \rho^2 k_B T_g \Delta\kappa_{Tg} \tag{11}$$

an expression is deduced for the "correlation length" $\xi \approx V^{1/3}$ in terms of quantities that are experimental measurable (*2,11*)

$$\xi = \left[\frac{(1-x)\Delta h^*}{RT^2}\right]^{2/3}\left\{\frac{k_B T_g \Delta\kappa_{Tg}}{\left\langle \Delta^2 \ln \tau \right\rangle}\right\}^{1/3} . \tag{12}$$

Here, the quantity $\left\langle \Delta^2 \ln \tau \right\rangle$ is a gauge of the "width of the distribution of relaxation times". This expression for the "correlation length" has been generalized (*12*) to include a non-equilibrium extension of the AG model described below.

Using the data in ref. 13, the "correlation length" is evaluated in the vicinity of the glass transition temperature T_g. One finds that $\xi(T_g)$ is in the range of 0.6 nm for boric oxide, while it is 0.7 nm for glycerol and 1.5 nm for PVAC (*11,14*). Thus, the "correlation length" does not diverge at or near T_g. Computer simulations are in agreement with this conclusion (*15-17*).

Inherent Structures: Dynamics in Basins. The relaxation time of glass-forming liquids is intricately linked to the topography of the potential energy hypersurface Φ (*1-3,7-8,15-17*). The absolute minima of the Φ hypersurface terrain is assumed to be the crystal. Since we are interested in supercooled liquids, the crystalline packings have been eliminated from Φ.

Steepest descent quenches have been utilized with considerable success to identify the local minima in model systems (*2,3,7*). What computer simulations reveal is that transitions to nearby minima are basically due to "localized" rearrangements in the volume of the sample (*2,3,7*). Further, even though a small set of the total number of the N particles participates in the "localized" motions, a number of these rearranging regions stretch across the volume of the sample (*2,3,7*). An implication of this observation is that there is an aggregation of transition states in regions intersecting the basins (*2,3,7*). In addition, although the heights of the various minima are of order of magnitude N, the value of Φ changes by a small amount as the phase point makes excursions from one local minimum to another minimum (*2,3,7*). The number of minima in the range $\phi + \Delta\phi$ scales as $\exp(N)\sigma_s(\phi)\Delta\phi$, where $\sigma_s(\phi)$ is the distribution of minima and ϕ is the depth of the local minima per particle (*2,3,7*).

To describe the potential energy hypersurface, we introduce a variable $\theta_\alpha(\Gamma_o)$ that is one if the configuration point Γ_0 is located within "cell" α and is zero otherwise (*2,18-20*). The collection of $\{\theta_\alpha(\Gamma_o)\}$ introduces a complete basis set in the configuration space. The phase space distribution $g(X,t)$ enables us to obtain the probability that at time t the configuration point Γ_o is in "cell" labeled α *(2,18-20)*

$$P_a(t) = \int dX\theta_\alpha(X)g(X,t). \tag{13}$$

At initial time the distribution function $g(X,0) = g_{equil}(X)Q(\theta)$; here $Q(\theta)$ is a function of the "cell" like variable. The evolution of $P_\alpha(t)$ is derived by projection operator techniques. If the system looses memory of where it was at initial time, then *(2,18-20)*

$$P_a(t) = \sum_{\alpha\neq\gamma}\left[K_{\alpha\gamma}P_\gamma(t) - K_{\gamma\alpha}P_\alpha(t)\right], \tag{14}$$

where $K_{a\gamma}$ - the probability per unit time for the system to execute a jump from the "cell" α to the "cell" γ is dictated by the principle of detailed balance *(2,7,20)*

$$K_{\alpha\to\gamma} = \left\{\frac{\langle M_\alpha(E)\rangle}{\langle M_\gamma(E)\rangle}\right\}^{1/2} A_{\alpha\to\gamma}(E). \tag{15}$$

The phase space volume of the "cell" α is indicated by $\langle M_\alpha(E)\rangle$ *(2,7,20)*. The matrix $A_{\alpha\to\gamma}(E)$ appearing in Eq. (15) is symmetric in "cell" variables.

Entropic forces. To elucidate the evolution of the system on the hypersurface, we introduce a set of intensive order parameters ζ. Examples of ζ are ϕ and the specific volume *(2,3,7)*. We consider the time dependence of the probability in the "space" of the order parameters *(7,20)*

$$P(\zeta_\alpha,t) = \exp(N\sigma(\zeta_\alpha))P_\alpha(t). \tag{16}$$

The distribution of the minima in the order parameter "space" $\sigma(\zeta_\alpha)$ is deduced from the total entropy of the system $S(\zeta_\alpha,E)$ and the volume in phase space of "cell" α *(7,20)*

$$\exp\left[S(\zeta_\alpha,E)/k_B\right] = \langle M_\alpha(E)\rangle\exp(N\sigma(\zeta_\alpha)). \tag{17}$$

In the limit the order parameters acquire continuous set of values, we obtain from Eqs. (14)-(17) a master equation *(7,20)*

$$\frac{dP(\zeta,t)}{dt} = \int d\zeta' A(\zeta',\zeta;E)\Big[\exp\{(S(\zeta,E) - S(\zeta',E))/2k_B\}P(\zeta',t)\Big)$$
$$- \exp\{(S(\zeta',E) - S(\zeta,E))/2k_B\}P(\zeta,t)\ \Big] \tag{18}$$

The transition matrix $A(\zeta',\zeta,E)$ is approximated by averaging $A_{\alpha\to\gamma}(E)$ over basin α located at ζ' and basin γ located at ζ, and then weighting it by the term $\exp\left(\left(\frac{N}{2}\right)[\sigma(\zeta)+\sigma(\zeta')]\right)$ *(7,20)*.

An expansion of the master equation in terms of system size is implemented since, as discussed earlier, the order parameters changes are approximately 1/N due to transitions between cells *(2,7,20)*. This leads to a Fokker-Planck equation in which the so-called "thermodynamic" force $\vec{F}$ is a direct consequence of "entropic barriers" *(2,7,20)*

$$\vec{F}(\zeta, \mathrm{E}) = (k_B T)^{-1} grad(S(\zeta,E)). \tag{19}$$

The mobility tensor $\vec{\mu}(\zeta,E)$ is given by *(7,20)*

$$\vec{\mu} = \frac{\beta}{2}\int d\zeta\, A_o(\zeta,\zeta-\zeta';E)(\Delta\zeta')^2, \tag{20}$$

where $A_o\left(\zeta,\frac{1}{2}(\zeta'-\zeta),\Delta\zeta';E\right) = A(\zeta,\zeta',E)$. Assuming linear response, the mobility (tensor) and the diffusion coefficient are related by the relation $\vec{D}(\zeta,E) = k_B T\vec{\mu}(\zeta,E)$.

We assume that the free-energy $F(\zeta)$ is an extremum. Variations of $F(\zeta)$ lead to the condition $\zeta^f = \zeta^f(E)$. The "thermodynamic" force $gradS(\zeta,E)$ vanishes on a surface defined by $\zeta^s = \zeta^s(E)$. The overlap of ζ^f and ζ^s at energy E imposes a restricted order parameter space: $\zeta^* = \zeta^*(E)$. The hypothesis here is that the characteristics of the relaxation rate on temperature in glass-forming liquids arise because the slowest mode (defined below) evolves towards equilibrium on this surface *(20)*.

We solve Eq. (18) in the "diagonal" approximation to the diffusion tensor and by expanding $P(\zeta,E,t)$ as $\sum_n a_n\psi_n(\zeta,E)\exp(-\omega_n t)$, where $\Psi_n(k,E)$ is the Fourier transform of $\psi_n(\zeta,E)$ in the order parameter "space". One obtains an integral equation for $\Psi_n(k,E)$ *(20)*

$$\left[D\left(\zeta^*, E\right)k^2 - \omega_n\right]\Psi_n(k,E) = \left\{\frac{V_\zeta}{(2\pi)^d}\right\}D\left(\zeta^*, E\right) \int \kappa(k-k';E)\Psi_n(k',E)dk' \tag{21}$$

V_ζ is the "volume" of the order parameter "space" of dimension d. A sum over the repeated index is implied in Eq. (21). In deriving Eq. (21) it is assumed that $D\left(\zeta^*, E\right)$ is parallel to $\kappa(\zeta,E) = \frac{\partial^2\left[S(\zeta,E)/k_B\right]}{\partial\zeta^2}$ -- the curvature of the "entropic barriers" at energy E and order parameters ζ.

Eq. (21) is remarkably similar to the Cooper pair equation in the BCS theory of superconductivity (*21,20*). The quantity $\Psi_n(k, E)$ is equivalent to the Cooper pair wave function in Fourier space (*21,20*). The electron pairs interact via an attractive interaction in the BCS theory (*21*). This term is analogous to the curvature $k(\zeta, E)$ of the "entropic barrier" (*21,20*). Stability conditions require that $\kappa\left(\zeta^*\right)$ be less than zero.

Relaxation Time and Cross-over Temperature. Our conjecture (*20*) is that the slowest mode evolves towards equilibrium on the order parameter surface $\zeta^* = \zeta^*(E)$. Since the Prigogine-Defay ratio is larger than unity for supercooled liquids, atleast two order parameters are required to describe the surface (*22,20*). The integral equation for $\Psi_n(\zeta, E)$ is solved explicitly. The relaxation time of the slowest mode is (*20*)

$$\tau_s \approx \tau_D\left(D^*\left(E,\zeta^*,\langle k^*\rangle\right)\right)\exp\left(\frac{B}{TS\left(\langle k^*\rangle,E\right)}\right). \tag{22}$$

Here k^* is a wave vector corresponding to the order parameter ζ^* and B is a function of T, k_B, V_ζ and $\langle\zeta^*\rangle^2$. If $S\left(\langle k^*,E\rangle\right) \approx s^*(E)f\left(\langle k^*\rangle\right)$, and $f\left(\langle k^*\rangle\right)$ is weakly temperature dependent, then the temperature dependence of the slowest mode is that predicted by Adam-Gibbs.

The term $A(\zeta,\zeta',E)$ appearing in Eq. (20) is linked to $K(\zeta,\zeta',E)$ - the probability per unit time; the later quantity is expressible in terms of the free-energy of "activation" $F^*(\zeta \to \zeta')$ and a frequency factor $\omega_o(\zeta)$

$$A(\zeta,\zeta',E) = \exp\left(-\beta/2\left[F(\zeta) - F(\zeta')\right]\right)\exp\left(-\beta/2\left[\sigma(\zeta) + \sigma(\zeta')\right]\right)$$
$$x\omega_o(\zeta)\left\langle\left\langle \exp - \beta/2F^*(\zeta \to \zeta')\right\rangle\right\rangle \tag{23}$$

The brackets $\langle\langle\ \rangle\rangle$ indicate an average over "cell" μ near ζ and "cell" α near ζ' (*7*). By expanding the free-energy and the configurational entropy terms in Eq. (23) on the "critical" surface $\zeta^*(E)$, and on substituting it back in Eq. (20), one obtains an expression for the diffusion coefficient $D^*(E)$. Further analysis leads to the existence of a cross-over temperature

$$T_x \approx \frac{\kappa_s}{k_B\kappa_u}. \tag{24}$$

Here, κ_s and κ_u are the curvature of the entropic and the potential energy barriers respectively. For temperature less than T_x the dominant mode of relaxation towards equilibrium is via "entropic" channels in configuration space (20).

Fixed Points, Topology, and Kauzmann Temperature. In this section we address the following question: Can one relate the characteristics of the "fixed points" associated with the Hamiltonian of a glass-forming liquid to the Kauzmann temperature T_2 and the polymorphism of the glassy state (1,24)?

To address this question we make the following observations. First, the conventional dogma is that different experimental routes should lead to almost the same thermodynamic glassy state. However, what has been recently discovered is that compression of a crystal at temperatures below T_g leads to high density glassy states (*24,25,26*). These glassy states on decompression change back to the low density vitreous ice, which itself may undergo a first-order phase transition to the high density phase (*24,25,26*). Further evidence of the so-called polymorphism phenomena in the glassy state comes from computer simulations of amorphous ice (*27*). Second, there are systems in which a fragile liquid is known to undergo a first-order phase transition to a strong liquid during quench (*1,24,28*). Examples of such systems include Y_2O_3-Al_2O_3 and silicon (*1,24,28*). Third, a strong liquid can be converted to fragile liquids by application of an external pressure (*1,24,25,26*). Fourth, as argued above, correlation length does not diverge near the glass transition temperature (*11-13*). This is in accord with experiments (*2,13*) as well as simulations (*15-17*). Finally, since the Prigogine-Defay ratio at T_k is larger than unity, at least two order parameters are required in addition to T and P (*22*).

The Hamiltonian of a glass-forming liquid under renormalization group transformations leads to Eqs. (9)-(10). The stationary points of these equations are the fixed points (*5,6*); they can be identified in principle. Based on our generalization of the Adam-Gibbs model, we make the hypothesis that the line of metastable states of the supercooled liquid (Figure 1) leads to an "unstable" fixed point D_*. By assumption, the line of metastable states is on a "critical" surface. Since T and two order parameters are required to describe the amorphous phase, the fixed point D_* is "twice unstable" (*5*). Further, since trajectories cannot leave a "critical" surface except at the fixed points (*5,6*), all trajectories on the "critical" surface flow away from D_* (Figure 1). The trajectory G_A on the "critical" surface flow towards a low temperature "unstable" fixed point P_A, while the trajectory G_P flows toward the "high" temperature fixed point. Off the "critical" surface, trajectory such as G_{P_o} flows towards the "stable" fixed point P_o - the "ground" state of the system (*8*). Thus, the fixed point D_* can be thought of as the "top" of a hill. G_A and G_P are thus ridge lines (*5,6*). The "unstable" fixed point P_A is a saddle point; it is unstable in one direction but stable in the orthogonal direction (*5,6*). There is a gully that runs from the saddle point to P_o.

Associated with D_*, there is a sub-domain of "relevant" interactions which would lead under a renormalization group transformation to a fixed point. The dimension of the domain D_* is smaller than the dimension of the parameter space of the Hamiltonian by two (*5,6*). This imposes topological constraints. To understand this point, one defines a "canonical surface" C. C is the set of values of the various interactions that one initially chooses in the model Hamiltonian (Figure 1). Since the fixed points are topological as well as random in nature with respect to the choice of the canonical surface C, "twice unstable" fixed point D_* is "elusive" (*5,6*). What this means is that the curve C is not likely to intersect the fixed point D_*.

The presence of the fixed point D_* is felt at high and at low temperatures. Since the renormalized trajectories are either into the high temperature or the low temperature phases, one may not be able to reach D_* in "equilibrium" experiments. In other words, the only way to be at D_* is to sit on "top" of the hill which is possible if one knows the initial coordinates. We identify the "twice unstable" fixed point D_* with the Kauzmann temperature.

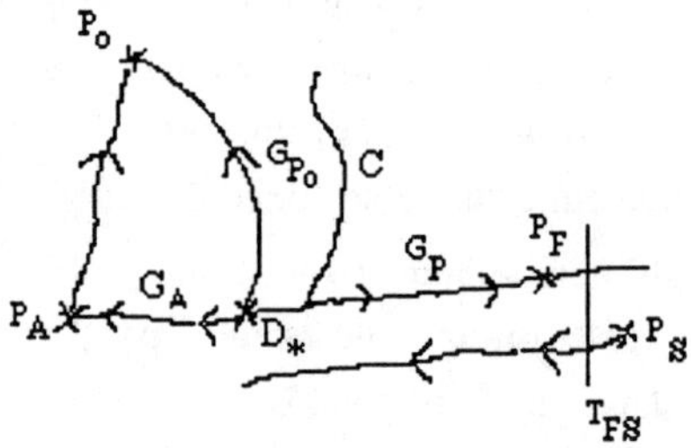

Figure 1 shows a schematic view of the topology of the renormalization group surface for glass-forming liquids. The fixed points D_*, P_A and P_F are on the "critical" surface. The fixed D_* is "twice unstable" and is identified with the Kauzmann temperature. The fixed point P_o denotes the "ground state" of the system. Polymorphism is described by the "discontinuity" fixed points P_F and P_S. The free-energy surfaces cross at a temperature T_{FS} and give rise to a first-order phase transition. The arrows indicate the directions of the renormalization group trajectories. P_A is a saddle point. Pressure effects have not included.

If Γ and Ω are assumed to be analytic at the "twice unstable" fixed point, then Eqs. (9)-(10) is solved as follows (*2,5,9*). The "effective couplings" are integrated until $L \approx \xi$. When $L \approx \xi$, the block correlation length and the free-energy are calculable by conventional methods. Eqs. (5) and (6) then allows us to construct the subsystem correlation length and the free-energy (*2,5,9*).

In the inherent structure picture of glass-forming liquids, regions of the configuration space have different topologies (*1,2,3*). The so-called "megabasins" are separated from other "megabasins" by potential energy barriers which are large compared to thermal fluctuations (*1,2,3*). The observed polymorphism is accounted for by a first-order phase transition from one "megabasin" to another on the potential energy hypersurface (*1,3*). Thus, there is a free-energy surface associated with each "branch" of the metastable states.

In a first-order phase transition there is a discontinuity in the order parameters for temperatures below a certain temperature; both the phases have a finite "correlation length" and are non-critical. However, in this case, one of the two phases, namely that corresponding to the fragile liquid is "critical". This implies the existence of "discontinuity" fixed points P_F and P_S on the free-energy surfaces (*29,30*). It is discontinuous in the sense that there is one fixed point on each free-energy branch of the coexisting phases (*31*). The "discontinuity" fixed points may be anywhere on the free-energy surfaces (*31*). The fixed point P_F on the free-energy surface is a critical "end point" since a "non-critical" phase coexists with a "critical" one. The two free-energy surfaces cross at a temperature T_{FS} leading to a first order phase transition. Consequently, the correlation length of the "critical" phase will be termed the "coherence length" (*30*). The later quantity measures "long-range order" in thermodynamic quantities that identifies the characteristics of the two phases (*30*). The trajectory on the free-energy surface of the "non-critical" phase flows towards a low temperature fixed point associated with the amorphous state of matter.

Concluding Remarks. The above formalism is capable of elucidating the roles played by the unstable and the stable modes (*32,33*) in normal mode analysis of transport coefficient such as self-diffusion. It will be potentially fruitful to investigate the relationship of this methodology to the "avoided" critical point (*34*), the scaling (*35, 36*) and the non-trapped and caged diffusive motion approaches to supercooled liquids *(37)*.

Acknowledgment. This work was supported in part by Boston College.

Literature Cited.

1. Angell, C. A. *Science* **1995**, 267, 1924.
2. Mohanty, U. *Adv. Chem. Phys.* **1994**, LXXXIX, 89.
3. Stillinger, F. *Science* **1995**, 267, 1935.
4. Adam, G.; Gibbs, J. H. *J. Chem. Phys.* **1965**, 43, 139.
5. Wilson, K.G. *Phys. Rev.* **1971**, B4, 3174.
6. Wilson. K. G.; Kogut, J. *Phys. Rep.* **1974**, 12, 75.
7. Stillinger, F. H. AT&T Bell Laboratories preprint, **1990**.
8. Gibbs, J. H.; DiMarzio, E. A. *J. Chem. Phys.* **1958**, 28, 373.
9. Mohanty, U. *Physica* **1990**, A162, 362.
10. Mohanty, U. *Physica* **1992**, A183, 579.
11. Mohanty, U. *J. Chem. Phys.* **1994**, 100, 5905. This corrects a minor error. The basic result remains unchanged, however.
12. Mohanty, U., Boston College, unpublished.
13. Moynihan, C. T.; Schroeder, J. Rensselar Polytechnic Institute preprint, **1993**.
14. Moynihan, C. T., Rensselar Polytechnic Institute, private communication, **1994**.
15. Ernst, R. M.; Nagel, S. R.; Grest, G. S. *Phys. Rev.* **1991**, B43, 8070.
16. Dasgupta, C.; Indrani, A. V.; Ramaswamy, S.; Phani, M. K. *Europhys. Lett.* **1991**, 15, 307.
17. Mountain, R. D.; Thirumalai, D. NIST and University of Maryland preprint, **1991**.
18. Zwanzig, R. *J. Stat. Phys.* **1983**, 30, 255.
19. Mohanty, U. *Phys. Rev.* **1986**, A34, 4993.
20. Mohanty, U.; Oppenhein, I.; Taubes, C. H. *Science* **1994**, 266, 425.
21. Cooper, L. *Phys. Rev.* **1956**, 104, 1189.
22. Gupta, P. K. *Rev. Solid. State Sci.* **1989**, 3, 221.
23. Wolf, G. H.; Wang, S.; Herbst, C. A.; Durben, D.J.; Oliver, W. J.; Kang, Z. C.; Halvorsen, C. in *High-Pressure Research: Application to Earth and Planetary Sciences*, eds. Manghnani,Y. S.; and Manghnani, M.H.; Terra Scientific/Am. Geophys. Union, Washington, pp. 503 (**1992**).
24. Angell, C. A. *Proc. Nat. Acad. Sci. USA.* **1995**, 92, 6675.
25. Mishima, O.; Calvert, L. D.; Whalley, E. *Nature* **1984**, 310, 393.

26. Mishima, O. *J. Chem. Phys.* **1994**, 100, 5910.
27. Poole, P.H.; Sciortino, F.; Grande, T.; Stanley, H. E.; and Angell, C. A.; *Phys. Rev. Lett.* **1994**, 73, 1632.
28. Aasland, S.; McMillian, P. F.; *Nature* **1994**, 369, 633.
29. Nienhuis, B.; Nauenberg, M. *Phys. Rev. Lett.* **1975**, 35, 477.
30. Fisher, M. E.; Berker, A. N. *Phys. Rev.* **1982**, B 26, 2507.
31. Stein, D. L.; and R. G. Palmer, R. G. *Phys. Rev.* **1988**, B38, 12035.
32. Cotterill, R.; Masden, J. *Phys. Rev.* **1988**, B33, 262.
33. Seeley, G.; Keyes, T.; Madan, B. *J. Chem. Phys.* **1991**, 95, 3847.
34. Kivelson, D.; Kivelson, S. A.; Zhao, X.; Nussinov, Z.; Tarjus, G. *Physica* **1995**, A219, 27.
35. Kirkpatrick, T. R.; Thirumalai, D.; Wolynes, P. G. *Phys. Rev.* **1989**, A40, 1045.
36. Souletie, J. Laboratoire Associe a l'Universite Joseph Fourier (C.N.R.S.) preprint, **1991**.
37. Odagaki, T. *Phys. Rev. Letts.* **1995**, 75, 3701.

Chapter 8

The Replica Approach to Glasses

Giorgio Parisi

Dipartimento di Fisica, Università di Roma La Sapienza, Istituto Nazionale di Fisica Nucleare, Sezione di Roma I, Piazzale Aldo Moro, 00185 Rome, Italy

Here we review the approach to glassy systems based on the replica method and we introduce the main ingredients of replica symmetry breaking. We explain why the replica method has been successful in spin glass and why it should be successful for real glasses.

The idea that glasses and spin glasses have something in common (beyond the name) is often stated in the literature. We can hope that the progresses that have been done in these last 20 years in the study spin glasses can be exported to glasses *(1–3)* . Many of these progresses employs the replica method with the assumption of spontaneous symmetry breaking.

The aim of this note is to describe the foundations of the replica method and to show why glasses are well inside its scope. It is organized as follows: in the first section we review the replica method and we introduce the formalism of replica symmetry breaking. In the second section we recall some mean field results that have been obtained for infinite range generalized spin glasses. In the third section we present some considerations on real glasses and in the final section we explain why the replica method should work for real glasses. Finally, in the Appendix we present a first computation of the properties of a soft sphere glass using the replica method.

The Replica Method

Why Replicas? If we deal with systems which order in a simple way at low temperature (i.e. crystal forming materials, ferromagnets), life is simple: we have to find the ground state and study the fluctuations around the ground state.

The situation is much more difficult when the for one reason or another (intrinsic randomness, chaotic dependance on the size of the system) the ground state is not known. In this case it is convenient to introduce replicas of the system. For example, in spin glasses, where the fundamental variables are Ising spins which are defined on the points of the lattice, one introduces n replicas of the same system *(1, 2)* , with Hamiltonian

$$H = \sum_{a=1,n} H[\sigma^a]. \tag{1}$$

Let us consider the case where at low temperature the system develops a spontaneous magnetization, which change sign from site to site with zero average, i.e.

$$< \sigma_a(i) >= m(i). \tag{2}$$

In this case the system average of the magnetization, i.e.

$$m \equiv \frac{\sum_{i=1,N} < \sigma_a(i) >}{N}, \tag{3}$$

is not informative. On the contrary if we look to two-replica quantity (for $a \neq b$), we find the rather interesting result:

$$q_{a,b} \equiv \frac{\sum_{i=1,N} < \sigma_a(i)\sigma_b(i) >}{N} = \frac{\sum_{i=1,N} m(i)^2}{N}. \tag{4}$$

In a nutshell, we use one replica to probe the properties the other replica. Correlation functions among different replicas (averaged over the system) contain information on the properties of the ground state of the system although they are not sufficient to determine the ground state.

Using this formalism we accomplish two goals:

- If we measure the correlations among replicas and not the ground state directly, we have quantities which should be more stable when we perturb the system.
- If we succeed to write down theoretical relations for the correlations among different replicas, we can use them to solve the model without having to compute the ground state.

The Breaking of Replica Symmetry. Things are more complicated if we have more than one ground state. For example, for Ising spins in the absence of a magnetic field there are always two possible values of the magnetization, which are obtained by changing the sign to all the spins.

Usually we cure this problem by adding a small magnetic field, but this does not always solve the problem, e.g. it does not solve the problem in the case of an antiferromagnet. In this case one finds that

$$< \sigma_a(i)\sigma_b(i) >= 0, \tag{5}$$

because the four (two for the replica a and two for the replica b) states give an exactly equal modulus and cancelling contribution.

The solution is simple: for example in the case $n = 2$ we consider the following Hamiltonian *(4–7)* :

$$H = H(\sigma_1) + H(\sigma_2) - \epsilon \sum_i \sigma_1(i)\sigma_2(i), \tag{6}$$

and we study its properties for small ϵ.

Now the Hamiltonian is invariant under the change of *both* the first and the second replica together. If there are two states, having magnetization $\pm m(i)$ one finds

$$\lim_{\epsilon \to 0^+} q(\epsilon) \quad = \quad - \lim_{\epsilon \to 0^-} q(\epsilon) = \frac{\sum_{i=1,N} m^2(i)}{N},$$

$$q(\epsilon) \equiv \frac{\sum_{i=1,N} < \sigma_1(i)\sigma_2(i) > \epsilon}{N}. \tag{7}$$

By adding a term which couples different replicas and observing if the correlations among replicas are continuous or discontinuous when this parameter changes sign, we can monitor the existence of one of more equilibrium states. Just because the equivalence among replicas is destroyed by an infinitesimal field where more than one equilibrium state is present, we say that in this situation replica symmetry is spontaneously broken *(1, 2)*.

It is evident that if we should consider only the case in which the equilibrium states differ by a simple symmetry transformation, the replica formalism would be correct, but may be not so useful. The real power of the replica formalism shows up when there are many equilibrium states of different free energy. In one of the most simple non trivial cases which can be studied by using this formalism, there are many equilibrium states and the probability of finding a state with free energy $F = \Delta F + F_0$ (F_0 is the free energy of the lowest lying state) increases as

$$P(\Delta F) \propto \exp(\beta m \Delta F), \tag{8}$$

where the parameter m characterize the way in which the replica symmetry is broken *(1, 2)*. This case is usually called one step breaking.

There is an algebraic formulation of the replica method in which one encodes the information coming from the structure of the states in correlations among replicas. The n replicas are divided in groups in such a way that correlations among replicas belonging in the same group are the same. For example in the previous case of one step replica symmetry breaking, one divides the n replicas in n/m groups of m replicas each. This construction may look rather artificial, but it turns out to be an extremely efficient way of describing in a compact way rather complex situations. Moreover, this formulation is well suited for doing computations *(1, 2)*. It may take some work to do the decoding and to find out which are the physical properties of the state distribution from the assumed form of replica symmetry breaking; however, this decoding is always possible and the structure of the probability distribution of the states mirrors the structure of the algebraic properties of the replicas.

Mean Field Results

In these recent years there have been much progress on the understanding of the behavior of glassy systems in the mean field approximation. The main results have been obtained for the following cases:

- Models with random quenched disorder have been well understood also from the dynamical point of view.
- Models with random quenched disorder, which display a glassy transition quite similar to that of the previous models. Some of the results obtained for systems with random quenched disorder have been extended also to these systems.

Let us see what happened in more detail.

Systems with Quenched Disorder. Generally speaking when the interaction range is infinite, the thermodynamical properties at equilibrium of a spin model can be computed analytically. When the system is disordered, we must use the replica method *(1, 2)*. A typical example of a model which can be solved with the replica method is a spin model with p spin interaction. The Hamiltonian we consider depends on some control variables J, which have a Gaussian distribution and play the same role of the random energies of the Random Energy Model (REM) *(10)* and by the spin variable σ. For $p = 1, 2, 3$ the Hamiltonians are respectively

$$H_J^1(\sigma) = \sum_{i=1,N} J_i\sigma_i; \quad H_J^2(\sigma) = \sum_{i,k=1,N}' J_{i,k}\sigma_i\sigma_k; \quad H_J^3(\sigma) = \sum_{i,k,l=1,N}' J_{i,k,l}\sigma_i\sigma_k\sigma_l\,, \tag{9}$$

where the primed sum indicates that all the indices are different *(8, 9)*. The N variables J must have a variance of $O(N^{(1-p)/2})$ in order to have a non trivial thermodynamical limit. The variables σ are usual Ising spins, which take the values ± 1. From now on we will consider only the case $p > 2$.

In the replica approaches one assumes that at low temperatures the phase space breaks into many valleys, (i.e. regions separated by high barriers in free energy). One also introduces the overlap among valleys as

$$q(\alpha, \gamma) \equiv \frac{\sum_{i=1,N} \sigma\alpha_i \sigma\gamma_i}{2N}, \tag{10}$$

where $\sigma\alpha$ and $\sigma\gamma$ are two generic configurations in the valley α and γ respectively.

In the simplest version of this method *(8, 9)* one introduces the typical overlap of two configurations inside the same valley (sometimes denoted by q_{EA}). Something must be said about the distribution of the free energies of the valleys. Only those which have minimum free energy are relevant for the thermodynamics. One finds that these valleys have zero overlap and have the following distribution of the *total* free energy F:

$$P(F) \propto \exp(\beta m(F - F_0)). \tag{11}$$

The two parameters, q and m give complementary information:

- The parameter q tells us to what extent two generic configurations in the same valley are similar one to the other; with this normalization q should go to 1 to zero temperature.
- The parameter m tells us how much the different valley differs in free energy. If we assume that the free energy differences remain finite when the temperature goes to zero, m is proportional to the temperature at small temperature.

The two parameters q and m are enough to characterize the behavior the system. Indeed the average value of the free energy density (f) can be written in a self consistent way as function of m and q ($f(q, m)$) and the value of these two parameters can be found as the solution of the stationarity equations:

$$\frac{\partial f}{\partial m} = \frac{\partial f}{\partial q} = 0. \tag{12}$$

The quantity q is of order $1 - \exp(-A\beta p)$ for large p, while the parameter m is 1 at the critical temperature, and has a nearly linear behavior at low temperatures.

The thermodynamical properties of the model are the same as is the REM *(10)* (indeed we recover the REM when $p \to \infty$ *(8)*): there is a transition at T_c with a discontinuity in the specific heat, with no divergent susceptibilities.

A very interesting finding is that if we consider the infinite model and we cool it starting at high temperature, there is a dynamical transition at a temperature $T_D > T_c$

(11) -(16) . At temperatures less than T_D the system is trapped in a metastable state with energy greater than the equilibrium energy. The existence of these *infinite* mean life metastable states is one of the most interesting results for these models. The correlation time (not the equilibrium susceptibilities) diverges at T_D and the mode-mode coupling become exact in this region.

Systems Without Quenched Disorder. We could ask how much of the previous results can be carried to models without quenched results. It has been found in the framework of the mean field theory (i.e. when the range of the interaction is infinite), that there a the partial equivalence of Hamiltonians with quenched and random disorder. More precisely it often possible to find Hamiltonians which have the same properties (at least in a region of the phase space) of the Hamiltonian without disorder (17) - (23) . An example of this intriguing phenomenon is the following.

The configuration space of our model is given by N Ising spin variables (3) . We have a line of line of length N and on each site there is an Ising variable $\sigma(i)$. The Hamiltonian is a four spin antiferromagnetic interaction

$$H = N^{-1} \sum_{i,k,j,l} \sigma(i)\sigma(k)\sigma(j)\sigma(l), \tag{13}$$

where the sum is restricted to the sites that satisfy the condition $i + k \equiv_N j + l$ (where $\equiv_N$ stand for congruent modulus N). This Hamiltonian has been proposed in the study of low autocorrelation sequences. With some work it can be rewritten as:

$$H = \sum_{i=1,N} (|B_i|^2 - 1)^2, \tag{14}$$

$$\text{where} \quad B_i = \sum_{k=1,N} R_{i,k}\sigma_k. \tag{15}$$

Here R is an unitary matrix, i.e.

$$\sum_{k=1,N} R_{i,k}\overline{R_{k,m}} = \delta_{i,m}. \tag{16}$$

which in this particular case is given by

$$R(k,m) = \frac{\exp(2\pi i\ km)}{N^{1/2}} \tag{17}$$

We could consider two different cases (3) :

- The matrix R is a random orthogonal matrix.
- The matrix R is given by eq (17).

The second case is a particular instance of the first one, exactly in the same way that a sequence of all zeros is a particular instance of a random sequence.

The first model can be studied using the replica method and one finds results very similar to those of the p-spin model we have already studied.

Now it can be proven that the statistical properties of the second model are identical to those of the first model, with however an extra phase. In the second model (at least for some peculiar value of N, e.g. N prime (3, 18)) there are configurations which have exactly zero energy. These configuration form isolated valleys which are separated from the others, but have much smaller energy and they have a very regular structure (like a crystal). An example of these configurations is

$$\sigma_k \equiv_N k^{(N-1)/2} \tag{18}$$

(The property $k^{(N-1)} \equiv 1$ for prime N, implies that in the previous equations $\sigma_k = \pm 1$). Although the sequence σ_k given by the previous equation is apparently random, it satisfies so many identities that it must be considered as an extremely ordered sequence (like a crystal). One finds out that from the thermodynamical point of view it is convenient for the system to jump to one of these ordered configurations at low temperature. More precisely there is a first order transition (like a real crystallization transition) at a temperature, which is higher that the dynamical one.

If the crystallization transition is avoided by one of the usual methods, (i.e. explicit interdiction of this region of phase space or sufficiently fast cooling), the properties of the second model are exactly the same of those of the first model. Similar considerations are also valid for other spin models *(19–21)* or for models of interacting particles in very large dimensions, where the effective range of the force goes to infinity *(22, 23)* .

We have seen that when we remove the quenched disorder in the Hamiltonian we find a quite positive effect: a crystallization transition appears like in some real systems. If we neglect crystallization, which is absent for some values of N, no new feature is present in system without quenched disorder.

These results are obtained for long range systems. As we shall see later the equivalence of short range systems with and without quenched disorder is an interesting and quite open problem.

Some Considerations on Real Glasses

Here we will select some of the many characteristics of glasses we think are important and should be understood. The main experimental findings about glasses that we would like to explain are the following:

- If we cool the system below some temperature (T_G), its energy depends on the cooling rate in a significant way. We can visualize T_G as the temperature at which the relaxation times become of the order of a hour.
- No thermodynamic anomaly is observed: the entropy (extrapolated at ultraslow cooling) is a linear function of the temperature in the region where such an extrapolation is possible. For a finite value of the cooling rate, the specific heat is nearly discontinuous. Data are consistent with the possibility that the true equilibrium value of the specific heat is also discontinuous at a temperature T_c lower than T_G. The difference of the entropy among the glassy phase and the crystal seems to vanish approximately at T_c.
- The relaxation time (and quantities related to it, e.g. the viscosity) diverges at low temperature. In many glasses (the fragile ones) the experimental data can be fitted as

$$\tau = \tau_0 \exp(\beta B(T)) \tag{19}$$

$$B(T) \propto (T - T_c)^{-\lambda} \tag{20}$$

where $\tau_0 \approx 10^{-13}s$ is a typical microscopic time, T_c is near to the value at which we could guess the presence of a discontinuity in the specific heat and the exponent λ is of order 1. The so called Vogel-Fulcher law *(25)* states that $\lambda = 1$. The precise value of λ is not well determined. The value 1 is consistent with the experimental data, but different values are not excluded.

Now the theoretical interpretation of these results is quite clear (quite similar phenomena happens also in spin glass models in the mean field limits, where they have been very carefully studied both numerically and analytically). When we go near to the glassy phase the system may be frozen in many different configurations whose number is exponentially large. The entropy density can thus be written as

$$S(T) = S_c(T) + \Sigma(T), \tag{21}$$

where $S_c(T)$ is the entropy inside each of these configurations (which is likely not too different from that of the crystal) and $\Sigma(T)$ is proportional to the logarithm of the total number of configurations (i.e. it is the configurational entropy, sometimes also called complexity *(24)*). The complexity vanishes linearly when T approaches T_c and remains zero at smaller temperature. This fact implies the presence of thermodynamic transition at $T = T_c$.

The viscosity is dominated by the hopping from one equilibrium to an other equilibrium configuration. The number of equilibrium configurations decrease when $T \to T_c$ and therefore the number of particles that are involved in each hopping process must increase. The free energy barriers for such a process increases at the same time and diverges at T_c; the precise way in which they diverge is a highly debated problem.

A consequence of this picture is that below T_c there number of different equilibrium configurations (which contribute to the partition function) should not anymore be exponentially large. In this region replica symmetry should be broken, i.e. an arbitrarily small force should be enough to keep together two different replicas for an arbitrary large time.

The replica formalism seems therefore well suited to capture the phase transition in glasses. At this moment it is not clear which is the best way in which this formalism should be used. We shall see in the next section a possible application.

Toward Real Glasses

It is interesting to find out if one can directly construct an approximation for the glass transitions in liquids. Some progress has been done in the framework of mean field theory in the infinite dimensional cases. Indeed the model for hard spheres moving on a sphere can be solved exactly in the high temperature phase when the dimension of the space goes to infinity in a suitable way *(22, 23)* .

One of the most interesting results is the suggestion that the replica method can be directly applied to real glasses. The idea is quite simple. We assume that in the glassy phase a finite large system may be in different valleys, labeled by γ. The probability distribution of the free energy of the valley is given by eq. (8). We can speak of a probability distribution because the shape of the valleys and their free energies depends on the total number of particles. Each valley may be characterized by the density

$$\rho(x)_\gamma \equiv < \rho(x) >_\gamma \tag{22}$$

In this case we can define two correlation functions

$$g(x) = \frac{\int dy < \rho(y)\rho(y+x) >_\gamma}{V}$$

$$g_R(x) = \frac{\int dy < \rho(y) >_\gamma < \rho(y+x) >_\alpha}{V}, \tag{23}$$

where for simplicity we have assumed that the density of the particles is 1.

A correct description of the low temperature phase must take into account both correlation functions. The replica method does it quite nicely: g is the correlation function inside one replica and g_R is the correlation function among two different replicas.

$$g(x) = \frac{\int dy < \rho_a(y)\rho_a(y+x >}{V}$$

$$g_R(x) = \frac{\int dy < \rho_a(x)\rho_b(y+x) >}{V} \text{ with } a \neq b, \tag{24}$$

where a and b label the replica and $\rho_a(x)$ is the density of particles of the replica a.

This would be more or less the traditional approach if we assume the existence of only one valley for the system. On the contrary if we assume that the state may be in many different valleys we have to say something about the free energy distribution of free energy of the valleys. This can be done using the replica method and using the variable m, defined in eq. (11), which characterize the probability distribution of the free energy of those valley having a value of the free energy near the minimum.

The problem is now to write closed equations for the two correlation functions g_R and g. Obviously in the high temperature phase we must have that $g_R = 1$ and the non-vanishing of $g_R - 1$ is a signal of entering into the glassy phase.

The first attempt in this direction was only a partial success *(26)* and it will be described in the appendix. A generalized hypernetted chain approximation was developed for the two functions g and g_R.

The replica formalism provides an automatic bookkeeping of all complications which would arise from the existence of many states. If one assumes a given form for replica symmetry breaking, it correspond to a given form for the probability distribution of the w. In the simplest case, called one step breaking, one divides the n replicas in n/m groups of m replicas and one assumes the following structure of correlation functions:

$$\begin{aligned} < \rho_a(x)\rho_a(y) > &= g(x-y) , \\ < \rho_a(x)\rho_b(y) > &= g_R(x-y) \text{ for } a \neq b \text{ in the same group ,} \\ < \rho_a(x)\rho_b(y) > &= 1 \text{ for } a \text{ and } b \text{ in different groups. .} \end{aligned} \tag{25}$$

Fortunately the results for the free energy density (per replica) do not depend on n so that we do not have to specify this (unphysical) parameter.

A non trivial solution was found at sufficient low temperature both for soft and hard spheres and the transition temperature to a glassy state was not very far from the numerically observed one. Unfortunately the value of the specific heat at low temperature is not the correct one (it strongly increases with decreasing temperature). Therefore the low temperature behavior is not the correct one; this should be not a surprise because an explicit computation shows that the corrections to this hypernetted chain approximation diverge at low temperature.

These results show the feasibility of a replica computation for real glasses, however they point in the direction that one must use something different from a replicated version of the hypernetted chain approximation. At the present moment it is not clear which approximation is the correct one, but I feel confident that a more reasonable one will be found in the near future.

We consider a system of N interacting particles in a volume V. We study the infinite volume limit in which $N \to \infty$ at fixed $\rho \equiv N/V$. The Hamiltonian is given by:

$$H(x) = \sum_{i \neq k} U(x_i - x_k). \tag{26}$$

We will consider a soft sphere case $U(x) = r^{-12}$. We will work at density one and we will introduce the parameter $\Gamma \equiv \beta^4$.

In this case the glass phase may be reached only with a very fast cooling rate, otherwise the system goes into the crystal phase. This difficulty may be easily removed by considering a binary mixture, but this problem is not relevant here.

The hypernetted chain (HNC) approximation consists in considering only a given class of diagrams in the virial expansion *(27)*. It gives a reasonable account of the liquid phase. We will consider here this approximation because it has the advantage of having a simple variational formulation. In the liquid phase, where the density is constant, the usual HNC equation (for the non replicated system) can be written as

$$g(x) = \exp\left(-\beta U(x) + W(x)\right), \tag{27}$$

where:

$$\begin{aligned} g(x) &= (1 + h(x)) =< \rho(x)\rho(0) > -\delta(x), \\ W(x) &\equiv \int \frac{d^d p}{(2\pi)^d} e^{-ipx} \frac{\mathbf{h}(p)^2}{1 + \mathbf{h}(p)}, \end{aligned} \tag{28}$$

and we denote by $\mathbf{h}(p)$ the Fourier transform of $h(x) \equiv g(x) - 1$.

This equation can be derived by minimizing with respect to $g(x)$ the following free energy per unit volume, in the space of functions of $|x|$:

$$\beta F = \int d^d x \, g(x)[\ln(g(x)) - 1 + \beta U(x)] + \int \frac{d^d q}{(2\pi)^d} L(\mathbf{h}(q)), \tag{29}$$

where $L(x) \equiv -\ln(1 + x) + x - x^2/2$.

The HNC equation gives a description of the liquid phase which is not perfect, but precise enough for our purpose: The energy (or equivalently the pressure), does not depart more than 15% from the correct value, and the correlation function is also well reproduced (see Fig. (1) *(26)*).

Ref. *(26)* proposed a bold generalization of the HNC equations for n replicas. The replicated free energy is now

$$\beta F = \int d^d x \sum_{a,b} g_{ab}(x)[\ln(g_{ab}(x)) - 1 + \beta U(x)\delta_{a,b}] + \mathrm{Tr} L(\mathbf{h}), \tag{30}$$

where $\mathbf{h}$ is now an operator both in x space and in replica space.

If one finds that at low enough temperatures there is a solution of the variational equations $\delta F/\delta g_{a,b} = 0$, where g_{ab} is non zero off the diagonal, replica symmetry is broken. In the case where g_{ab} is of the form shown in eq. (25), this equation can be used to compute the properties of the correlation function in the glassy phase. This approach amounts to a study of the density modulations in the glass phase at the level of the two point function. In the glass phase $\rho\gamma(x)$ becomes space dependent. However, as argued in the introduction, the necessity of averaging over the states γ forces us to study this x dependence at the level of the two point correlations. So our correlation g_R reflects the structure of $\rho\gamma(x)$ as a sum of peaks of unit weights, smoothed by the average over states.

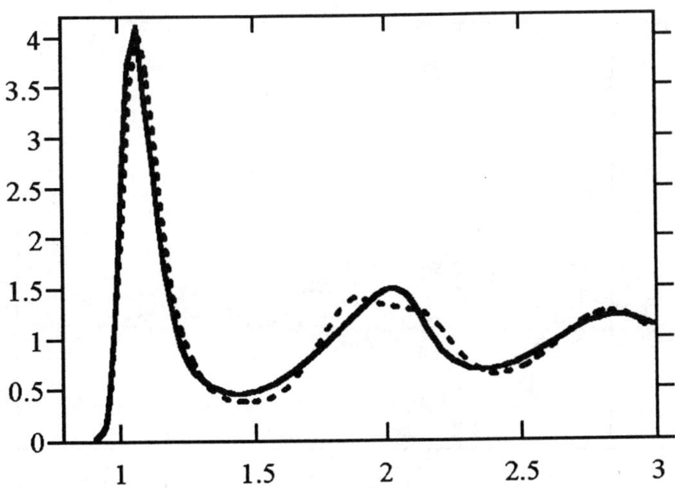

Figure 1: The correlation function of a system of soft spheres as a function of the distance at the dimensionless inverse density $\Gamma = 1.6$ corresponding to the glassy transition: numerical simulations (dashed line) and replica symmetric HNC equation (full line).

Within the one step breaking scheme (25), the free energy per replica is:

$$\beta F = \int d^d x \{g(x)[\ln(g(x)) - 1 + \beta U(x)] - (1-m) g_R(x)[\ln(g_R(x)) - 1]\}$$
$$- \int \frac{d^d q}{(2\pi)^d} \left(\frac{1}{2}\mathbf{h}(q)^2 - \frac{1}{2}(1-m)\mathbf{h}_R(q)^2 - \mathbf{h}(q) \right.$$
$$\left. + \frac{1}{m}\ln[1 + \mathbf{h}(q) - (1-m)\mathbf{h}_R(q)] - \frac{1-m}{m}\ln[1 + \mathbf{h}(q) - \mathbf{h}_R(q)]. \right) \quad (31)$$

Two transitions can be found: the static and the dynamical transition.

- The static transition is identified as the temperature (or density) at which there exists a non trivial solution to the replicated HNC equations :

$$\frac{\delta F}{\delta g(x) = 0} \quad \delta F/\delta g_R(x) = 0 \quad (32)$$

and

$$\partial F/\partial m = 0, \quad (33)$$

for $m \in [0,1]$ (in fact we must minimize the free energy with respect to g, but maximize with respect to g_R and m).
- The dynamical transition may be characterized as the highest temperature at which there is a non trivial solution of the two stationarity equations $\delta F/\delta g(x) = 0$ and $\delta F/\delta g_R(x) = 0$ at $m = 1^-$. The corresponding equations are obtained by substituting $m \to 1$ in the first two equations of (32). The equation for g is identical to the usual HNC equation (28), while g_R is a solution of $g_R(x) = \exp(W_R(x))$, with W_R can be extracted from the second equation of (32) at $m = 1$.

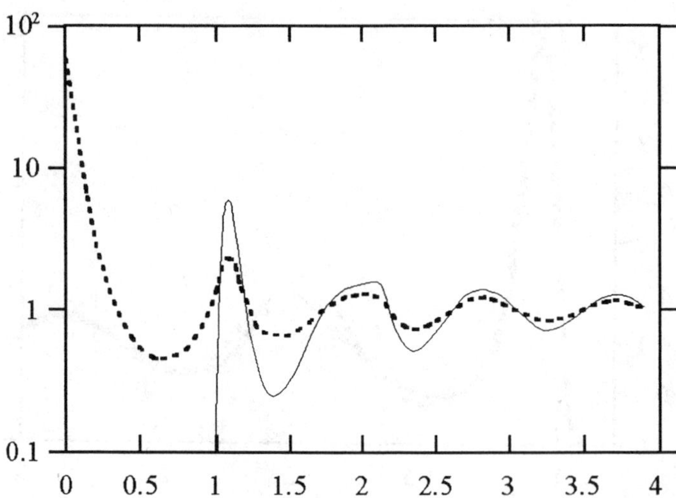

Figure 2: The correlations $g(r)$ (full line) and $g_R(r)$ (broken line) as functions of the distance, for soft spheres at the density where replica symmetry is broken (e.g. $\gamma = 2.15$).

If we look for the numerical solution of the replicated HNC equations, we find a dynamical transition at $\Gamma \simeq 2.05$, and a static replica symmetry breaking solution at $\Gamma \simeq 2.15$. In numerical simulations the glass transition is found at a smaller value of Γ, namely $\Gamma = 1.6$. In the glass phase, the correlation function $g_1(r)$ is essentially a smoothed form of the function $g(r)$ plus an extra contribution at short distance which has integral near to 1 (see Fig.2). This form seems very reasonable: considering the definition (2) of g_R, we see that it basically characterizes the average over γ of the product $\rho\gamma(x)\rho\gamma(y)$, which is precisely expected to have this kind of peak structure.

In spite of this nice form for g_R, this solution has some problems. A first one is found on the value of the energy. Although there is a discontinuity in the specific heat at T_c, it is extremely small and the final effects on the internal energy are more or less invisible. The specific heat remains extremely large. Moreover the value of m has a very unusual dependence on the temperature. In all the known models with one step replica symmetry breaking, the breakpoint m varies linearly with T at low temperatures. Here we have a very different behavior.. We have also computed the dynamical internal energy and found that it differs from the equilibrium one by an extremely small account.

If we consider the qualitative behavior of the correlation functions, we find a reasonable form, on the other hand the energy in the glassy phase turns out to be quite wrong. This computation should be considered only as a first step and work is in progress to find another approximation (different from the replicated HNC) which should be similar in spirit, but technically more sound.

Literature Cited

[1] Mézard, M.; Parisi, G.;Virasoro, M.A. *Spin Glass Theory and Beyond*, World Scientific: Singapore, 1987.
[2] Parisi, G. *Field Theory, Disorder and Simulations*, World Scientific: Singapore, 1992.
[3] Marinari, E.; Parisi, G.; Ruiz-Lorenzo, J.J. *Numerical Simulations of Spin Glass Systems*, cond-mat 9701016.
[4] Caracciolo, S.; Parisi, G.; Patarnello, S.; Sourlas, N. *Europhys. Lett.* **1990,** *11,* 783.
[5] Kurchan, J.; Parisi, G.; Virasoro, M. A. *J. Physique* **1993,** *3,* 18.
[6] Parisi, G. *Gauge Theories, Spin Glasses and Real Glasses*, Talk presented at the Oskar Klein Centennial Symposium, cond-mat 9411115.
[7] Franz, S.; Parisi, G. *J. Phys. I (France)* **1995**, *5,* 1401.
[8] Gross, D. J.; Mézard, M. *Nucl. Phys. B* **1984,** *240,* 431.
[9] Gardner, E. *Nucl. Phys. B* **1985,** *257,* 747.
[10] Derrida, B. *Phys. Rev. B* **1981,** *24,* 2613.
[11] Kirkpatrick, T. R.; Thirumalai, D. *Phys. Rev. Lett.* **1987,** *58,* 2091.
[12] Kirkpatrick, T. R.; Thirumalai, D. *Phys. Rev. B* **1987,** *36,* 5388.
[13] Kirkpatrick, T. R.; Thirumalai, D.;Wolynes, P.G. *Phys. Rev. A* **1989,** *40,* 1045.
[14] Crisanti, A.; Horner, H.; Sommers, H.-J. *Z. Phys. B* **1993,** *92,* 257.
[15] Cugliandolo, L.; Kurchan, J. *Phys. Rev. Lett.***1993,** *71,* 173.
[16] Bouchaud, J.-P.; Cugliandolo, L.; Kurchan, J.; Mézard, M. cond-mat 9511042.
[17] Marinari, E.; Parisi, G.; Ritort, F. *J.Phys.A (Math.Gen.)* **1994,** *27,* 7615; *J.Phys.A (Math.Gen.)* **1994,** *27,* 7647.
[18] Borsari, I.; Graffi S.; Unguendoli, F. *J.Phys.A (Math.Gen.)*, to appear and *Deterministic Spin Models With a Glassy Phase Transition*, cond-mat 9605133.
[19] Franz, S.; Hertz, J. *Nordita Preprint*, cond-mat 940807.
[20] Marinari, E.; Parisi, G.; Ritort, F. *J. Phys. A (Math. Gen.)* **1995,** *28,* 1234.
[21] Marinari, E.; Parisi, G.; Ritort, F. *J. Phys. A (Math. Gen.)* **1995,** *28,* 327.
[22] Cugliandolo, L.; Kurchan, J.; Monasson, E.; Parisi, G. *Math. Gen.* **1996,** *29,* 1347.
[23] Cugliandolo, L.; Kurchan, J.; Parisi, G.; Ritort, F. *Phys. Rev. Lett.* **1995,** *74,* 1012.
[24] Niewenhuizen, Th. M. *Complexity as the Driving Force for Glassy Transitions*, cond-mat 9701044 and references therein.
[25] Vogel, H. *Phys. Z*, **1921**, *22*, 645; Fulcher, G.S. *J. Am. Ceram. Soc.,* **1925,** *6,* 339.
[26] Mézard, M.; Parisi, G cond-mat 9602002.
[27] See for instance Hansen, J. P.;Macdonald, I. R. *Theory of Simple Liquids*; Academic: London, 1986, or Temperley, H.N.V.;Rowlinson, J.S.; Rushbrooke, G.S. *Physics of Simple Liquids*; NorthHolland: Amsterdam 1968.

Chapter 9

Growing Length Scales in Supercooled Liquids

Raymond D. Mountain

Physical and Chemical Properties Division, National Institute of Science and Technology, Gaithersburg, MD 20899

Molecular dynamics simulations of strongly supercooled liquids are used to identify a length scale that increases with the degree of supercooling. Some of the consequences for the theory of the strongly supercooled state of glass forming liquids implied by this particular length scale are examined. Some new ways of characterizing supercooled states are suggested.

In this chapter, we examine some of the evidence that has been developed from molecular dynamics simulations about the ways supercooled liquids differ from equilibrium liquids. Our attention is mainly on the nature of the "order" that develops as the liquid is supercooled. The notion of the existence of a length scale that increases with deceasing temperature enters several theories of glass formation and the supercooled liquid.(*1*) The concept of an increasing length goes back at least to the introduction by Adam and Gibbs of the concept of "cooperative rearranging regions" in supercooled liquids.(*2*) These regions are purported to grow with decreasing temperature. Unfortunately, an unambiguous identification of such regions has not been reported. As will be demonstrated below, it is possible to identify a "hydrodynamic length", $\mathcal{L}$, that increases with the degree of supercooling.(*3*) It is not obvious how this length would pertain to a theory and probably this length is not directly related to cooperative regions nor to simple correlation measures. The growth with decreasing temperature of this unconventional length scale indicates that significant differences exist between the equilibrium liquid and the supercooled liquid. These differences appear to be associated with some sort of inhomogeneity that develops as the temperature decreases. The type of inhomogeneity is not known at present.

The message contained in this article has two parts. The first is that the order in a supercooled liquid involves inhomogeneities that involve many particles. The second is that the characterization of the inhomogeneities that develop in supercooled liquids requires a statement of the observation time as well as of the spatial extent of the correlations. This will be illustrated using a two-dimensional model of a glass forming liquid mixture. This simple model has the advantage that it is easy to examine the configuration of the atoms visually. One type of inhomogeneity that develops in this system is identified using trajectory plots of the atoms. A detailed characterization of this inhomogeneity is a topic for future research.

The hydrodynamic length provides an example of how molecular dynamics simulations can contribute to the process of developing a coherent understanding of a complex subject such as strongly supercooled liquids. A great deal of effort has been devoted to examining the predictions of mode coupling theory.(*4*) This topic will not be examined here as the theory seems to apply only to the high temperature region where inhomogeneities are not significant.

In this article, the temperature is used as the important variable when defining the state of the fluid. It is also possible to consider the case where the temperature is held fixed and the pressure is varied.(*5–7*) In that case, one would speak of overpressing rather than supercooling. Although there are some subtle differences that can occur when the pressure is the variable defining the state, the distinction between the two paths in thermodynamic space will be ignored here.

The remainder of the chapter is organized as follows. First, a small amount of material needed to understand the basis for the hydrodynamic length is presented. Some of the previous molecular dynamics studies of lengths in supercooled liquids are described in the section on largely unsuccesful searches for growing lengths. Next, a two-dimensional Lennard-Jones mixture is described. This glass forming mixture is used to illustrate how one obtains the hydrodynamic length that is introduced in the section on transverse current correlations. The chapter concludes with a discussion of the possible significance of the existence of this length that increases with the degree of supercooling of the mixture.

Some preliminaries

Molecular dynamics simulations are extensively discussed in several books(*8–10*) so only details of this method that are pertinent to this chapter will be introduced here. The molecular level formulation of collective quantities is typically made in terms of spatial Fourier transforms of the number density and the momentum density. The molecular level theory of transport coefficients is formulated in terms of these densities.(*11*) The Fourier transform of the number density at wavevector **q** is

$$n_{\mathbf{q}} = \frac{1}{\sqrt{N}} \sum_{j=1}^{N} \exp(i\mathbf{q} \cdot \mathbf{r}_j)$$

and the Fourier transform of the momentum density is

$$\mathbf{J}_{\mathbf{q}} = \frac{1}{\sqrt{N}} \sum_{j=1}^{N} \mathbf{p}_j \exp(i\mathbf{q} \cdot \mathbf{r}_j).$$

Here N is the number of particles, and $\mathbf{r}_j$ and $\mathbf{p}_j$ are the position and the momentum of particle j. The density-density time correlation,

$$F(q,t) = \langle n_{\mathbf{q}}(t_0) n_{-\mathbf{q}}(t+t_0) \rangle,$$

is known as the intermediate scattering function and is observable in scattering measurements.(*12*) It is also a quantity that can be generated in a molecular dynamics simulation. The $\langle ... \rangle$ notation indicates an average over time origins, t_0.

The corresponding quantity for the current,

$$C(q,t) = \langle \mathbf{J}_{\mathbf{q}}(t_0) \mathbf{J}_{-\mathbf{q}}(t+t_0) \rangle,$$

can be resolved into three distinct correlation functions. The interesting ones are the longitudinal current correlation function, $C_L(q,t)$, the correlation function of the momentum components parallel to the wavevector **q**, and the transverse current correlation function, $C_T(q,t)$, the correlation function of the momentum components perpendicular to the wavevector **q**. The remainder of the current-current correlation function describes cross correlations between the parallel and perpendicular components of the momentum and is of no interest for the present discussion. The longitudinal current correlation function is proportional to the second time derivative of $F(q,t)$ and thus is also an experimentally observable quantity. The transverse current correlation function will be discussed further here. Direct observation of the transverse current correlation function is possible only through simulations.

Largely unsuccessful searches for growing lengths

There have been several molecular dynamics studies directed toward identifying a length scale that increases with the degree of supercooling. In these studies, a correlation length is associated with the decay of spatial correlations of some order parameter. The challenge is to find a suitable order parameter. To date, no order parameter that drives the glass transition has been found. However, some other interesting features have been reported.

The first study that we mention found no indication in the relaxation of the intermediate scattering function, $F(q,t)$, of a length that changed with the degree of overpressing.(*6*) Instead, a change in the relaxation process was observed. At high temperatures (or low densities) the relaxation involves fluidlike, continuous motion of the atoms. As the degree of supercooling increases, the relaxation mechanism changes to one of atoms hopping over barriers. This change occurs well before the glass transition is reached and is consistent with the picture provided by extended mode coupling theory.(*4*)

The relaxation of near neighbor pairs and the relaxation of bond orientational correlations of neighbors were examined for a strongly supercooled mixture.(*13*) No indication of any length scale in these relaxation processes was found as the liquid was supercooled. Essentially the same results were reported for another mixture.(*14*) These authors also examined a four-point spatial correlation function and compared it to a product of two-point correlation functions. No difference was found beyond first neighbor distances, indicating that no growing length exists for these correlations.

There are some simulation results that suggest that some sort of length develops. The first result is that a specific component of the stress-stress correlation function develops a range greater than the near neighbor distance at a temperature well above the glass transition temperature.(*15*) This observation was made for a model of liquid iron and has not been reported for other systems. It would be interesting to see if this type of order exists in other glass forming systems.

A second result is that in binary mixtures of Lennard-Jones particles with not too different sizes, a specific type of spatial ordering develops above the glass transition.(*5*) This ordering is described in terms of the "inherent structure" of the potential energy landscape.(*16*) At high temperatures, there are many local minima in the potential surface and the inherent structure is independent of the temperature. In the strongly supercooled liquid region, the system becomes localized in a single minimum on the potential surface. This minimum is one where dense packing is preferred. This process of settling into a single minimum is the structural relaxation mechanism. The resulting order is described as a cluster of face sharing and interpenetrating icosahedra. This cluster tenuously extends over the volume of the system. This is a rather different picture of what constitutes "order". Instead of order in terms of an order parameter defined at the atomic level, order is found between groups of atoms and can be viewed as a cluster of clusters.

A two-dimensional example

The example system that we use to illustrate the discussion of the hydrodynamic length is a two-dimensional Lennard-Jones mixture. The ratio of diameters of the two components is 1.5/1. This size difference provides the frustration of crystallization needed to reach the glassy state.(*17*). There are 100 large type 1 particles and 400 small type 2 particles. The interactions are specified to be

$$\phi_{\alpha\beta}(r) = 4\epsilon\left[(\sigma_{\alpha\beta}/r)^{12} - (\sigma_{\alpha\beta}/r)^{6}\right],$$

with σ_{11} = 1.5, σ_{12} = 1.25, and σ_{22} = 1, so σ_{22} sets the length scale. The energy scale is set by ϵ = 1, and the mass of type 2 particles is unity. The mass of the larger particles is taken arbitratily to be 2.2. The time unit

$$\tau = \sqrt{m_2 \sigma_{22}^2/\epsilon}$$

is obtained by writing the equations of motion in terms of these units. Temperatures are reported in units of ϵ/k_B, the ratio of the energy unit to Boltzmann's constant.

This system has been used in a previous study of the strongly supercooled state.(*18*) The results reported here are for a different set of supercooled states. The 500 particles were placed in a square cell with sides of length L = $24.5\sigma_{22}$ subject to periodic boundary conditions in both directions. The equations of motion were integrated using the Beeman algorithm(*19, 20*) with a time step $\Delta t = 0.005\tau$. A total of six states at this density were simulated. The temperature varied from a high temperature liquid state with T = 3.18 down to a solid, glassy state with T = 0.69.

Transverse current correlation functions

Transverse current correlation functions are the vehicle to be used to introduce a hydrodynamic length using the method introduced in (*3*). First, the transverse current is that part of the momentum density that is perpendicular to the wavevector. In the small-q limit, where a hydrodynamic description of the fluctuations in the liquid is applicable,

$$C_T(q,t)/C_T(q,0) = \exp(-\eta q^2 t/\rho),$$

where η and ρ are the shear viscosity and the density of the liquid.

At larger values of the wavevector, the liquid begins to exhibit viscoelastic behavior rather than hydrodynamic behavior. If the density is sufficiently high, and temperature is sufficiently low, the transverse current correlation function will exhibit oscillatory behavior. A consequence of this is that the power spectrum of the transverse current fluctuations, $C_T(q,\omega)$, which is the the real part of the time Fourier transform of $C_T(q,t)$, will have a maximum at a frequency greater than zero. The position of the maximum in the power spectrum, $\omega_m(q)$, is the propagation frequency of a transverse momentum fluctuation, sometimes known as a shear wave.

An example of this type of analysis is shown in Figures 1–3 for the two-dimensional Lennard-Jones liquid mixture discussed previously. Figure 1 shows $C_T(q,\omega)$ for the T = 1.65 state with $q = 2\pi/L$. Figure 2 is a plot of $\omega_m(q)$ vs q for this state. There is a range of q with linear dispersion typical of a propagating mode. This region is followed by a nondispersive region at larger values of q that is not shown in Figure 2. The nondispersive region reflects the dominance of single particle momentum correlations at large q values and indicates that the collective nature of the transverse current fluctuations is largely absent at these values of the wavevector.

A hydrodynamic length is obtained from the dispersion curve of the transverse modes by extrapolating $\omega_m(q)$ to zero. The intercept of the dispersion curve on the q-axis determines the smallest value of wavevector, q_{min} for which a propagating transverse mode exists. That in turn determines the largest wavelength of these modes and that wavelength is the hydrodynamic length, $\mathcal{L}$,

$$\mathcal{L} = 2\pi/q_{min}.$$

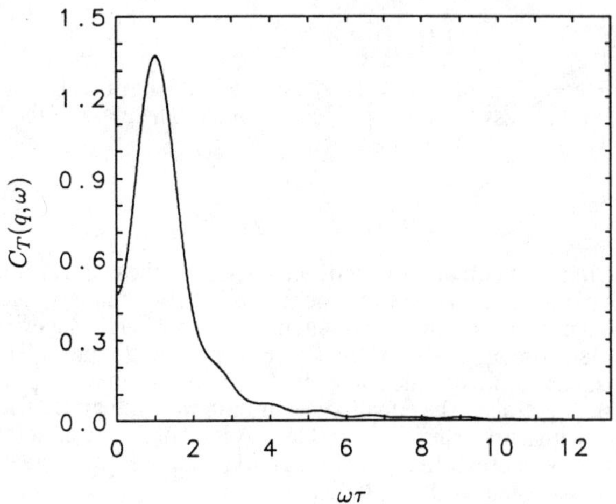

Figure 1. The power spectrum of the transverse current correlation function for the T = 1.65 state of the two-dimensional mixture.

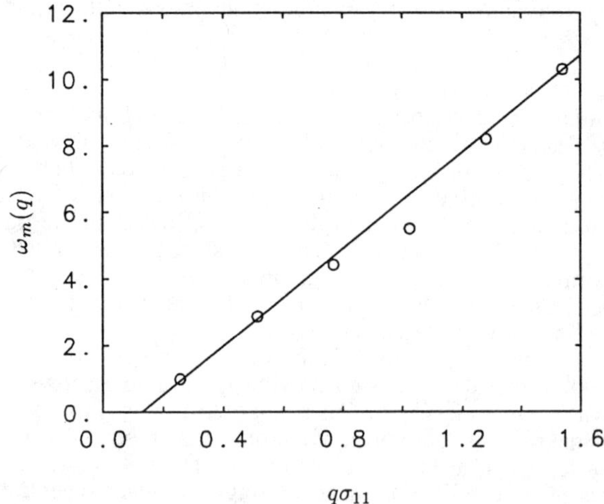

Figure 2. The dispersion relation for the transverse current correlations for the T = 1.65 state of the two-dimensional mixture. The solid line provides the extrapolation to $\omega_m(q) = 0$.

The temperature variation of $\mathcal{L}$ is displayed in Figure 3 with $\mathcal{L}$ in units of the simulation cell size L. It should be noted that the explicit values of $\mathcal{L}$ probably are system size dependent, particularly those that are larger than L itself. However, the important point is that this length increases rapidly with decreasing temperature. This demonstrates that there is a growing length scale for supercooled liquids. This property of the hydrodynamic length has been noted in many different systems.(*21, 22*)

Discussion

The existence of a growing hydrodynamic length has some implications about the type of order that develops as a liquid is supercooled. The existence of propagating transverse current correlations means that there are regions in the fluid that exhibit solid-like elastic behavior for time intervals of sufficient duration that oscillations in the transverse current can occur. It is not necessary that these regions are permanent features. In fact, they cannot be permanent when there is linear growth with time of the mean-square displacement of particles for long times. This is illustrated in Figure 4 where the time dependence of the mean-square displacement of the particles is displayed for the T =1.65 state. The lifetime of these transient regions is on the order of 1–2 τ for the two-dimensional states that are examined in Figures 1–2.(*22*) A similar time interval was found for a three-dimensional mixture. This indicates that the significant change is in the growth of $\mathcal{L}$ and not in the lifetimes of the solid-like regions.

The two-dimensional mixture can provide some insight into the type of ordering that leads to propagating transverse modes. Figure 5 contains a series of trajectory plots for the state examined in Figures 1–2. The plots are made by placing a small circle at the position of each particle at each time step for some specified interval. The light gray part of the trajectory represents postitions with a solid-like displacement from the initial position of the particle while the black part of the trajectory represents the liquid-like portion of the trajectory. For Figure 5, the boundary between solid-like and liquid-like displacements was set to occur when the square of the displacement of a particle from its initial position was 0.1. As may be observed from Figure 4, this is approximately the magnitude of the displacement where the onset of the linear portion of the mean-square displacement occurs. The time origin for each of the trajectories is identical. The upper left-hand trajectory is for the interval 0.5τ. For this time interval the system appears almost totally solid-like. The upper right-hand trajectory is for the longer interval 1τ. Note that some regions of the system are beginning to exhibit more fluid-like behavior. The lower left-hand trajectory is for the interval 2τ. The regions of fluid-like behavior are seen to be expanding at the expense of the solid-like regions. Even so, there are regions that are principally solid-like with an extent of several particle diameters. The final trajectory is for the relatively long interval, 4τ. The fluid-like regions are now seen to extend across much of the system.

This sequence of trajectory plots indicates that the fluid-like regions tend to nucleate and grow over an interval of several τ rather than occuring more-or-less uniformly over the system. This sort of behavior is what one might reasonably expect to be needed for the propagation of transverse modes with wavelengths of several particle diameters. This behavior also illustrates the point that the inhomogeneity that develops in this supercooled liquid has both a spatial and temporal component. The trajectory plots also suggest that the solid-like regions are fairly compact while the fluid-like regions are tenuous and develop along the boundaries of the solid-like regions. A more quantative analysis of these regions has not been made. One approach to the characterization of these regions is found in (*18*).

The existence of the hydrodynamic length, $\mathcal{L}$ and the growth of this length with the degree of supercooling is a general feature of supercooled liquids. This feature indicates the presence of a particular type of inhomogeneity in the liquid that is revealed only when local structure is examined over "short" time intervals. This has consequences for the theory of the strongly supercooled liquids that are only now being explored.

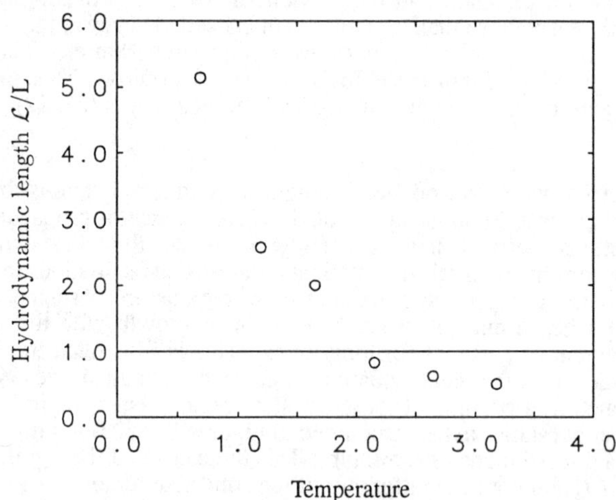

Figure 3. The temperature dependence of the hydrodynamic length, $\mathcal{L}$, for the two-dimensional mixture. $\mathcal{L}$ is in units of the size of the simulation cell, L. The points above the horizontal line indicate that the longest wavelength propagating transverse excitation excedes the size of the simulation cell.

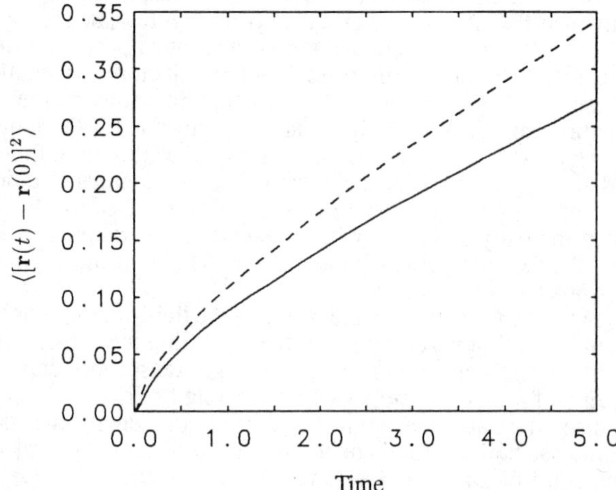

Figure 4. The time dependence of the mean-square displacement of the particles of the two-dimensional mixture for the T = 1.65 state becomes linear for long times. The solid line represents the type 1 particles and the dashed line represents the type 2 particles.

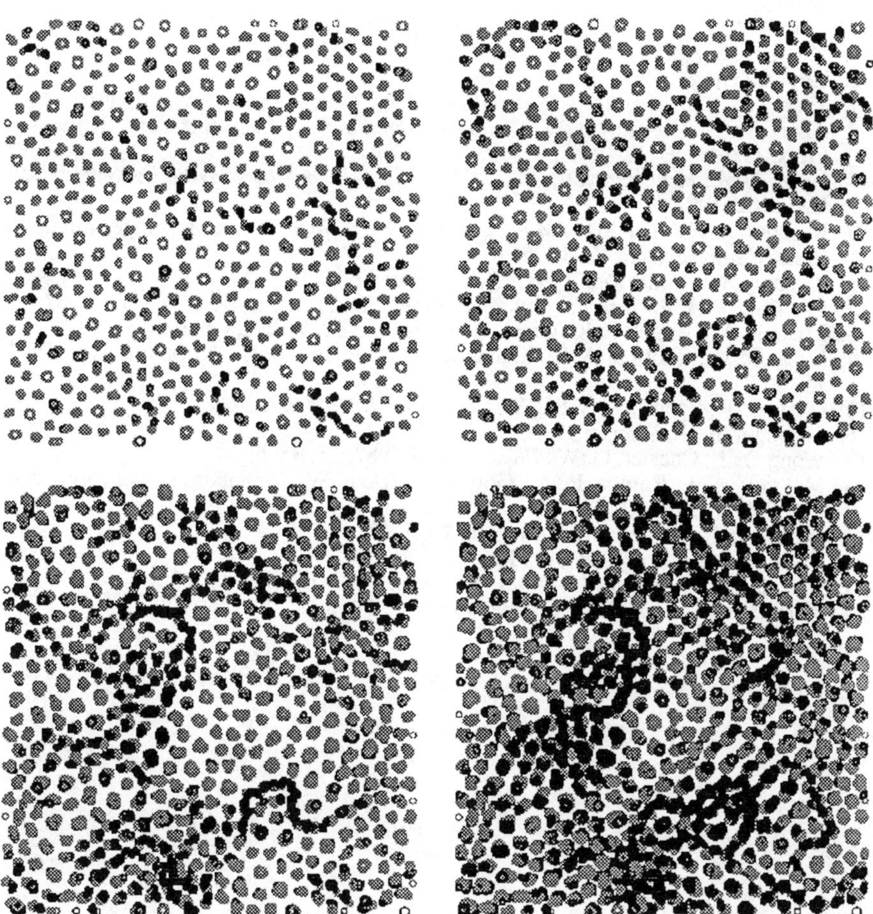

Figure 5. Trajectory plots for the T = 1.65 state of the two-dimensional mixture. The black portions of the plots indicate fluid-like mobility and the gray portions of the plots indicate solid-like mobility. The duration for each plot is indicated in the text.

Literature cited

1. Sethna, J. P.; *Europhys. Lett.* **1989**, *6*, 529.
2. Adam, G; Gibbs, J. H.; *J. Chem. Phys.* **1965**, *43*, 139.
3. Jacucci, G; McDonald, I. R. In *Liquid and Amorphous Metals*; Lüscher, E.; Coufal, H., Eds.; Sijthoff & Noordhoff, Alphen ann den Rijn, The Netherlands, 1980; pp 143-157.
4. Götze, W.; Sjögren, L; *Rep. Prog. Phys.* **1992**, *55*, 241.
5. Jónsson, H.; Andersen, H. C.; *Phys. Rev. Lett.* **1988**, *60*, 2295.
6. Ullo, J.; Yip, S.; *Phys. Rev. A* **1989**, *39*, 5877.
7. Shumway, S. L.; Clarke, A. S.; Jónsson, H.: *J. Chem. Phys.* **1995**, *102* 1796.
8. Allen, M. P.; Tildesley, D. J. *Computer Simulation of Liquids*; Oxford University Press: Oxford, U. K., 1987.
9. Haile, J. M.; *Molecular Dynamics Simulation: Elementary Methods*: Wiley: New York, NY, 1992.
10. Rapaport, D. C.; *The Art of Molecular Dynamics Simulation*; Cambridge University Press; Cambridge, U. K., 1995.
11. Zwanzig, R.; *Annu. Rev. Phys. Chem.* **1965**, *16*, 67.
12. Egelstaff, P. A. *An Introduction to the Liquid State*; Academic Press, NY, 1967.
13. Ernst, R. M.; Nagel, S. R.; Grest, G. S.; *Phys. Rev. B* **1991**, *43*, 8070.
14. Dasgupta, C.; Indrani, A. V.; Ramaswamy, S.; Phani, M. K.; *Europhys. Lett.* **1991**, *15*. 307.
15. Chen, S.-P.; Egami, T.; Vitek, V.; *Phys. Rev. B* **1988**, *37*, 2440.
16. Stillinger, F. H.; *Science* **1995**, *267*, 1935.
17. Wong, Y. J.; Chester, G. W.; *Phys. Rev. B* **1987**, *35*, 3506.
18. Mel'cuk, A. I.; Ramos, R. A.; Gould, H.; Klein, W.; Mountain, R. D.; *Phys. Rev. Lett.* **1995**, *75*, 2522.
19. Schofield, P.; *Comp. Phy. Comm.* **1973**, *5*, 17.
20. Beeman, D.; *J. Comp. Phys.* **1976**, *20*, 130.
21. Barrat, J.-L.; Klein, M. L.: *Annu. Rev. Phys. Chem.* **1991**, *42*, 23.
22. Mountain, R. D.; *J. Phys. Chem.* **1995**, *102*, 5408.

Chapter 10

Shear Viscosity and Diffusion in Supercooled Liquids

Frank H. Stillinger

Bell Laboratories, Lucent Technologies, Inc., 600 Mountain Avenue, Room 1D–359, Murray Hill, NJ 07974

Under normal liquid conditions (at equilibrium or when moderately supercooled), molecular rotational and translational diffusion adhere closely to the simple Stokes-Einstein-Debye hydrodynamic model. Deeply supercooled "fragile" liquids present a striking exception, with translational diffusion alone occurring up to $10^2 - 10^3$ times "too fast". This anomaly rests upon the statistics of fluctuating local viscosity in the fragile glass-forming medium. The simplest version of this picture identifies large fluidized domains that spontaneously appear and disappear in a dynamically inert matrix. Plausible and self-consistent values characterizing these fluctuations are cited for orthoterphenyl (OTP) at its glass transition.

Liquids that supercool easily and form glasses have presented many conceptual puzzles, whose resolution has led to deeper understanding of these substances at the molecular level. This paper focusses on one of those puzzles that mainly concerns the so-called "fragile" glass formers, identified by the markedly non-Arrhenius temperature dependences of their shear viscosity $\eta(T)$ and various mean relaxation times $\tau(T)$ (*1, 2*). Specifically, the need to reconcile separate measurements of translational and of rotational diffusion rates in fragile liquids just above their glass transition temperature T_g leads inevitably to a molecular picture of the medium that is strongly inhomogeneous in both space and time.

Translational and rotational diffusion constants, D_{trans} and D_{rot}, measure rates of increase with time of the mean-square positional and angular displacements for tagged molecules:

$$\begin{aligned} \langle(\delta\mathbf{r})^2\rangle &\cong 6D_{\text{trans}}\delta t\,, \\ \langle(\delta\theta)^2\rangle &\cong 4D_{\text{rot}}\delta t\,. \end{aligned} \tag{1}$$

It has been traditional to assess measured values of these diffusion constants by means of the elementary Stokes-Einstein-Debye (SED) hydrodynamic model of a Brownian sphere with radius R embedded in a homogeneous, incompressible, viscous medium (*3*)–(*6*). Assuming that "sticking" boundary conditions apply in this model, one has

$$\begin{aligned} D_{\text{trans}} &= k_BT/6\pi\eta(T)R\,, \\ D_{\text{rot}} &= k_BT/8\pi\eta(T)R^3\,, \end{aligned} \tag{2}$$

where k_B is Boltzmann's constant. These expressions apply both to the case of a pure liquid (where the "sphere" represents the same chemical species as that composing the surrounding viscous medium), as well as to the cases of selected "probe" molecules of various sizes dissolved in the viscous liquid of interest.

The SED expressions 2 enjoy surprising success at representing measured diffusion rates for a wide variety of substances in their equilibrium liquid ranges, and under modest amounts of supercooling. The temperature variations of D_{trans} and D_{rot} track that of $T/\eta(T)$, and the implied hydrodynamic radii R from equation 2 have reasonable values for the molecules examined. Furthermore, the SED model retains its validity under elevated pressure conditions (*7*).

Recent experiments on deeply supercooled fragile liquids have revealed strong violations of the simple SED model that produces equation 2 (*8*)–(*11*). The second of equation 2 continues to describe rotational diffusion down to T_g (at which point η may be approximately 15 orders of magnitude larger than its melting temperature value); however measured translational diffusion rates below about $1.2\,T_g$ display positive deviations from the first of SED expressions 2 that can grow to an enhancement factor of 10^2 or more (*8*)–(*11*).

Even though the selective breakdown of the SED description at first sight seems paradoxical, the large magnitude of the phenomenon at T_g suggests that an explanation need only be straightforward, not subtle. Indeed, such an explanation is available (*12, 13*). The following Section II lays the groundwork for resolving the puzzle by presenting some general remarks about the "rugged potential energy landscape" that describes intermolecular interactions in fragile liquids. This leads naturally into Section III, a mesoscopic description of the fragile glass-forming medium in terms of a fluctuating local viscosity field. Section IV then shows how certain combinations of length and time scale parameters characterizing this field create a selective bottlenecking scenario that decouples the rotational and translation diffusion processes, just as experiment demands. Section V concludes the presentation with a discussion of several issues, including stretched-exponential (KWW) relaxation, and connections between the SED model violations and the Adam-Gibbs theory of relaxation processes in glass formers (*14*).

I Potential Energy Landscape

Molecular interactions in a glass-forming substance, and consequently its thermal and dynamical behaviors, are determined by a potential energy function $\Phi(\mathbf{r}_1 \ldots \mathbf{r}_N)$. Here the $\mathbf{r}_i$ represent coordinates of the individual molecules, N in number, and generally will include internal degrees of freedom (orientation, conformation) as well as center position. The total number of variables for a macroscopic material sample may be some multiple of Avogadso's number, and the geometry of the Φ hypersurface in this huge multidimensional configuration space is incredibly complicated. Yet some simple features of the "Φ-scape" and their implications for the SED puzzle can be extracted from basic measurements.

Orthoterphenyl (OTP) has become one of the most widely studied fragile glass formers. Plazek, Bero, and Chay (*15*) have tabulated its shear viscosity $\eta(T)$ over a wide temperature range, from well above its melting point ($T_m = 329\text{K}$) to the vicinity of its glass transition ($T_g \cong 240$K). One can represent this data in the following manner:

$$\eta(T) = \eta_0 \exp[F^*(T)/k_B T] \tag{3}$$

where η_0 is temperature independent, and F^* is an activation free energy for viscous flow. Formal activation energy E^* and activation entropy S^* then follow:

$$\begin{aligned} E^*(T) &= \partial[F^*(T)/T]/\partial(1/T)\,, \\ S^*(T) &= -\partial F^*(T)/\partial T\,. \end{aligned} \tag{4}$$

One finds that E^* is approximately 4.3 kcal/mole in the hot OTP liquid regime ($T \cong$ 500K), but increases dramatically to approximately 86 kcal/mole at T_g. Over the same range S^*/k_B increases by about 140. Similar results would be found for other fragile liquids.

The large increases in E^* and S^* as T declines stem from the strongly non-Arrhenius character of $\eta(T)$ for OTP, and indicate a change in kinetic mechanism for molecular rearrangements. In view of the fact that the heat of fusion of OTP is 4.103 kcal/mole (*16*), the above numbers suggest that while molecular motions in high-temperature shear flow can be resolved into shifts of single molecules or small groups of molecules, the corresponding motions near T_g entail concerted rearrangements of much larger groups.

Figure 1 translates this notion into a schematic view of the multidimensional Φ-scape. The topography shown in 1(a) represents a portion of the multidimensional configuration space inhabited by the system at high temperature; the texture is uniformly rough with modest barriers separating neighboring minima. By contrast, 1(b) illustrates a nonuniform topography that applies to the region inhabited at low temperature; modest barriers still exist between neighboring minima, but over a larger distance scale a substantial degree of "cratering" appears. The low-temperature preference is to reside near the low-potential-energy crater bottoms, but in order for the system to relax structurally to even lower energy, it is necessary to exit the initial crater and search for an even deeper one. The activation energy E^* in this low-temperature scenario involves the net rise required to move between craters, substantially larger than the mean barrier height separating neighboring single minima. The low-temperature activation entropy measures the extent of configuration space explored between successively inhabited craters, evidently quite large in magnitude.

II Local Mobility Variations

The structural relaxation process indicated in Figure 1(b) for cold fragile liquids requires unbundling a large local region in the medium, and repacking the loosened molecules in an alternative pattern to yield an equally low or lower potential energy. The kinetic sequence that accomplishes this process temporarily transforms an initially nearly rigid region into a more mobile, or fluid, state and then back to a rigid condition. In a macroscopic system, many locales of varying size and lifetime can simultaneously experience this scenario. Consequently one could describe the state of the glass-forming material in conventional 3-space by means of a position and time dependent mobility field. An equivalent description for the present analysis utilizes a space and time dependent local viscosity $\eta_{\text{loc}}(\mathbf{r}, t)$. Autocorrelation functions of this scalar field identify characteristic mean lengths and lifetimes for the spontaneous structural fluctuations.

Full characterization of the statistical properties of the fluctuating field η_{loc} is unnecessary for present purposes. Instead, it suffices just to distinguish regions of high and of low molecular mobility, and to assign mean values of η_{loc} to these regions:

$$\begin{aligned} \eta_{loc}(\mathbf{r}, t) &\cong \eta_0 \quad \text{(high mobility region)}, \\ &\cong \eta_\infty \quad \text{(low mobility region)}. \end{aligned} \tag{5}$$

A similar binary view of the local kinetics, stated in terms of translational diffusion rates and their fluctuations, appears in a Zwanzig study (*17*). If the cutoff criterion between "high mobility" and "low mobility" is biased in favor of the latter, regions of the former type will be rare and disconnected, as schematically illustrated in Figure 2.

III Selective Bottlenecking

We are now in a position to explain the gross violation of the SED expression for D_{trans}, while that for D_{rot} remains approximately valid, in the case of fragile liquids near T_g. As mentioned earlier, only a simple picture should be required to rationalize this strong qualitative effect, and so we use the simplifying assumption that η_∞ is substantially infinite compared to η_0, and to the macroscopic measurable viscosity η for the material. Therefore we need only to be concerned with four parameters that represent mean-value properties of the individual fluidized domains that spontaneously appear and disappear in the glass-forming medium. These are (a) the domain internal viscosity η_0, (b) domain mean volume v_0, (c) domain appearance rate per unit volume of the medium r_0, and (d) the mean domain lifetime t_0. One must keep in mind that these parameters are all temperature dependent, presumably increasing toward well-defined limits as T declines to T_g.

The macroscopic rate of shear relaxation in a viscoelastic medium is determined by the Maxwell relaxation time (*18*)

$$\tau_s = \eta(T)/G_\infty, \tag{6}$$

where G_∞ is the (essentially) temperature-independent high-frequency shear modulus. Imagine that near T_g the system were in a uniform state of shear stress. The dominating inert and rigid matrix is incapable of relaxing that stress. However, a fluctuation that suddenly converts a portion of that inert matrix into a fluidized domain permits its unbundled molecules to rearrange within its volume $\cong v_0$ so as to relieve the imposed stress there. Other locales must await the stochastic appearance of their own fluidizing fluctuations to experience similar stress reduction. Consequently, appearance rate r_0 and volume v_0 control overall stress relaxation. This is equivalent to recognizing that the Maxwell time $\tau_s(T)$, and hence $\eta(T)$ itself, will be proportional to the combination

$$\tau_s(T), \eta(T) \propto (r_0 v_0)^{-1}, \tag{7}$$

and so the temperature dependence of this combination becomes that of τ_s and η. As T_g is approached from above, τ_s increases to the maximum time normally available for experiments, approximately $10^3 - 10^4$s.

With the four nominally independent parameters combined in different ways, several characteristic times can be identified. In fact it is a separation of these time scales that

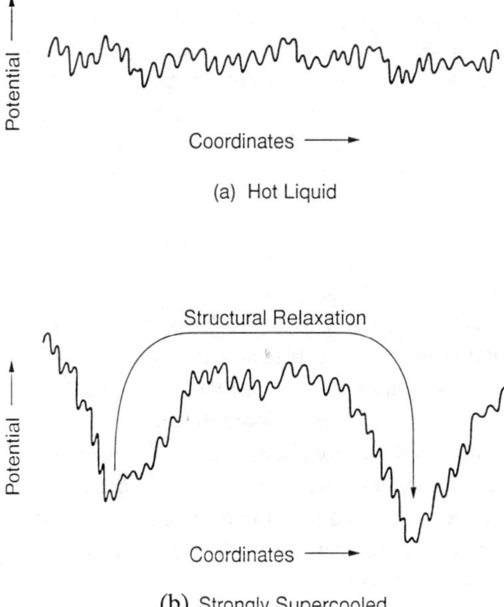

Figure 1. Schematic views of the rugged potential energy landscape for a fragile glass-forming liquid. Part (a) represents the uniformly rough topography in the region of the coordinate space occupied by the system when it is a hot liquid above its thermodynamic melting point. Part (b) indicates the deeply cratered topography in the region visited by the strongly supercooled fragile liquid.

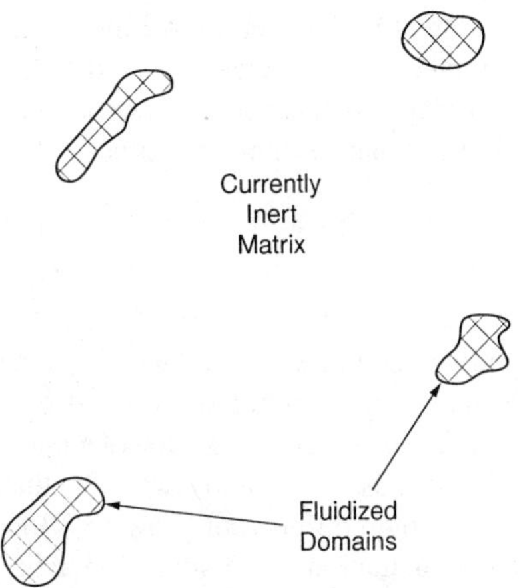

Figure 2. Binary classification of an amorphous glass-forming medium into high-mobility fluidized domains, and a relatively inert, low-mobility matrix. The criterion for distinction between the two has been selected so that the former are rare and disconnected.

provides a mechanism for violating the SED results 2. Specifically, consider the case where t_0 substantially exceeds the rotational relaxation time within a fluidized domain, while the mean domain volume v_0 is sufficiently large so that intradomain translational diffusion can span only a small fraction of the domain diameter in t_0. Assuming that the SED expressions 2 with η_0 hold within fluidized domains, these statements convert into the following strong inequalities (*12*):

$$\frac{4\pi\eta_0 R^3}{k_B T} \ll t_0 \ll \frac{\pi\eta_0 v_0^{2/3} R}{k_B T}. \tag{8}$$

The physical significance of this time-scale ordering is that once a domain appears its interior molecules rotationally relax during an early small portion of its lifetime, so that the dominant remainder of t_0 has no further effect, and thus is "dead time." By contrast, the entire t_0 contributes to overall translational diffusion.

The picture just advocated requires both rotational and translational Brownian motion to rely on spontaneous appearance and later disappearance of fluidized domains, but this stochastic process exerts quite distinct bottlenecking influences on the two. When translated into expressions for the respective diffusion constants, this picture of selective bottlenecking leads to the following (*12*):

$$\begin{aligned} D_{\text{rot}}(T) &\cong r_0 v_0/2\,, \\ D_{\text{trans}}(T) &\cong k_B T r_0 v_0 t_0/6\pi\eta_0 R\,. \end{aligned} \tag{9}$$

The earlier equation 7 implies that D_{rot} should have the same temperature dependence (non-Arrhenius) as η^{-1}, in agreement with the second of equation 2. However, D_{trans} exhibits the additional factor t_0/η_0, which provides the mean to enhance D_{trans} by the required two or more orders of magnitude near T_g.

Insufficient experimental data is currently available for any fragile substance to assign η_0, v_0, r_0, and t_0 at all relevant temperatures. However it is possible to find reasonable estimates that are consistent with experimental measurements and with the time-scale separation 8. Table I, taken from reference (*13*), displays such a set of estimates for OTP at its T_g; this set produce an enhancement factor of 10^2 for translational diffusion compared to the SED value, and exhibits the time-scale separation demanded by equation 8.

IV Discussion

Legitimate concerns might well be raised about the use of four mean value parameters (η_0, v_0, r_0, t_0) to describe fluctuating, distributed, attributes of fluidized regions. For example, the right members of equation 7 and of the first of equation 9 respectively contain the combinations $(r_0 v_0)^{-1}$ and $r_0 v_0$. If r_0 and v_0 were fluctuating variables, then generally the average $\langle (r_0 v_0)^{-1}\rangle$ would differ from $\langle r_0 v_0\rangle^{-1}$. However in the present context we expect anticorrelated fluctuations, with small domains spontaneously appearing more frequently than larger ones. Mathematically it is easy to demonstrate that if two random variables fluctuate with perfect anticorrelation such that their product is a constant, then indeed

$$\langle (xy)^{-1}\rangle \equiv \langle xy\rangle^{-1}\,. \tag{10}$$

Table I: Fluidized domain parameters and related quantities for OTP at T_g [a,b]

$\eta_0(\mathrm{P})$	6.10×10^6
$v_0(\text{Å}^3)$	1.70×10^5
$r_0(\mathrm{cm}^{-3}\mathrm{s}^{-1})$	2.11×10^{15}
$t_0(\mathrm{s})$	0.426
n_0 [c]	500
ϕ_0 [d]	1.53×10^{-4}
$R(\text{Å})$	3.5
$4\pi\eta_0 R^3/k_B T_g(\mathrm{s})$	9.92×10^{-2}
$\pi\eta_0 v_0^{2/3} R/k_B T_g(\mathrm{s})$	6.21

[a] Values taken from ref. 13.
[b] $T_g = 240\mathrm{K}, \quad \eta(T_g) = 4 \times 10^{12}\mathrm{P}$.
[c] n_0 = mean number of molecules in v_0.
[d] ϕ_0 = volume fraction of fluidized domains.

While it is too much to expect fluidized domain characteristics to follow this behavior precisely, it may be appropriate to suggest that rough conformity obtains, sufficient to justify the simple level of description utilized here.

Deeply supercooled glass-forming liquids seem universally to exhibit stretched-exponential relaxation functions (*2, 19*):

$$\phi(t) \cong \exp\{-[t/\tau(T)]^\beta\}, \quad 0 < \beta < 1\,. \tag{11}$$

Strong experimental evidence now exists not only that glassy media are dynamically heterogeneous, but that such heterogeneity underlies stretched exponential relaxation (*20*)–(*22*). The fluctuating field $\eta(\mathbf{r}, t)$ is the means for describing this situation in the present context, and its inverse controls the local (in space and time) rate of decay of structure. Consequently, near the glass transition we must have the mathematical connection:

$$\left\langle \exp\left\{K \int_0^t [\eta(\mathbf{r}, t')]^{-1} dt\right\}\right\rangle = \exp\{-[t/\tau(T)]^\beta\}\,, \tag{12}$$

where K is a positive constant dependent on the property whose relaxation is examined, and where $\langle \ldots \rangle$ indicates a spatial average over all $\mathbf{r}$ in the system volume. Experimental determinations of β and τ thus provide partial information about $\eta(\mathbf{r}, t)$.

Translational diffusion rates for probe molecules of various sizes have been studied experimentally as a means to determine the average size of dynamical heterogeneities — fluidized domains in the present analysis (*11, 23*). Certainly a spherical probe whose volume far exceeds v_0 would be expected to conform to the SED norm, though probes utilized so far are likely not in this size range. Although this large-probe strategy has merit, considerable caution needs to be exercised in its interpretation. Uncertainties exist about how foreign probes in a glass-forming medium might locally perturb the dynamics, and the larger the probe the larger such a perturbation would tend to be. Nevertheless, the resulting estimates of the mean linear dimension of inhomogeneities is in the range of several nanometers for OTP at T_g (*20*), which is in rough agreement with Table I.

The Adam-Gibbs theory of cooperative relaxation in glass-forming liquids has maintained prominence in the field over the more than thirty years since its inception (*14*). That approach connects the temperature-dependent mean relaxation time to calorimetric entropy, using the concept of "cooperatively rearranging regions." It is tempting, though probably not correct, to identify these Adam-Gibbs regions with the fluidized domains of the present paper. By themselves, the four parameters used here to characterize the fluidized domains do not suffice to determine configurational entropy, so at least one further postulate would be required to connect to the Adam-Gibbs viewpoint. One class of theoretical models suggests in fact that the Adam-Gibbs theory cannot be generally valid (*24*); whether it properly describes a restricted class of theoretical models and real substances remains an open issue.

For additional perspectives on the phenomena discussed in this paper, the reader may wish to consult the presentations (*25*) and (*26*).

References

[1] C. A. Angell, *J. Non-Cryst. Solids* **102**, 205 (1988).

[2] R. Böhmer, K. L. Ngai, C. A. Angell, and D. J. Plazek, *J. Chem. Phys.* **99**, 4201 (1993).

[3] A. Einstein, *Ann. Phys. (Leipzig)* **17**, 549 (1905); **19**, 371 (1906).

[4] L. D. Landau and E. M. Lifshitz, *Fluid Mechanics*, 2nd ed. (Pergamon, New York, 1987), pp. 63–67.

[5] P. Debye, *Polar Molecules* (Dover, New York, 1929), pp. 77–86.

[6] N. E. Hill, W. E. Vaughan, A. H. Price, and M. Davies, *Dielectric Properties and Molecular Behavior* (Van Nostrand Reinhold, London, 1969), pp. 60–63.

[7] J. Jonas, D. Hasha, and S. G. Huang, *J. Phys. Chem.* **84**, 109 (1980).

[8] F. Fujara, B. Geil, H. Sillescu, and G. Fleischer, *Z. Phys. B* **88**, 195 (1992).

[9] M. T. Cicerone and M. D. Ediger, *J. Phys. Chem.* **97**, 10489 (1993).

[10] E. Rössler and P. Eiermann, *J. Chem. Phys.* **100**, 5237 (1994).

[11] M. T. Cicerone and M. D. Ediger, *J. Chem. Phys.* **104**, 7210 (1996).

[12] F. H. Stillinger and J. A. Hodgdon, *Phys. Rev. E* **50**, 2064 (1994).

[13] F. H. Stillinger and J. A. Hodgdon, *Phys. Rev. E* **53**, 2995 (1996).

[14] G. Adam and J. H. Gibbs, *J. Chem. Phys.* **43**, 139 (1965).

[15] D. J. Plazek, C. A. Bero, and I.-C. Chay, *J. Non-Cryst. Solids* **172-174**, 181 (1994).

[16] S. S. Chang and A. B. Bestul, *J. Chem. Phys.* **56**, 503 (1972).

[17] R. Zwanzig, *Chem. Phys. Letters* **164**, 639 (1989).

[18] J. Jäckle, *Rep. Prog. Phys.* **49**, 171 (1986).

[19] K. L. Ngai, *Comments Solid State Phys.* **9**, 127 and 141 (1979).

[20] M. T. Cicerone and M. D. Ediger, *J. Chem. Phys.* **103**, 5684 (1995).

[21] A. Heuer, M. Wilhelm, H. Zimmermann, and H. W. Spiess, *Phys. Rev. Letters* **75**, 2851 (1995).

[22] B. Scheiner, R. Böhmer, A. Loidl, and R. V. Chamberlin, *Science* **274**, 752 (1996).

[23] G. Heuberger and H. Sillescu, *J. Phys. Chem.* **100**, 15255 (1996).

[24] F. H. Stillinger, *Physica D* **000**, 0000 (1997).

[25] G. Tarjus and D. Kivelson, *J. Chem. Phys.* **103**, 3021 (1995).

[26] D. N. Perera and P. Harrowell, *J. Chem. Phys.* **104**, 2369 (1996).

Chapter 11

Phase Separation in Silicate Melts: Limits of Solubilities

L. René Corrales

Environmental Molecular Sciences Laboratory, Pacific Northwest National Laboratory, Richland, WA 99352

A statistical mechanical theory for silica melts is used to investigate the phase equilibrium behavior of binary silicates. The theory couples a Flory-type lattice model with a set of chemical equilibrium reactions that together capture the interplay between the solvation of a metal oxide into the silica network and the rearrangement of the network structure that lead to phase separation. The theory produces two-phase coexistence curves with interesting features characteristic of being in the proximity of a higher-order critical point. The theory contains the qualitative behavior and essential features of simple binary silicate melts.

Silicate melts are known to form multiple phases,[1] where each phase has an inherent preference to solvate specific metal ions and, thus, form a distinct network structure. Partitioning of the metal oxide components, along with the characteristic network structure associated with the partitioning, lead to phases with different mechanical, thermodynamic and chemically reactive properties.[2,3] The partitioning occurs in response to metal oxides forming specific coordinations about a central silicon, where the preferred coordination differs with each type of metal oxide. In simple binary systems, the incompatibility occurs when specific coordinated sites cannot be neighbors, due to enthalpy and entropy constraints that force a rearrangement of the network structure to minimize the free energy of the system. The interplay between the solvation of the metal oxide into the network, the formation of the preferred coordinated states, and the optimization of the network structure lead to miscibility gaps in a glass melt.[4]

Equilibrium chemical reactions govern the solvation of the metal oxide into the silica network, and determine their preferred coordination about a central silicon that define the silicate species. In principle, each type of metal oxide has a preference for the formation of specific silicate species whose concentration and distribution vary as a function of concentration and temperature. Thus, the equilibrium distribution of the silicate species is regulated by the affinity of reactions that are defined in terms of the species activities. In such a complex system, it is the net sum of reactions that regulates the species distribution that in turn drive the phase equilibrium behavior. This occurs because of structural incompatibilities between silicate species that favor aggregation of similar silicate species and partitioning of dissimilar silicate species, thus forming more than one phase. Effectively, the enthalpy drives the formation of silicate species, and the mixing entropy drives the phase separation. The interplay of chemical reactions and phase equilibrium is known to lead to interesting and important phase equilibria and higher-order critical phenomena.[5]

Statistical mechanics provides useful approaches to study the phase coexistence and speciation of silicate melts. The phase transition can be viewed as being driven by the entropy for a fixed species distribution. The mixing entropy is determined using Flory-Huggins theory[6] (a combinatorial method) where the species interactions provide the favorable and disfavorable interactions that lead to phase separation.[7] Speciation can be studied using a structural thermodynamic model where a set of reactions dictate the species formation and from which species activities can be determined.[8] The activities define equilibrium constants that contain both enthalpy and entropy contributions of forming the species in the network. A comprehensive description of the relationship between speciation and thermodynamic properties can be provided by combining a combinatorial approach for the entropy with a chemical equilibrium theory for the speciation that is described below.[4]

For a binary silicate system, a set of chemical reactions are written that lead to the incorporation of a metal oxide into the silica network and, hence, define five possible silicate species. The net effect is to depolymerize the network by transforming a bridging oxygen (BO) bond into a nonbridging oxygen (NBO) bond by the addition of the metal oxide. The BO bond is considered to be a covalent bond and the NBO bond is an ionic bond, held together by Coulombic interactions between the anionic oxygen atoms and the metal cations. Each of the five silicate species is defined by the possible arrangements of BO and NBO bonds about a silicon atom. Excess metal oxide species, or those that have not reacted, occupy interstitial sites. In the following work, the excess metal oxide do not aggregate to form a separate phase and only tetrahedral coordinated silicate species are considered.

Description of Theory

The solvation of a metal oxide into the silica network consists of transforming a bridging oxygen (BO) bond to a nonbridging oxygen bond (NBO) via insertion of a M_2O that contributes an O^{2-} to the system and forms two O^- upon incorporation into the silicate lattice[8,9]

$$\equiv\text{Si—O—Si}\equiv + \text{M}_2\text{O} \leftrightarrow \begin{array}{l}\equiv\text{Si—O}^-\ \text{M}^+ \\ \text{M}^+\ \text{O}^-\text{—Si}\equiv\end{array}$$

The chemical reaction is considered to be in equilibrium. A silicon can be four-fold coordinated by a combination of BO and NBO bonds, which correspond to a covalent oxygen and a pair of ionic oxygens, respectively. The Q species notation[10] is used to describe each of the possible combinations of bonds. It must be made clear that a BO or a NBO are shared between silicon tetrahedra. Hence, in the following definitions of the Q species a half-BO corresponds to sharing a covalent oxygen, and a half-NBO corresponds to a single ionic oxygen. Thus, a Q_4 site is made up of four half-BO bonds, a Q_3 site is made up of three half-BO and one half-NBO bonds, a Q_2 site is made up of two half-BO and two half-NBO bonds, a Q_1 site is made up of one half-BO and two half-NBO bonds, and a Q_0 site is made up of four half-NBO bonds. The equilibrium reactions that describe the solvation of metal oxide to produce any of the Q species from the Q_4 species are written as:

Eq. 1

$$Q_4 + \tfrac{1}{2} M_2O \overset{K_3}{\longleftrightarrow} Q_3$$

$$Q_4 + M_2O \overset{K_2}{\longleftrightarrow} Q_2$$

$$Q_4 + \tfrac{3}{2} M_2O \overset{K_1}{\longleftrightarrow} Q_1$$

$$Q_4 + 2M_2O \overset{K_0}{\longleftrightarrow} Q_0$$

The chemical equilibrium constants are given by

Eq. 2

$$-kT \ln K_i = \Delta G_i = \Delta H_i - T\,\Delta S_i$$

All cross reactions are possible whose chemical equations and chemical equilibrium constants are linear combinations of Eq. 1.[9] The cross reactions lead to a migration or diffusion of the NBO along the networks and to a redistribution of the Q species as a function of composition and temperature. On a tetrahedral lattice, the ability to redistribute the bond types allows sampling of the distribution via Monte Carlo simulations and bond exchange probabilities.

Assigned to each species are their conjugate activities z_i, which act as statistical weights for each of the Q_i species. Thus, the chemical equilibrium constants are written in terms of the species activities or the species concentrations, which for convenience are also defined by Q_i.

Eq. 3

$$K_i = \frac{z_i}{z_4 z_m^{(4-i)/2}} = \frac{Q_i}{Q_4 n_m^{(4-i)/2}}$$

where i = {3,2,1,0}, n_m is the number concentration of free metal oxide, and $z_k = \exp(\mu_k / kT)$ for k = {0,1,2,3,4,m} are the activities of each of the species. The statistical weights define the model such that a suitable partition function as a sum over all sites with a single Q species having the statistical weight of that species type can be expressed. Such a partition function for the Q species distribution located on a tetrahedral lattice is

Eq. 4

$$Y = \sum_Q z_4^{Q_4} z_3^{Q_3} z_2^{Q_2} z_1^{Q_1} z_0^{Q_0} z_m^{n_m} \Gamma[Q_4, Q_3, Q_2, Q_1, Q_0]$$

where the sum is taken over all Q sites of the lattice. The configurational contributions of the ways of arranging each of the species on the lattice is given by $\Gamma[Q_4,Q_3,Q_2,Q_1,Q_0]$. However, although the Q species define sites on a lattice and an appropriate phenomenological free energy can be derived in terms of the concentration and chemical equilibrium constants for their formation, the order parameter of the system is not clearly defined.

The natural order parameter of the system is the degree to which the network has depolymerized, or the number concentration of broken bonds. Alternatively, the degree to which the system is polymerized can be considered. However, because the chemical reactions are defined from the fully polymerized network, the first definition of the order parameter is appropriate. Calculating the deterministic set of equations for phase equilibrium from a site model representation of the free energy, in terms of the NBOs as the order parameter, is neither straightforward nor necessarily the correct approach. Instead, the motivation is to arrive at a partition function, and thereby a free energy, in terms of a corresponding bond model. Details of this transformation can be found in Reference 4.

The activities are defined in terms of the Q species that occupy the sites of a tetrahedral lattice. The transformation from the bond to the site model requires defining the sites in terms of the bond types. This is done by determining the probabilities of forming each of the specific Q sites in terms of the bond types, namely BO and NBO bonds. The probability of forming a Q_4 site from n_{bo} BO bonds is n_{bo}^4. The probability of forming a Q_3 site from n_{nbo} NBO and n_{bo} BO bonds is $n_{bo}^3 n_{nbo}$. Similarly, the probability of forming a Q_2 is $n_{bo}^2 n_{nbo}^2$, the probability of forming a Q_1 is $n_{bo} n_{nbo}^3$, and the probability of forming a Q_0 is n_{nbo}^4. Moreover, the asymmetric species have ways of being formed that are uniquely different from one another in that they lead to different network connectivities, and so they must be counted. The Q_3 and Q_1 species each have four distinguishable ways of being formed on a site, and the Q_2 species has six distinguishable ways of being formed on a site.

The mass conservation equations for the bond model is given by the total number of bonds in the system

Eq. 5

$$N = n_{bo} + n_{nbo}$$

The conservation equation for the metal oxide is given by

Eq. 6

$$N_m = n_m + 2N^3 n_{nbo}$$

where N_m is the total number of metal oxide in the system with each NBO bond contributing a full metal oxide. A partition function for the bond model can be formulated from Eq. 4 by substituting the above probabilities to form each Q species from the bond types. By using Eq. 5, 6 and 3 the partition function for the bond model takes on the following form

Eq. 7

$$Z = z_4^{N^4} z_m^{N_m} \sum_N (K_3)^{4N^3 n_{nbo}} \left(\frac{K_2}{K_3^2}\right)^{6N^2 n_{nbo}^2} \left(\frac{K_1 K_3^3}{K_2^3}\right)^{4N\, n_{nbo}^3} \left(\frac{K_0 K_2^6}{K_1^4 K_3^4}\right)^{n_{nbo}^4} \Gamma[N, n_{nbo}]$$

where the summation is over all N bonds and $\Gamma[N,n_{nbo}]$ is the number of ways of arranging the bonds on a tetrahedral lattice consisting of N bonds, given they must be arranged in such a way as to form combinations of specific Q sites.

The configurational contribution, given by Γ, is calculated by determining the number of ways of arranging each of the sites on a lattice, keeping in mind that sites are formed by the probabilities of forming each site form the bond types. The number of ways of arranging sites on a lattice is determined following the Flory method for placing polymers on a lattice.

Eq. 8

$$\Omega = \frac{N^4!\,4^{4n_{nbo}n_{bo}^3}\,6^{6n_{nbo}^2 n_{bo}^2}\,4^{4n_{nbo}^3 n_{bo}}}{\left(4n_{nbo}n_{bo}^3\right)!\left(6n_{nbo}^2 n_{bo}^2\right)!\left(4n_{nbo}^3 n_{bo}\right)!\,n_{nbo}^4!\left(N^4 - 4n_{nbo}n_{bo}^3 - 6n_{nbo}^2 n_{bo}^2 - 4n_{nbo}^3 n_{bo} - n_{nbo}^4\right)!}$$

The configurational entropy of the system is calculated by taking the log of the number of ways of arranging the sites on a lattice and simplifies to

Eq. 9

$$S = k \ln\Omega = 4\left[N^3 n_{nbo} \ln\frac{n_{nbo}}{N} + N^3\left(N - n_{nbo}\right)\ln\frac{\left(N - n_{nbo}\right)}{N}\right]$$

An approximate free energy for the bond model can be written by considering the logarithm of Eq. 7 and including the configurational entropy given by Eq. 9 to arrive at

Eq. 10

$$G=-N^4kT\ln z_4-N_mkT\ln z_m+4N^3n_{nbo}B_1+6N^2n_{nbo}^2B_2+4Nn_{nbo}^3B_3+n_{nbo}^4B_4$$
$$+4N^3n_{nbo}\,kT\ln\frac{n_{nbo}}{N}+4N^3\left(N-n_{nbo}\right)kT\ln\left(\frac{N-n_{nbo}}{N}\right)$$

where

Eq. 11

$$B_1=-kT\ln K_3=\Delta G_3$$
$$B_2=-kT\ln\left(\frac{K_2}{K_3^2}\right)=\Delta G_2-2\Delta G_3$$
$$B_3=-kT\ln\left(\frac{K_1K_3^3}{K_2^3}\right)=\Delta G_1-3\Delta G_2+3\Delta G_3$$
$$B_4=-kT\ln\left(\frac{K_0K_2^6}{K_1^4K_3^4}\right)=\Delta G_0-4\Delta G_1+6\Delta G_2-6\Delta G_3$$

To understand how to extract useful thermodynamic properties, the isothermal differential of the free energy is expressed as in standard thermodynamics

Eq. 12

$$dG_T=\mu_4dQ_4+\mu_3dQ_3+\mu_2dQ_2+\mu_1dQ_1+\mu_0dQ_0+\mu_mdn_m$$
$$=4\mu_{bo}dn_{bo}{}^4+(3\mu_{bo}+\mu_{nbo})d\left[4n_{nbo}n_{bo}{}^3\right]+(2\mu_{bo}+2\mu_{nbo})d\left[6n_{nbo}^2n_{bo}{}^2\right]$$
$$+(\mu_{bo}+3\mu_{nbo})d\left[4n_{nbo}^3n_{bo}\right]+4\mu_{nbo}dn_{nbo}^3+\mu_mdn_m$$

where μ_i for i = {1,2,3,4} are the chemical potentials of the Q species, μ_m is the chemical potential for the free metal oxide, and μ_{nbo} and μ_{bo} are the chemical potentials corresponding to the NBO and BO bonds. In Eq. 12, the transformation from the Q site activities and species concentrations to the bond activities and site formation probabilities is made. By using Eqs. 5 and 6, expanding out the polynomial terms and rearranging Eq. 12 simplifies to

Eq. 13

$$dG_T=4\mu_{bo}\,dN^4+4\Delta\,d\left[N^3n_{nbo}\right]+\mu_mdN_m$$

where

Eq. 14

$$\Delta=\mu_{nbo}-\mu_{bo}-\tfrac{1}{2}\mu_m$$

This form of the isothermal differential of the free energy reveals the conditions to satisfy both chemical equilibrium and phase equilibrium in the bond model. From the definition of Δ, it represents the negative affinity of reaction of creating a NBO from a BO and a metal oxide. To satisfy chemical equilibrium, Δ must be equal to zero in all phases that are in phase equilibrium. The phase equilibrium conditions are then defined by μ_{bo} and μ_m being equal in all phases.

For ease of calculation, the differential free energy is expressed in terms of dN and dn_{nbo}

Eq. 15

$$dG_T = 4N^3 \mu\, dN + 4N^3 \Delta\, dn_{nbo} + \mu_m dn_m$$

where

Eq. 16

$$\mu = 4\mu_{bo} + 3\frac{n_{nbo}}{N}\Delta$$

From these equations, μ_{bo} is given by

Eq. 17

$$\mu_{bo} = \mu - 3\frac{n_{nbo}}{N}\Delta \equiv \frac{n_{nbo}}{N}\frac{\partial G}{\partial n_{nbo}} - G$$

The latter equality indicates the chemical potential of the bridging oxygen bond can also be obtained from the common tangent construction.

Calculating the indicated partial differentials in Eq. 15 on the free energy Eq. 10, Δ is

Eq. 18

$$\frac{\Delta}{kT} = \frac{B_1}{kT} + 3x\frac{B_2}{kT} + 3x^2\frac{B_3}{kT} + x^3\frac{B_4}{kT} + \ln\left(\frac{x}{1-x}\right)$$

where $x = n_{nbo}/N$. By determining μ and using Eq. 17, μ_{bo} is

Eq. 19

$$\frac{4\mu_{bo}}{kT} = \left[-\ln z_4 - 6x^2\frac{B_2}{kT} - 8x^3\frac{B_3}{kT} - 3x^4\frac{B_4}{kT} + 4\ln(1-x)\right]$$

One of the conditions for phase equilibrium is given by Eq. 19.

The second equilibrium condition, determined by μ_m being equal in each phase, does not lead to an explicit expression in terms of the field parameters, the B_is, and the order parameter, n_{nbo}. To this end, Δ can be used, recognizing that it contains the chemical potential of the free metal oxide, μ_m, as well as the chemical potential of the metal oxide incorporated into the network, μ_{nbo}. Thus, $\Delta + \mu_{bo} = \mu_{nbo} - 1/2\ \mu_m$ is used for the second phase equilibrium condition. This phase equilibrium condition must be coupled with the chemical equilibrium condition that requires $\Delta = 0$ also be satisfied in each of the phases in equilibrium.

Alternatively, Δ is used for Maxwell's construction or to satisfy the equal area rule. Thus, Δ can be used for the second phase equilibrium condition while simultaneously ensuring that it is equal to zero in all phases that are in equilibrium. Both approaches lead to equivalent results, and require that one of the field parameters, namely B_1, be searched and solved in a self-consistent manner.

Thus, two-phase equilibrium of the system is determined by searching for the solution of $\mu_{bo}(x', T) = \mu_{bo}(x'', T)$ and $\Delta(x', B_1, T) = \Delta(x'', B_1, T) = 0$ at fixed values of B_2, B_3, B_4, and T. The entire coexistence curve is then determined by varying T and solving for x' and x''.

The critical point condition is given by the second and third partial derivative of the free energy with respect to the order parameter be equal to zero and is identical to

Eq. 20

$$\left(\frac{\partial \Delta/kT}{\partial x}\right)_{T_c,x_c} = \left(\frac{\partial^2 \Delta/kT}{\partial x^2}\right)_{T_c,x_c} = 0$$

These conditions are easily calculated from Eq. 18. The parameters B_2 and B_3 can be determined using the critical temperature and composition of a system while keep B_4 fixed. The role of B_4 is to introduce an asymmetric stretch in the coexistence curve, and so in principle can also be fit to available data.

Using the definitions of the equilibrium constants in terms of the activities of each species given by Eq. 3, the average number density of each Q species is calculated from Eq. 10 using

Eq. 21

$$\frac{\partial G/N^4kT}{\partial(-\ln z_i)} = \langle q_i \rangle$$

The total number of silica units is given by

Eq. 22

$$X_{SiO_2} = \sum_i q_i = (1-x)^4 + 4x(1-x)^3 + 6x^2(1-x)^2 + 4x^3(1-x) + x^4 = 1$$

Using Eq. 6 along with

Eq. 23

$$\frac{\partial G/N^4kT}{\partial(-\ln z_m)} = \langle x_m \rangle = \frac{N_m}{N^4} - 2x$$

where $x_m = n_m/N^4$, the total number of metal oxide units in the system is given by

Eq. 24

$$X_m = \frac{N_m}{N^4} = x_m + 2x$$

Thus the total mass conservation of this system is given by

Eq. 25

$$N_T = X_m + X_{SiO_2}$$

The apparent fraction of metal oxide in the system is

Eq. 26

$$\chi_m = \frac{x_m + 2x}{1 + x_m + 2x}$$

and the apparent mole fraction of silica in the system is

Eq. 27

$$\chi_{SiO_2} = \frac{1}{1 + x_m + 2x}$$

Phase Diagrams

Two-phase coexistence curves are determined for a fixed value of the parameter B_4= -4.99 and of the critical temperature T_c=1000 K, corresponding to B_4/kT_c = -0.43. The critical composition is varied from the dilute side to the high concentration side of the metal oxide fraction. The parameters B_2 and B_3 are determined from the critical composition and critical temperature using the equations for the critical point obtained from Eq. 19. The resulting phase diagrams are shown in Figures 1-5.

In Figure 1a, the critical point sits on the dilute side at χ_{M20}= 0.44 and corresponds to x = 0.3 for the number concentration of NBO bonds, as defined above. This value of B_4 shows only a slight asymmetric stretch of the coexistence curve on the right branch. The shape of the coexistence curve is similar to those observed in binary silicates containing Na_2O.

In Figure 1b, the critical point is at χ_{M20}= 0.55, corresponding to x = 0.4. Note the drastic change in the shape of the right-hand branch of the coexistence curve, that emphasizes the asymmetric stretch with a shoulder to the right of the critical point. A similar coexistence curve is observed in binary silicates containing CaO or MgO.

The phase coexistence curve shown in Figure 1c is significantly flatter in appearance than all the other coexistence curves for the same values of B_4. The critical composition occurs at χ_{M20}= 0.66, corresponding to x = 0.5, which is identical to the critical composition of the nonsymmetric tricritical point that is known to exist for this theory. The parameter values lie very close to this higher order critical point and is the reason why the coexistence curve has such a pronounced flatness.

In Figure 1d, the critical point occurs at χ_{M20}= 0.77, corresponding to x = 0.6. This coexistence curve nearly mirrors that in Figure 2, with its shoulder to the left of the critical point.

In Figure 1e, the critical point occurs at χ_{M20}= 0.87, corresponding to x = 0.7. This coexistence curve has a pronounced asymmetry on the left branch compared to that in Figure 1a.

Discussion

This theory shows a rich assortment of phase behavior that includes three-phase equilibrium, critical endpoints and a nonsymmetric tricritical point, that have not been shown here. The nonsymmetric tricritical point is where three phases that are in equilibrium simultaneously coalesce to form a single phase. The critical endpoints is where two of the three phases have coalesced to a critical point that remains in equilibrium with the third phase.

Experimentally determined phase coexistence curves for binary silicates have focused on two-phase coexistence curves. This work shows the theory is capable of capturing the wide behavior, in terms of the shapes of the coexistence curves, seen in binary silicate melts. The shapes of the curves are strongly coupled to the proximity of the parameters to the nonsymmetric tricritical point, as well as to the position of the critical composition of the system.

Although a number of approximations have been used, that includes the use of Flory theory for computing the mixing entropy, a more severe approximation has been to keep the free metal oxide off the lattice. Incorporating the metal oxide into the lattice will result in a mixing entropy contribution to the system, that will be significant in the high concentration regime of the phase diagram. Additionally, it would be desirable to turn on the interaction energy of the metal oxide with the other components of the system, that can lead to aggregation of a pure metal oxide phase and to favoring further uptake of the metal oxide into the network.

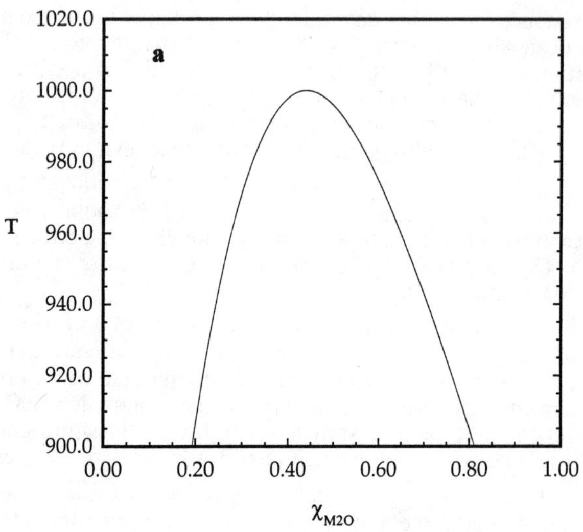

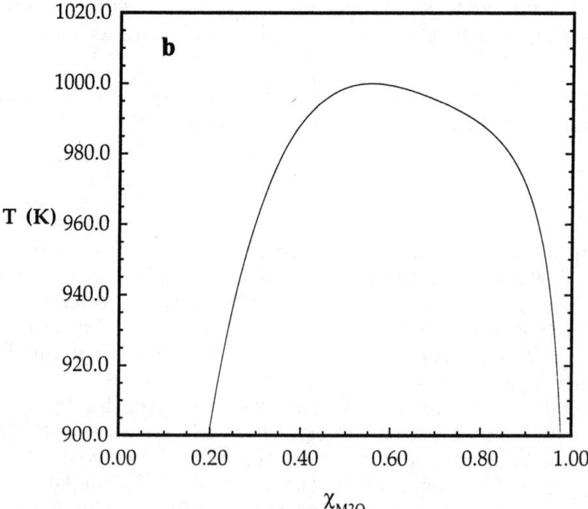

Figure 1. Phase coexistence curves at T_c = 1000 K at critical compositions of a) χ_{M2O} = 0.44, b) χ_{M2O} = 0.55, c) χ_{M2O} = 0.66, d) χ_{M2O} = 0.77, and e) χ_{M2O} = 0.87. The series of coexistence curves pass through the vicinity of a nonsymmetric tricritical point that leads to the asymmetric broadening of the coexistence curves.

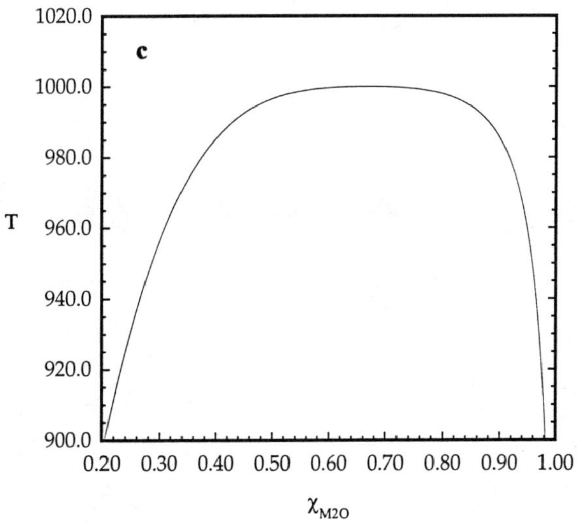

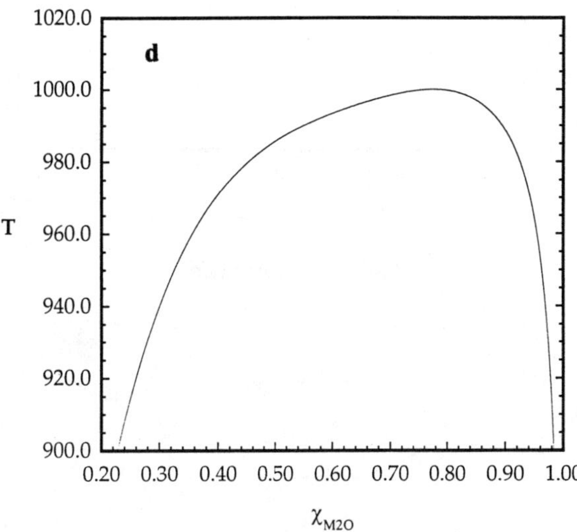

Figure 1. Continued.

Continued on next page

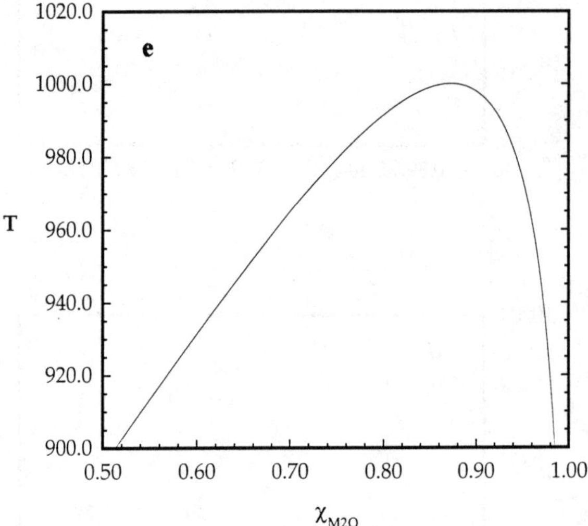

Figure 1. Continued.

The proximity of the coexistence curves, presented here, to the nonsymmetric tricritical point leads to the broadening asymmetric behavior. In general, the coexistence curves of binary silicate systems are observed to occur in the dilute regime. In the high concentration regime other phase transformations may occur, such as metal oxide aggregation. Due to temperature constraints, most phase separation of silicates is observed in the solid phases. Thus, phase domains may be glass or crystalline depending upon quench history and the propensity for a phase to crystallize from the melt. Phases rich in metal oxide may be expected to crystallize thereby driving a phase separated system into a single phase. Such phenomenon must be further explored.

Expansion of this theoretical approach to include multiple network forming and network modifying components can be carried out, although the analytic solutions may become intractable. However, solutions can in principle be obtained using Monte Carlo simulation techniques.

Acknowledgements

This work was performed under the auspices of the Division of Chemical Sciences, Office of Basic Energy Sciences, U.S. Department of Energy under Contract DE-AC06-76RLO 1830 with Batelle Memorial Institute that operates the Pacific Northwest National Laboratory under Grant No. DE-FG06-89ER-75522 with the U.S. Department of Energy.

References

[1] R. H. Doremus, Glass Sience, (John Wiley & Sons Inc., New York), 2nd Edition, 1994.
[2] D. R. Uhlmann and A.G. Kolbeck, Phys. Chem. Glasses, 1976, **17**, 146
[3] R.J. Charles, Phys. Chem. Glasses, 1969, **10**, 169.
[4] L.R. Corrales and K.D. Keefer, J. Chem. Phys., 1997, **106**, 6460.
[5] L.R. Corrales and J.C. Wheeler, J. Chem. Phys., 1989, **91**, 7097.
[6] P. Flory, J. Chem. Phys., 1942, **10**, 51; 1944, **12**, 426.
[7] R.J. Araujo, J. Non-Cryt. Solids, 1983, **55**, 257;
P.J. Bray, R.V. Mulkern and E.J. Holupka, J. Non-Cryst. Solids, 1985, **75**, 37.
[8] W.G. Dorfeld, Phys. Chem. Glasses, 1988, **29**, 179.
[9] R. Dron, J. Non-Cryst. Solids, 1982 **53**, 267.
[10] G. Englehardt, H. Jancke, D. Hoebbel and W. Weiker, Z. Anorg. Alg. Chem., 1975, **418**, 17.

Experimental Advances in Supercooled Liquids

Chapter 12

Higher Order Time-Correlation Functions from Fluctuating Energy Landscape Models: Comparison with Reduced Four-Dimensional NMR Spectroscopy

G. Diezemann, G. Hinze, R. Böhmer, and H. Sillescu

Institut für Physikalische Chemie, Johannes Gutenberg Universität, Staudinger Weg 7, D–55099 Mainz, Germany

Recent reduced 4D-NMR experiments on supercooled ortho-terphenyl near its calorimetric glass transition temperature reveal heterogeneous reorientational dynamics. The dynamics of these heterogeneities is modeled in the framework of different energy landscape models that account for environmental fluctuations. Molecular reorientation rates are modeled by thermally activated processes with activation energies that fluctuate in time. Fluctuations of the barriers to reorientation are described in terms of transitions among states characterizing the different activation energies. It is shown that quite different scenarios for these transitions yield results in qualitative agreement with experimental data. Apart from the time honored Anderson and Ullman model no other locally and globally connected scenario that we have considered can be ruled out on the basis of our present experiments.

The relaxation of supercooled liquids near their calorimetric glass transition shows a number of characteristic features [for reviews see (1,2)]. Apart from the strong temperature dependence of the primary relaxation (α- relaxation) the pronounced non-exponential decay of typical time correlation functions and the non-Debye form of the corresponding susceptibilities as observed e.g. in dielectric experiments are subject of controversial discussions (3,4 and references therein).

Typically, time correlation functions related to the α-relaxation can very well be fitted to the Kohlrausch function:

$$F_{KWW}(t) = F_0 \exp\{-(t/\tau_0)^{\beta}\} , \qquad (1)$$

where F_0 is an amplitude, τ_0 a characteristic decay time, and $\beta \leq 1$ a measure for the deviation from exponential decay (β=1). There are two extreme scenarios which might be viewed as the microscopic origin for the occurrence of a non-exponential time correlation function. Consider the macroscopic ensemble (the sample as a whole) as consisting of a large number of independent sub-ensembles. Then the time correlation function F(t) of the total ensemble is given by a superposition of the time correlation functions of the sub-ensembles, i.e. $F(t) = \Sigma f_i(t)$.

In the *homogeneous* case the time correlation function, $f_i(t)$, of each sub-ensemble is assumed to be of the Kohlrausch form, say, with the same values for τ_0 and β. The superposition then results in the form of equation 1 with a sum of amplitudes yielding F_0.

The other extreme scenario, the *heterogeneous* case is associated with the assumption of a Debye form ($\beta=1$) for the time correlation functions of the sub-ensembles, i.e. $f_i(t) = g_i \exp(-t/\tau_i)$, where g_i is the statistical weight and τ_i the characteristic correlation time. The macroscopically observed response function is then recovered by assuming an appropriate continuous distribution of correlation times [meaning that the weights g_i are replaced by $g(\tau)$].

Note that apart from their spectral characterization nothing is said about the sub-ensembles. In models for the α-relaxation it often is assumed that the supercooled liquid consists of 'regions' or 'domains' of different sizes, packing fractions, etc. (2,3,5) which are characterized by different correlation times. Only within the framework of such models it is possible to assign different correlation times to different regions in the sample and thus to relate heterogeneity in the time correlation functions to spatial heterogeneity.

Since most experiments measure two time correlation functions in the linear response regime, it is not possible to distinguish between the above described scenarios. [An exception is provided by deuteron NMR spin lattice relaxation experiments near the glass transition (6), which show that at least part of the relaxation-spectrum is heterogeneous.] What is needed in order to distinguish between homogeneous and heterogeneous relaxation is a method that allows to select a (slow) sub-ensemble and monitor its subsequent relaxation. What can we expect from such a selection to be different in both scenarios? In the homogeneous case such a selection just is impossible due to the fact that there is only a single characteristic time scale in the system. In the heterogeneous case, on the other hand, such a spectral selection should be possible and the relevant question then is whether a selected slow sub-ensemble remains slow or changes its correlation time in the course of time. In such a case the question immediately arises as to how long this re-equilibration takes.

Up to now there exist three types of experiments which are able to provide such a spectral selection and a means to observe the subsequent re-equilibration. These are a reduced 4D-NMR experiment (7), the non-resonant spectral hole burning technique (8) and an optical deep bleach experiment (9). In Ref. 7 the principal idea was employed for the first time. Later on further multidimensional NMR studies on polymeric (10) but also on low molecular (11) viscous liquids were carried out. We note that all of the experiments performed up to now have shown that a spectral selection is possible. This means that it is possible to assign different characteristic decay times to different sub-ensembles. All experiments clearly demonstrate that the dynamics of the selected sub-ensemble change in the course of time, which means that the correlation time distribution is not static as assumed in the extreme heterogeneous scenario.

Furthermore all three types of experiments mentioned above allow to measure the rotational relaxation of the molecules. Thus the result of changing reorientational rates means that the rotational motion of the molecules cannot be modeled by a Markovian stochastic process. One way of modeling such a non-Markovian process consists in describing it as the projection from a higher dimensional Markov process, as discussed by Sillescu (12).

In the present paper we briefly describe the specific four-time correlation function measured in the reduced 4D-NMR experiment and present some results of our experiment on supercooled ortho-terphenyl (o-TP) (11). Here, we compare the predictions of fluctuating energy landscape models as used to mimic the exchange between different reorientational rates to the experimental results. Finally, we discuss the significance of the model calculations and draw some conclusions.

Reduced 4D-NMR

Instead of describing the experimental procedure in detail, we briefly summarize the points relevant for our present purpose. The reader is referred to Ref. 12 for more detailed information. All experiments are taken from Ref. 11. They were performed on o-TP at a temperature of about 254 K (11 K above the calorimetric glass transition temperature), for experimental details see Ref. (11).

To start with, consider the three pulse experiment depicted in Figure 1a. After the first pulse a second one at a time t_p is applied, which stores the phase ωt_p of the spins. Here ω denotes the quadrupolar frequency of the deuterons (which depends solely on the orientation of the considered C-D bond relative to that of the external magnetic field). After a mixing time t_{fil} a third pulse is applied. The amplitude of the two-time (2t) echo occurring with a delay of t_p after the third pulse is conveniently written as

$$F_2(t_p,t_{fil}) = \langle \exp(i\omega(0)t_p)\exp(-i\omega(t_{fil})t_p)\rangle, \tag{2}$$

where $\langle\cdots\rangle$ denotes a thermal average. The cosine and sine parts of $F_2(t_p,t_{fil})$ (denoted by $F_2^{cc}(t_p,t_{fil})$ and $F_2^{ss}(t_p,t_{fil})$, respectively) are obtained by appropriate choice of the lengths and the phases of the pulses. The idea of the reduced 4D-NMR experiment is to use the 2t-echo experiment as a dynamical filter. Writing equation 2 in the form $\langle \cos((\omega(0)-\omega(t_{fil}))t_p)\rangle$ shows that only molecules which have not changed their orientation during t_{fil} appreciably contribute to the echo. On the other hand, choosing t_p long enough to ensure that each angular jump of the molecules leads to a total loss of correlation allows to adjust t_p in such a way that $F_2(t_p,t_{fil})$ measures the correlation function of angular jumps. In o-TP it is known that the mean jump angle is approximately 10^0 (13). Therefore the value of $t_p = 25\,\mu s$ chosen in all experiments is sufficient to treat each angular jump as being responsible for a total loss of correlation. The observed $F_2(t_p = 25\,\mu s, t_{fil})$ at T = 254 K can be fitted excellently to a Kohlrausch function (see equation 1) with the parameters $\tau_0 = 17$ ms and

$\beta = 0.42$, yielding an average inverse angular jump rate of 49.7 ms. It has to be noted that the rotational correlation function of the second Legendre polynomial, obtained from $F_2^{ss}(t_p \rightarrow 0, t_{fil})$, decays with a mean rotational correlation time of 540 ms (13), i.e. a factor of 10 slower than the angular jump correlation function.

In Figure 1b the principle of the four-time (4t) echo experiment is shown. The 2t-echo serves as a filter which can only be passed by those molecules which have not performed an angular jump during the filter time t_{fil}. After a mixing time t_{rex} the same filter is applied again.

This means that only molecules which remained slow during the time interval t_{rex} pass also the second filter and contribute to the 4t-echo. Since four evolution times are involved in the pulse sequence one indeed is concerned with a four time correlation function. However, here we measured the 4t-echo as a function of t_{rex} for fixed $t_p = 25\,\mu s$ and t_{fil}. The same experiment was then repeated for different values of t_{fil} (cf. Ref. 11). Out of the possible 4t-echo functions in the present context we consider only the following one

$$F_4(t_{rex}) \propto \cos[(\omega(0) - \omega(t_{fil}))t_p]\cos(\omega(t_{rex} + t_{fil})t_p)\cos(\omega(t_{rex} + 2t_{fil})t_p) \quad , \qquad (3)$$

cf. Refs. 11 and 12. This particular 4t-echo function has the advantage that in case that random reorientational jumps govern the rotational dynamics it depends solely on the change of the reorientational jump rates but not on the orientational relaxation itself. Since we have chosen $t_p = 25\,\mu s$ large enough for each jump to lead to a total loss of correlation, we can use the model of random reorientational jumps even though we know the mean jump angle to be 10^0.

In Figure 2, $F_4(t_{rex})$ functions are presented for different values of the filter time. Additionally shown are fits of $F_4(t_{rex})$ to the Kohlrausch like expression (solid lines)

$$F_4(t_{rex}) = F_4^{\infty} + (1 - F_4^{\infty})\exp\{-(kt_{rex})^{\beta_4}\} \qquad (4)$$

yielding the results shown in Figure 3. Since we are not going to discuss the plateau values F_4^{∞} further in the present context, they are not shown in Figure 3. We just mention that all model calculations tend to very similar under-estimates of F_4^{∞} for large t_{fil} (cf. Figure 4a in Ref. 11). The general quality of the fits is good, as can be inferred from Figure 2.

Note that the decay of the 4t-echo becomes longer as t_{fil} increases. At first glance this result suggests that molecules with smaller angular jump rates also change these rates more slowly than molecules which reorient faster. As will become clear below, no strict correlation between angular jump rates and exchange rates is necessary in order to interpret the experimental data.

Fluctuating energy landscape models

Sillescu (12) has worked out a two-state model of environmental fluctuations and applied it to the description and interpretation of the various 4t-echo functions. In Ref.

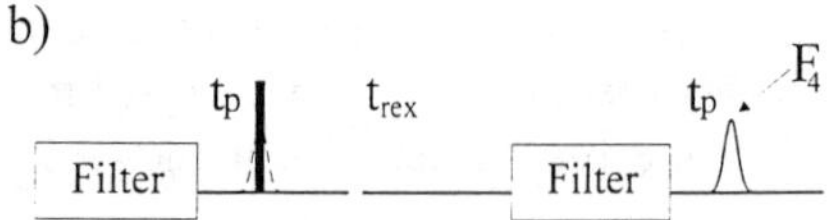

Figure 1. a) The two time stimulated echo pulse sequence. The right trace shows the interpretation as a dynamical filter. b) The reduced 4D-NMR pulse sequence: Only molecules which passed the first filter *and* have not changed their angular jump rates during the rate exchange interval contribute to the four time echo (Adapted from Ref. 11).

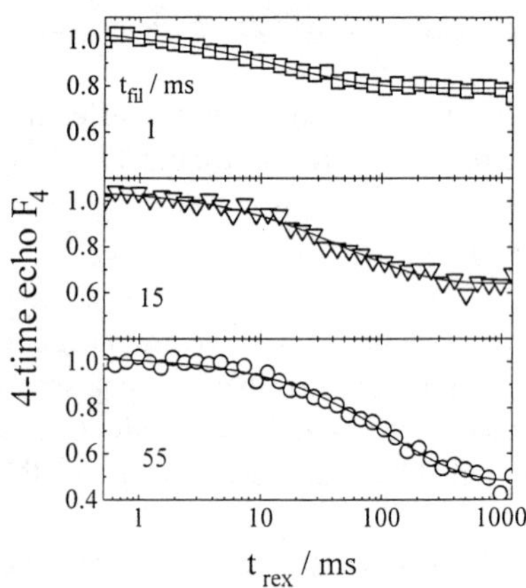

Figure 2. The normalized 4t-echo decay as a function of the rate exchange interval t_{rex} for various filter times t_{fil}.

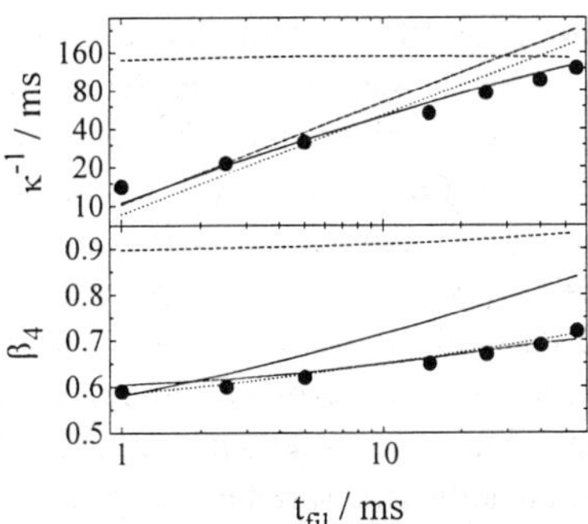

Figure 3. Results of fits of the four time echo decays to equation 4. Symbols. experimental data (11). The lines are from different model calculations with the parameters given in the text: dotted lines: Anderson-Ullman model; dashed lines: generalized Anderson-Ullman model; dot-dashed lines: random energy model; full lines: generalized Anderson-Ullman model with gamma distribution for the energies.

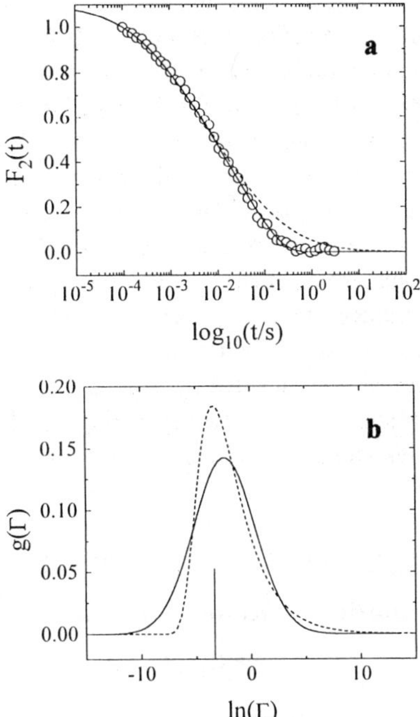

Figure 4. a) The two time echo function versus time; open symbols represent experimental results (11), full line: calculation according to the Anderson-Ullman model, see text, broken line: calculations with vanishing exchange rate. b) Decay rate distribution functions for the calculated 2t-echoes as shown in a) The distributions are normalized to an area of unity.

11 we performed calculations for different multi-state models for which distributions of angular jump rates and exchange rates were considered without recourse to specific models for the physics of the environmental fluctuations. In order to model the behavior seen in Figure 3 in this chapter we consider a class of models for environmental fluctuations which can be viewed as energy landscape fluctuation models and compare the calculated $F_4(t_{rex})$ functions to the experimental ones.

The idea underlying all models of environmental fluctuations is to account for the non-Markovian behavior of the rotational dynamics. This means that in the simplest case one assumes that the angular jump rates Γ relevant for the present discussion are functions of an additional variable ε, i.e. $\Gamma = \Gamma(\varepsilon)$. As ε characterizes the dynamical state of the molecules we denote it as a *dynamical variable*.. Different dynamical states are characterized by different values of the variable ε. In a physical picture of a viscous liquid one may view ε as a local packing density, an energy (potential energy, thermal energy, free energy), a local free volume, etc. (3,14,15). For molecules of high energy (corresponding e.g. to low local density) one expects a low barrier to reorientation and vice versa. Furthermore, in a given dynamical state the angular jump rate is assumed to be fixed. Changes in the angular jump rates are modeled by transitions between different dynamical states. This means if ε changes its value to ε', there is an associated transition from $\Gamma(\varepsilon)$ to $\Gamma(\varepsilon')$. Different models differ by the definition of the transition rates $\kappa(\varepsilon',\varepsilon)$ from state ε to state ε' and in the assumed functional dependence of the angular jump rates $\Gamma(\varepsilon)$.

This class of models for the non-Markovian rotational dynamics allows to treat the stochastic process (Ω,ε) [(orientation, dynamical state)] as a composite Markov process (12,16). [Note that the projection of the process (Ω,ε) onto the stochastic process Ω is no longer Markovian.] The most prominent consequence of this procedure is that the dynamics of the stochastic process (Ω,ε) can be described by a master equation (16). This means that the probability for finding a molecule in orientation Ω and in dynamical state ε, (Ω,ε), at time t, conditional on (Ω_0,ε_0) at time 0 obeys a set of linear differential equations.

It is exactly this probability, denoted by $P_{1|1}(\Omega,\varepsilon;t|\Omega_0,\varepsilon_0)$ which is needed in the calculations of the 2t- and 4t-echoes. Remembering that the Larmor frequency is a function of molecular orientation, $\omega = \omega(\Omega)$, we have for example for the cosine part of the 2t-echo

$$
\begin{aligned}
F_2^{cc}(t_p,t_{fil}) &= \langle\cos(\omega(\Omega(0))t_p)\cos(\omega(\Omega(t_{fil}))t_p)\rangle \equiv \langle\cos(\omega(\Omega_1)t_p)\cos(\omega(\Omega_2)t_p)\rangle \\
&= \iint d\varepsilon_1\, d\varepsilon_2 \iint d\Omega_1\, d\Omega_2\, \cos(\omega(\Omega_1)t_p)\cos(\omega(\Omega_2)t_p)P_1(\Omega_1,\varepsilon_1)\times \qquad (5)\\
&\qquad \times P_{1|1}(\Omega_2,\varepsilon_2;t_{fil}|\Omega_1,\varepsilon_1)
\end{aligned}
$$

where $P_1(\Omega_1,\varepsilon_1)$ denotes the a priori probability, cf. (12). All other correlation functions can be written in a similar way. The master equation for the conditional probabilities reads as:

$$\frac{\partial}{\partial t} P_{1|1}(\Omega,\varepsilon;t|\Omega_0,\varepsilon_0) = \iint d\Omega' d\varepsilon' \, \Pi(\Omega,\varepsilon|\Omega',\varepsilon') P_{1|1}(\Omega',\varepsilon';t|\Omega_0,\varepsilon_0) \tag{6}$$

where $\Pi(\Omega,\varepsilon|\Omega',\varepsilon')$ denotes the transition rate from $(\Omega',\varepsilon') \rightarrow (\Omega,\varepsilon)$. In principle a number of models can be constructed by assuming different functional forms for the transition rates. Here, we restrict the analysis to the case for which rotational random jumps govern the Ω-dependence and assume that transitions $\varepsilon \rightarrow \varepsilon'$ have no effect on the orientations. Then, with these restrictions different models differ solely in the definition of the transition rates $\kappa(\varepsilon',\varepsilon)$. Even under these constraints many different scenarios are possible as well. The choices we will discuss in the following belong to the most simple ones.

Next, the $P_{1|1}$ are expanded into Wigner rotation matrix elements, which in a discrete version read (12):

$$P_{1|1}(\Omega,\varepsilon_k;t|\Omega_0,\varepsilon_i) = \sum_{Lmn} C_{ki}^{(L)}(t) \frac{2L+1}{8\pi^2} D_{mn}{}^{(L)}(\Omega) \, D_{mn}{}^{(L)}(\Omega_0)^* \tag{7}$$

The expansion coefficients depend on the rank L of the matrices but are independent of m and n and again obey a master equation:

$$\frac{\partial}{\partial t} C_{ki}^{(L)}(t) = -(1-\delta_{L,0}) \Gamma_k C_{ki}^{(L)}(t) + \sum_n \kappa_{kn} C_{ni}^{(L)}(t) \tag{8}$$

where Γ_k and κ_{nk} are abbreviations for $\Gamma(\varepsilon_k)$ and $\kappa(\varepsilon_n,\varepsilon_k)$, respectively. It is obvious from equation 8 that the angular jump rates are independent of L, reflecting the well known fact that experimental methods, which measure rotational correlation functions of different values L lead to the same results in case of large angular jumps. (This, of course, is not true in case of rotational diffusion.) Furthermore, the time evolution of the coefficients $C_{ki}^{(0)}(t)$ is independent of the angular jump rates Γ_k.

For a calculation of the 2t- and 4t-echoes one proceeds in the following way. One diagonalizes the matrix defined implicitly (from the Γ_k and κ_{nk}) in equation 8 and represents the expansion coefficients $C_{ki}^{(L)}(t)$ in terms of the eigenvalues and eigenvectors of that matrix. From the results of that procedure the conditional probabilities $P_{1|1}$ and subsequently the correlation functions according to equation 5 and similar expressions are obtained. Instead of going into computational details here (see Ref. 12) we want to discuss the results for different models governing the dynamics of the dynamical variable ε and the functional dependence of the angular jump rates on ε. In terms of the $C_{ki}^{(L)}(t)$ coefficients the relevant 2t- and 4t-echo functions are approximately given by (cf. equations 2 and 3):

$$F_2(t_{fil}) \cong \sum_{ij} p_i^{eq} C_{ji}^{(L)}(t_{fil}) \tag{9}$$

$$F_4(t_{rex}) \propto \sum_{ijkl} p_i^{eq} C_{ji}^{(L)}(t_{fil}) C_{lk}^{(L)}(t_{fil}) C_{kj}^{(0)}(t_{rex}) \tag{10}$$

Here, p_i^{eq} denotes an equilibrium probability, i.e. the equilibrium population of states characterized by ε_i. Note, that the dependence on the 'rate exchange interval' t_{rex} of this particular 4t-echo function is determined solely via the time dependence of the $C_{ki}^{(0)}(t_{rex})$. These, however, do not depend on the angular jump rates at all, as can be seen easily from the master equation, equation 8. This means, that during t_{rex} only the exchange between different angular jump rates is observed. Of course, this holds exactly only if the model of rotational random jumps is applicable.

One very simple environmental fluctuation model has been formulated by Anderson and Ullman (17) in an attempt to explain the asymmetric form of the apparent distribution of rotational rates as observed in dielectric experiments on viscous liquids. These authors assumed that molecular reorientations depend on the local free volume at the site of the molecule considered. If the molecule is in an element of large free volume, the angular jump rate is assumed to be large. This is a physical reasonable assumption, since the barrier to reorientation should be relatively low. The temporal fluctuations in free volume were described as an Ornstein-Uhlenbeck process (16). The master equation for the probability $P(\varepsilon,t)$ of finding a molecule in a specific element of free volume in this case is a Fokker-Planck equation

$$\partial P/\partial t = D\,\partial^2 P/\partial\varepsilon^2 + \gamma\,\partial(\varepsilon P)/\partial\varepsilon$$

where $P \equiv P(\varepsilon,t)$ and ε denotes the deviation of free volume from the average value. The equilibrium probability $P^{eq}(\varepsilon) = P(\varepsilon,\infty)$ is given by a Gaussian with width $\sigma = (D/\gamma)^{1/2}$. As functional form for the dependence of the angular jump rates on ε Anderson and Ullman chose $\Gamma(\varepsilon) = \Gamma_0 \exp(\pm\varepsilon^C)$, where the plus sign holds for positive values of ε and the minus sign for negative ones. The exponent 'c' was arbitrarily chosen between 1 and 2.

In view of later generalizations it is suggestive to reinterpret the 'free volume' as 'energy'. This means we choose ε as an energy variable. A molecule in a state of 'large' free volume is thus assumed to be in a region of low local density and thus of high thermal energy and vice versa.

Next, we have to specify a functional form for the angular jump rates. It seems natural to choose the angular jump rates according to an Arrhenius form:

$$\Gamma(\varepsilon) = \Gamma'_\infty \exp\{-E_R(\varepsilon)/T\} \tag{11a}$$

where $E_R(\varepsilon)$ is a (dimensionless) activation barrier against reorientations and T is the (dimensionless) temperature. We do not assume the energies ε and $E_R(\varepsilon)$ to be the same. Rather we will take them to be proportional to one another in the sense that molecules in a more 'liquid-like' region (large ε) experience a lower barrier to rotation $E_R(\varepsilon) = E_R^0 - \chi_\Gamma \varepsilon$. Thus,

$$\Gamma(\varepsilon) = \Gamma_\infty \exp(\chi_\Gamma \varepsilon / T) \tag{11b}$$

This form for $\Gamma(\varepsilon)$ with Γ_∞ and χ_Γ as adjustable parameters will be used throughout in the calculations.

For numerical calculations the Fokker-Planck equation has to be written in a discrete form. One then ends up with equation 8, where the transition rates (in a canonical form, i.e. obeying detailed balance) are given by [cf. (18)]:

$$\kappa_{ik} = \frac{D}{(\Delta_\varepsilon)^2} \exp\left\{-(\varepsilon_i^2 - \varepsilon_k^2)/(4\sigma^2)\right\} \delta_{i,k\pm 1} \tag{12}$$

Here, Δ_ε denotes the spacing between two adjacent values of ε and the Kronecker symbol means that transition rates between values of ε which are more than one 'step' apart from each other vanish, i.e. the exchange is 'diffusive'. The Gaussian equilibrium probability [$p_i^{eq} \propto \exp\{-\varepsilon_i^2/(2\sigma^2)\}$] results in a logarithmic Gaussian distribution $g(\Gamma)$ of angular jump rates by virtue of equation 11 and a width which is related to the width of p_i^{eq} by $\sigma_\Gamma = (\chi_\Gamma / T)\sigma$.

In Figure 4a the 2t-echo according to equation 9 is shown together with the experimental data for σ=2.8, χ_Γ=1.0 and D/Γ_∞=0.5 (Γ_∞ can be chosen arbitrarily since it merely is a scale factor and the temperature is set to T=1). Numerically, we took N values ε_1 to ε_N with N typically being 50. We diagonalize the matrix defined by equation 8 with the transition rates defined in equation 12, calculate the time dependence of the coefficients $C_{ki}^{(L)}(t)$ and subsequently compute the 2t- and 4t-echoes. N=50 was checked to be sufficiently large by comparing the results to calculations with larger N values. The dashed line in Figure 4a shows $F_2(t_{fil})$ without exchange (D=0). The tailing in this curve is characteristic for a logarithmically symmetric distribution of decay rates. The stronger decay at longer times which is a hallmark for a Kohlrausch function is achieved by finite exchange. Figure 4b shows this effect on the distribution of decay rates. The full curve is the log-Gaussian with σ=2.8 and the dashed one the effective distribution including exchange, corresponding to a Kohlrausch distribution for β=0.42. The solid line marks the exchange rate D. The effect of the exchange is readily seen in skewing the 'bare' distribution.

Using the same parameters, we then calculated the 4t-echoes according to equation 10 for filter times varying between 1 and 55 ms. Instead of showing the curves we present the results of fits to equation 4 as the dashed lines in Figure 3. It is clearly seen that the simple Anderson-Ullman model is not able to reproduce any of the features seen in the experimental data. The reason for this is quite simple. The increases in κ^{-1} and β_4 with filter time t_{fil} show that there is a broad distribution of exchange rates κ_{ik} in the system. In the Anderson-Ullman model, on the other hand, there exists only a single exchange rate, D. The deviations in β_4 from unity (the value one naively would expect for a single rate) solely stem from the requirement of detailed balance $\kappa_{ik}\, p_k^{eq} = \kappa_{ki}\, p_i^{eq}$.

In order to develop a more realistic model let us consider the physical content of the Anderson-Ullman model in more detail. Since we interpret ε as an energy, the exchange rate D should be understood as an escape rate, i.e. $D = D_\infty \exp(-E_{esc}/T)$.

Then the interpretation of the Anderson-Ullman model is equivalent to saying that there exists a unique barrier height in the energy landscape in which the depth of the valleys correspond to different dynamical states. This, of course, is highly unplausible for the energy landscape of a complex system like a viscous liquid. However, this consideration leads us to an obvious generalization of the Anderson-Ullman model. Instead of a single escape energy one expects a distribution of escape energies, i.e. a density of energy states which is not given by a delta function. The Anderson-Ullman model might thus be seen as the simplest model which disregards the finite width of the density of states and merely considers the mean. We therefore introduce a density of states (DOS), $n(\varepsilon)$ and assume the exchange rates $D(\varepsilon)$ to be modified according to $D(\varepsilon)=D_\infty \exp(-(E^0_{esc}-\varepsilon)/T)$. This means that we have a distribution of escape rates depending on energy. The probability $P_{1|1}$ then has to be calculated from a generalized Fokker-Planck equation. Here, we again us a discrete version (equation 8) with the exchange rates (see equation 12) according to:

$$\kappa_{ik}=\frac{\kappa_\infty}{(\Delta_\varepsilon)^2}\exp(\varepsilon_k/T)\exp\left\{-(\varepsilon_i^2-\varepsilon_k^2)/(4\sigma^2)\right\}\delta_{i,k\pm 1} \tag{13}$$

where $\kappa_\infty = D\exp(-E^0_{esc}/T)$ to be consistent with equation 12. Additionally, the equilibrium probabilities are now given by:

$$p_i^{eq}=\frac{1}{\sqrt{2\pi}\sigma}\exp(-(\varepsilon_i-\bar{\varepsilon})^2/(2\sigma^2)) \quad ; \quad \bar{\varepsilon}=-\sigma^2 T \tag{14}$$

for a Gaussian DOS since statistical mechanics requires $p_i^{eq}\propto n(\varepsilon_i)\exp(-\varepsilon_i/T)$.

As before, we calculate F_2 and F_4 according to equations 9 and 10. As parameters we have chosen (T=1), σ=2.3, χ_Γ=1.7 (σ_Γ=3.9) and $\kappa_\infty/\Gamma_\infty$=0.26. The results for the 2t-echo look similar to the ones for the Anderson-Ullman model (except for a more pronounced tailing of the 2t-echo without exchange due to the broader distribution of bare angular jump rates) and are therefore not shown. The $F_4(t_{rex})$ functions obtained this way were fitted using equation 4. The results of the fits are shown as the dotted lines in Figure 3. It is seen that this slightly modified Anderson-Ullman model qualitatively reproduces the features seen in the experimental data. The experimental values of β_4 are met almost exactly, the characteristic decay times change somewhat too strongly with increasing filter time.

Since the extended Anderson-Ullman model shows the prominent characteristic features observed experimentally, the next thing we do is to ask whether the four time correlation function measured in this specific experiment allows to distinguish between different models for the transition rates κ_{ik} among the states of different energy. In the two models considered so far the exchange proceeds only among states of similar energy. This means that a small angular jump rate becomes larger only gradually in the course of time via many transitions towards fast reorientations. If transitions take place in a complex energy landscape there is no reason to assume that these transitions are restricted to energy minima of only slightly different depth. One might alternatively expect that if an escape out of one minimum has taken place one may end up in any other minimum. Stated in other terms, one would like to be able to distinguish between models that differ in the connectivity of the dynamical states. In this perspective, the diffusive scenario is viewed as a nearest neighbor random walk among the dynamical states biased by the ε-dependent escape rates, while the random scenario corresponds to a globally connected model, where each dynamical state is connected to any other.

Thus, in order to see whether the information content of the reduced 4D-NMR experiment is sufficient to distinguish between different models, we next consider a random energy model (19). This model, originally introduced as a simple model for spin glasses, has been applied to the study of protein folding kinetics (20) and also for modeling transport in viscous liquids (21).

We choose a Gaussian DOS as in the generalization of the Anderson-Ullman model and assume the following form for the transition rates between states of different energies [cf. (21,22)]:

$$\kappa_{ik} = \kappa_\infty n(\varepsilon_i)\exp(\varepsilon_k / T) \quad ; \quad n(\varepsilon_i) = \frac{1}{\sqrt{2\pi}\sigma}\exp(-(\varepsilon_i)^2 / (2\sigma^2)) \qquad (15)$$

This means we consider the transition $\varepsilon_k \rightarrow \varepsilon_i$ as an escape out of the state ε_k and the destination state ε_i is chosen randomly, i.e. according to the probability of its existence given by the DOS $n(\varepsilon_i)$. The equilibrium population of the states is again given by equation 14. The distribution of angular jump rates is taken to be a log-Gaussian as discussed in context with the Anderson-Ullman model. Also for this model we calculated the 2t- and 4t-echoes and the fit results according to equation 4 are represented by the dot-dashed lines in Figure 3. The parameters chosen are σ=2.0, χ_Γ=1.7 (σ_Γ=3.4), $\kappa_\infty / \Gamma_\infty$=0.13 and T=1.

It is seen from Figure 3 that the results of the random energy model are very similar to those of the generalized Anderson-Ullman model. This means that even though the dynamics of the transitions among different states is quite different in these two models, this does not show up in the results of the specific four time correlation function observed in reduced 4D-NMR. We only mention here that also other choices for the κ_{ik} yield results similar to those presented in Figure 3. What seems to be most important for the behavior of the 4t-echo is the finite width of the distribution of exchange rates rather than the exact scenario of how the angular jump rates fluctuate due to exchange. The fact that the exchange is more efficient in 'speeding up' slowly reorienting molecules in the random energy model is merely reflected in the correspondingly smaller scale κ_∞ as compared to the generalized Anderson-Ullman model. However, this also does not serve as a criterion to distinguish different scenarios, because if the attempt frequency Γ_∞ is chosen to give reasonable values, then the same holds for κ_∞ in both models.

As a final example let us consider the case in which the DOS is not given by a Gaussian. We consider the generalized Anderson-Ullman model with the exchange rates given by equation 13, but the DOS is chosen according to an asymmetric distribution. As an example we choose a gamma distribution (16):

$$n(\varepsilon) \propto \varepsilon^{\nu-1} \exp(-a\varepsilon) \quad ; \quad (a>0, \nu>0) \tag{16}$$

Consequently the equilibrium probability changes from a Gaussian to

$$p_i^{eq} \propto n(\varepsilon_i)\exp(-\varepsilon_i / T). \tag{17}$$

The use of an asymmetric DOS gives rise to an associated change in the distribution of angular jump rates $\Gamma(\varepsilon_i)$.

Calculations with ν=8, a=0.045, χ_Γ=1.15, $\kappa_\infty / \Gamma_\infty$=31.0 and T=1 are represented as the solid lines in Figure 3. It is seen that in this case the experimental values of κ^{-1} are reproduced almost exactly, whereas the agreement with the experimental β_4 becomes somewhat poorer.

Concluding Discussion

By performing reduced 4D-NMR experiments we were able to show that the reorientational dynamics in supercooled o-TP is of a heterogeneous nature. This heterogeneity is, however, not static but dynamic in the sense that slowly reorienting molecules become faster in the course of time and vice versa. By measuring the filter time dependence of the 4t-echo it becomes obvious that it is not possible to ascribe a well defined lifetime to the heterogeneities. Instead, it appears that slowly reorienting molecules seem to change their angular jump rates more slowly than the quickly reorienting molecules do, but see also the discussion in Ref. (11). The life time of the heterogeneities is similar to the time scale of reorientations. In the deep bleach experiments on dye-molecules in supercooled o-TP Cicerone and Ediger (9) found that the mean life time of the heterogeneities is much longer (about a factor of 10^2-10^3) than the reorientational correlation times. This is different from what we have found. It has to be noted, however, that the Cicerone-Ediger experiment was performed at lower temperatures than ours.

Since the dependence of the 4t-echo upon the rate exchange time t_{rex} is determined solely by the dynamics of the exchange between different angular jump rates (at least under the given experimental conditions) it becomes obvious from the increase in the time scale κ^{-1} and in the stretching parameter β_4 that there exists a distribution of exchange rates.

In an attempt to model the observed behavior we considered different fluctuating energy landscape models. We started with a time honored model due to Anderson and Ullman, which in its original form contains only a single exchange rate. This allows to mimic the shape of the two-time echo function but is not capable to reproduce the experimental 4t-echoes. Therefore, we proposed a generalization of the original Anderson-Ullman model by assuming a Gaussian distribution of escape energies. This 'diffusive' scenario already yielded satisfactory results. Since the dynamics in this model consists of transitions between states of very similar energy only, the model incorporates a very restricted connectivity similar to the situation encountered in a nearest neighbor random walk. In another model we assumed all dynamical states to be connected (globally connected model) and obtained results very similar to the locally connected case. Thus, it is not possible from the specific four time correlation function measured in the discussed experiment to discriminate between these different scenarios for the dynamics of the heterogeneuos reorientations in o-TP.

It has to be pointed out that the Arrhenius form for the dependence of the angular jump rates on the energy variable (equation 11), is a rather restrictive one and may be lifted if desired. However, in an energy landscape picture it is physically appealing to attribute a lower barrier to reorientation to those molecules with a higher (thermal) energy.

Literature Cited

1. Angell, C.A.; *Science* **1995**, *267* , 1924
2. Mohanty, U.; *Adv. Chem. Phys.* **1995**, *89* , 89
3. Chamberlin, R.V.; *Europhys. Lett.* **1996**, *33* , 545
4. Phillips, J.C.; *Rep. Prog. Phys.* **1996**, *59* , 1133
5. Adam, G; Gibbs, J.H.; *J. Chem. Phys.* **1965**, *43* , 139
6. Schnauss, W.; Fujara, F.; Hartmann, K.; Sillescu, H.; *Chem. Phys. Lett.* **1990**, *166* , 381
7. Schmidt-Rohr, K.; Spiess, H.W.; *Phys. Rev. Lett.* **1991**, *66* , 3020
8. Schiener, B; Böhmer, R; Loidl, A.; Chamberlin, R.V.; *Science* **1996**, *274* , 752
9. Cicerone, M.T.; Ediger, M.D.; *J. Chem. Phys.* **1995**, *103* , 5684
10. Heuer, A.; Wilhelm, M.; Zimmermann, H.; Spiess, H.W.; *Phys. Rev. Lett.* **1995**, *75* , 2851
11. Böhmer, R.; Hinze, G.; Diezemann, G.; Geil, B.; Sillescu, H.; *Europhys. Lett.* **1996**, *36* , 55
12. Sillescu, H.; *J. Chem. Phys.* **1996**, *104* , 4877
13. Chang, I.; Fujara, F.; Geil, B.; Heuberger, G.; Mangel, T.; Sillescu, H.; *J. Non-Cryst. Solids* **1994**, *172-174* , 248; Geil, B. to be published
14. Goldstein, M.; *J. Chem. Phys.* **1969**, *51* , 3728
15. Stillinger, F. H.; *Science* **1995**, *267* , 1935
16. van Kampen, N.G.; *Stochastic Processes In Physics And Chemistry* ; North-Holland, Amsterdam 1992
17. Anderson, J. E.; Ullman, R.; *J. Chem. Phys.* **1967**, *47* , 2178
18. Agmon, N.; Hopfield, J.J.; *J.Chem. Phys.* **1983**, *78* , 6947
19. Derrida, B.; *Phys. Rev. Lett.* **1980**, *45* , 79
20. Saven, J. G.; Wang, J.; Wolynes, P. G.; *J. Chem. Phys.* **1994**, *101* , 11037
21. Dyre, J. C.; *Phys. Rev.* **1995**, *B51* , 12276
22. Shakhnovich, E. I.; Gutin, A. M.; *Europhys. Lett.* **1989**, *9* , 569

Chapter 13

High-Frequency Dielectric Spectroscopy of Glass-Forming Liquids

P. Lunkenheimer, A. Pimenov, M. Dressel, B. Gorshunov[1], U. Schneider, B. Schiener[2], and A. Loidl

Experimentalphysik V, Universität Augsburg, Universitätsstrasse 2, D–86135 Augsburg, Germany

We give an overview of our recent results from broadband dielectric spectroscopy on various glass forming liquids as glycerol, $[Ca(NO_3)_2]_{0.4}[KNO_3]_{0.6}$, $[Ca(NO_3)_2]_{0.4}[RbNO_3]_{0.6}$, propylene-carbonate, and Salol. For the first time the dielectric loss has been investigated systematically in the crossover regime from the α-relaxation to the far-infrared (FIR) response. For all materials investigated we observe a minimum in $\varepsilon''(\nu)$ in the GHz-THz range with a significantly sublinear high-frequency wing. This behavior cannot be explained by a simple transition from the α-relaxation peak to the FIR bands but has to be attributed to additional fast processes. We compare our results to the predictions of the mode coupling theory of the glass transition.

Stimulated by recent theoretical approaches, the fast dynamics in glass-forming liquids came into the focus of interest. Various scenarios have been proposed to describe or predict the dynamic susceptibility at high frequencies in the GHz-THz region [*1-5*]. Maybe the most controversially discussed approach is the mode coupling theory (MCT) of the glass transition [*1*] which makes distinct predictions for the high-frequency region. The relevant frequency range up to now has been investigated mainly by neutron and light scattering experiments (see, e.g., [*6-11*]). Only recently, we were able to extend the frequency range of dielectric experiments on glass-forming liquids up to 370 GHz [*12,13*] using backward wave oscillators (BWO) as coherent sources of electromagnetic radiation [*14*]. These experiments, which have been performed on glycerol [*12*], $[Ca(NO_3)_2]_{0.4}[KNO_3]_{0.6}$ (CKN) [*13*], and $[Ca(NO_3)_2]_{0.4}[RbNO_3]_{0.6}$ (CRN) [*13*], for the first time revealed the existence of a relatively broad minimum in the

[1]Permanent address: Institute of General Physics, Russian Academy of Sciences, Moscow, Russia

[2]Current address: Institut für Physikalische Chemie, Johannes Gutenberg Universität of Mainz, Mainz, Germany

frequency dependence of the dielectric loss, $\varepsilon''(\nu)$, at frequencies above GHz. Its functional form and temperature dependence cannot be explained by a simple crossover from the structural (α-)relaxation to the FIR response but is indicative of additional fast processes prevailing in these glass-forming liquids. In the present paper we will give a review of our dielectric results obtained up to now including data on glycerol in an extended frequency range up to 950 GHz and first results on propylene carbonate (PC) and Salol. Our findings are compared to the predictions of the mode coupling theory (MCT) of the glass transition [*1*].

Experimental Details

Broadband dielectric measurements involve the use of various techniques. At frequencies 10 µHz $\leq \nu \leq$ 1 kHz measurements were performed in the time domain using a spectrometer which is based on a design described by Mopsik [*15*]. The autobalance bridge HP4284 was used at 20 Hz $\leq \nu \leq$ 1 MHz. For the radio-frequency and microwave range (1 MHz $\leq \nu \leq$ 10 GHz) a reflectometric technique was employed [*16*] using the HP4191 impedance analyzer and the HP8510 network analyzer. In addition, at frequencies 100 MHz $\leq \nu \leq$ 40 GHz data were taken in transmission with the HP8510 network analyzer using waveguides and coaxial lines filled with the sample material. Various ovens, closed-cycle refrigerators, and He-cryostats have been utilized to cover the different temperature ranges.

In the most relevant frequency range around some 100 GHz, quasi-optical measurements were performed by utilizing the submillimeter spectrometer "Epsilon" [*14*]. The Mach-Zehnder setup of the interferometer allows the determination of both components, the transmission coefficient and the change in phase upon passing of the electromagnetic wave through the sample. Five different backward wave oscillators were employed to cover the frequency range from 60 GHz up to 950 GHz; the signal was detected by a Golay cell and amplified using lock-in technique. The liquids were put in specially designed cells made of polished stainless steel with thin plane-parallel glass windows with a typical diameter of 15 mm; depending on the range of frequency and temperature the thickness of the sample cell was varied between 1 mm and 30 mm. The cell was placed at the end of a cold finger of a continuous flow He^4-cryostat allowing to perform the experiments down to 10 K. High temperature measurements were carried out in a custom-made oven up to 500 K. The data were analyzed by using optical formulas for multilayer interference [*17*] with the known thickness and optical parameters of the windows in order to get the real and imaginary part of the dielectric susceptibility of the sample as a function of frequency at various temperatures.

In general it is more difficult to obtain reliable absolute values of the real part of the dielectric susceptibility due to stray capacitances at low frequencies and due to phase measurement errors in the submillimeter range. Therefore the data presented in this article are restricted to the dielectric loss.

To cover the complete frequency range, a single $\varepsilon''(\nu)$ curve at a given temperature combines results from different experimental setups. For the measurements at $\nu \leq$ 40 GHz there are some uncertainties of the absolute values originating from an ill-defined geometry of the samples or parasitic elements. Therefore it often was necessary to shift the log ε" values of the measurements at the lower frequencies ($\nu \leq$ 40 GHz) with respect to the high-frequency results ($\nu \geq 60$ GHz) in order to construct a smooth ε"(ν) curve. It is important to note, that ε"(ν,T) from each experimental setup is shifted by one gauge factor only, which depends neither on frequency nor on temperature. In addition, the results at $\nu \leq 1$ MHz, obtained with the autobalance bridge and the results at $\nu \geq 60$ GHz, obtained with the quasi-optical spectrometer, provide very precise absolute values of ε" and were used to scale the data in the other frequency ranges, if necessary.

Results and Discussion

Glycerol. Glycerol ($T_g \approx 190$ K) is a relatively strong [*18*] hydrogen-bonded glass former with a fragility parameter [*19*] of m ≈ 53. From neutron and light scattering experiments [*9-11*] the imaginary part of the dynamic susceptibility, χ", has been determined. At frequencies of some 100 GHz χ"(ν) exhibits a minimum. These results have been compared to the predictions of the MCT and deviations from the simplest scaling laws of MCT have been found. This finding was attributed to additional vibrational excitations contributing at the high-frequency wing of the minimum and leading to a relatively steep increase of χ"(ν) towards the microscopic excitation bands [*9-11*]. These vibrational contributions which give rise to a peak, commonly called Boson-peak, seem to be most pronounced in strong glass formers as glycerol [*20*]. In a recent work [*21*] it has been shown that using a more sophisticated evaluation within MCT it is possible to describe the scattering data on glycerol, including the Boson peak. In earlier dielectric experiments the relaxation dynamics in glycerol has been studied over almost 16 decades in frequency, up to 40 GHz [*3, 22-24*]. A minimum could not be detected and clear deviations to the light and neutron scattering results were obtained. However, it has been pointed out very early by Wong and Angell [*5*] that a minimum dielectric loss must exist in the crossover region between structural relaxation and the FIR resonances. They predicted that at low temperatures a plateau should develop due to a constant (i.e. frequency independent) loss.

Figure 1 shows the dielectric loss, ε"(ν), for frequencies up to 950 GHz. At the highest frequencies investigated a minimum shows up. At lower frequencies, the data reveal a well developed α-peak which, close to the maximum, can be described by the Fourier transform of the Kohlrausch-Williams-Watts (KWW) function, $\Phi_0 \exp[-(t/\tau)^\beta]$, with the stretching exponent β and the relaxation time τ (dashed lines) [*12,24*]. The exponent β increases with temperature varying between 0.65 at 180 K and 0.85 at 330 K. The relaxation time τ increases significantly stronger than thermally activated and can be parameterized according to a Vogel-Fulcher (VF) law with a VF temperature close to 130 K [*3,24*]. At temperatures, T > 260 K, the high-frequency wing of the loss peak reveals only one power-law which smoothly connects to the low-frequency side of the minimum (Figure 1). At temperatures below 260 K, an additional power law develops at high frequencies with exponents clearly smaller than β which significantly differs from the KWW fit [*24*]. At these temperatures, the α-response can be described perfectly well using Nagel's universal scaling ansatz [*3*].

In the inset of Figure 1 the dielectric results for 295 K are compared with neutron and light scattering data [*10,11*]. As the scattering results give no information about the absolute value of χ'' the datasets have been scaled. Independent of the scaling procedure the $\chi''(\nu)$ from the scattering experiments increases significantly stronger towards high frequencies if compared to the dielectric results. It seems that there are additional contributions, most probably of vibrational origin which couple only weakly to the dipolar reorientations determining the dielectric response. This additional density of states, whose contribution to $\chi''(\nu)$ is usually termed Boson peak, seems to significantly contribute to density correlations especially in strong liquids, as has been worked out in detail by Sokolov *et al.* [*20*]. The inset of Figure 1 suggests that these contributions are most pronounced in the neutron scattering results and to a lesser extent in the light scattering results. While in neutron scattering experiments a good coupling to the density fluctuations is achieved, it recently has been found that light scattering experiments are much more influenced by the coupling to orientational fluctuations [*25*].

In Figure 2 we analyze the minimum region in terms of a simple crossover from the structural relaxation to the FIR response. Figure 2a shows the dielectric loss vs. frequency at 273 K, replotted from Figure 1. The high frequency wing of the α-response follows a power law with an exponent of -0.63. The increase of $\varepsilon''(\nu)$ towards the FIR bands is assumed to follow a power law with an exponent of 1, a behavior which is common to a variety of glasses [*26*]. In the region of the minimum, the two contributions are assumed to be superimposed. Depending on the strength and the resonance frequency of the FIR peak, two possible scenarios are obtained (Figure 2a). In scenario 1 (dashed line) the minimum susceptibility is fixed at the experimentally observed value. However, the frequency of the minimum cannot be described correctly. In scenario 2 (dash-dotted line) the minimum frequency is fixed at the experimentally observed value. But now the minimum is too deep and additional processes have to be considered in order to describe the experimental curve. If one introduces an additional constant loss contribution [*5*], it is possible to obtain good fits to the data. However, to take account of the whole data set (Figure 1), a *temperature dependent* constant loss is necessary, as demonstrated in Figure 2b. The temperature dependence of the constant loss is shown in the inset. Up to now there is no theoretical foundation for such a behavior. But from these considerations it becomes obvious that an additional process at high frequencies has to be taken into account. Fast processes are considered by the coupling model (CM) [*2*] and the MCT [*1*].

Following the predictions of the CM, for large β values the fast process becomes less prominent and a minimum cannot be expected as a consequence of the fast process alone. In order to describe the minimum observed in the neutron and light scattering experiments on glycerol [*10,11*] and also in the results of a molecular dynamics simulation of *ortho*-terphenyl [*27*] within the framework of the CM, Roland *et al.* [*27,28*] took into account additional vibrational contributions. But then again the problem of the temperature dependent constant loss is encountered.

In the following we compare our data to the MCT predictions. Both the minimum and the transition region to the α-process can be described by the MCT using the interpolation [*1*]:

$$\varepsilon''(\nu) = \varepsilon_{min}\,[a(\nu/\nu_{min})^{-b} + b(\nu/\nu_{min})^{a}]/(a+b) \qquad (1)$$

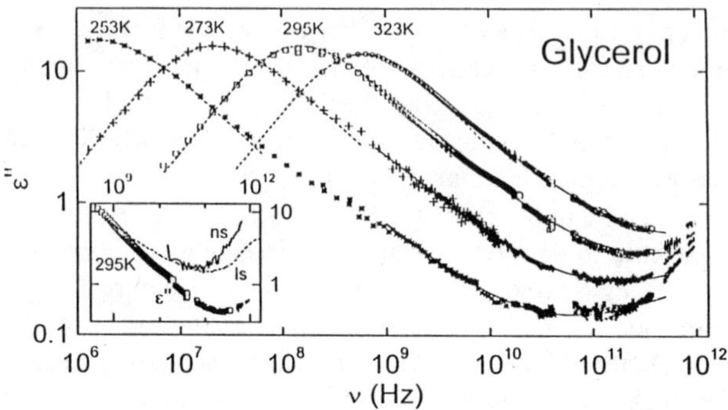

Figure 1. Frequency dependence of the dielectric loss of glycerol for four temperatures. The dashed lines are fits using the KWW function. The solid lines have been calculated with the MCT expression, Eq. (1), with a = 0.325 and b = 0.63 which corresponds to an exponent parameter λ = 0.705. In the inset the dielectric data for 295 K are compared to the susceptibility from light (ls, 293 K) [*10*] and neutron scattering (ns, 293 K) experiments [*11*]. The scattering data sets have been vertically shifted to give a comparable intensity of the α-process

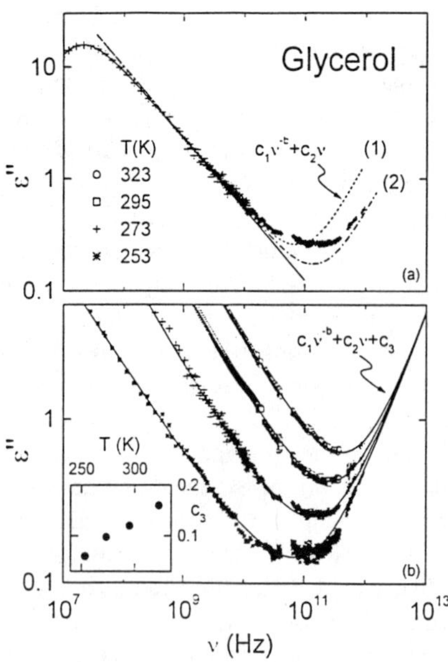

Figure 2. (a) Frequency dependence of the dielectric loss in glycerol for 273 K. The solid line is a fit using the Cole-Davidson ansatz with β_{CD} = 0.63 [*12*]. The dashed and dash-dotted lines have been calculated using the expression indicated in the figure with b = 0.63. (b) $\varepsilon''(\nu)$ for various temperatures. The solid lines are fits by the expression indicated in the figure with b = 0.63 and $c_2 = 4.4 \times 10^{-13}$ for all temperatures except for 253 K where b = 0.56. The resulting temperature dependence of c_3 is shown in the inset.

The exponents a and b describe the high and low frequency wing of the minimum, respectively, which are identical for all temperatures and constrained by the exponent factor $\lambda = \Gamma^2(1\text{-a})/\Gamma(1\text{-2a}) = \Gamma^2(1\text{+b})/\Gamma(1\text{+2b})$ where Γ denotes the Gamma function. This restricts the exponent a to values below 0.4, i.e. a significantly sublinear increase of $\varepsilon''(\nu)$ at frequencies above the minimum is predicted. ε_{min} and ν_{min} are the height and the position of the minimum, respectively. We want to emphasize that Eq. (1) is only an approximation of $\varepsilon''(\nu)$ near the minimum and valid only above the critical temperature T_c. In addition, it does not take account of the Boson peak contributions. However, it is useful for a first comparison of model predictions and experimental results. The solid lines in Figure 1 are obtained by Eq. (1) with $\lambda = 0.705$ ($a = 0.325$, $b = 0.63$) [*12*]. While the data are well described up to the minimum frequency, at the highest frequencies investigated we find clear discrepancies between data and fit. This may well be due to the remainders of the excess vibrational contributions observed in the neutron- and light-scattering experiments and an analysis including the Boson peak contribution [*21*] is necessary to describe the data up to the highest frequencies. However, from Figure 1 it seems reasonable that position and height of the minimum determined from the fits is only weakly influenced by these contributions. An analysis of the critical behavior near T_c neglecting the Boson peak is performed in [*12*]. We want to emphasize that the deviations at high frequencies were interpreted as indications for systematic deviations from MCT predictions by Sokolov *et al.* [*20*]. This behavior has been documented for many glass forming systems [*20*] and was ascribed to a coupling of vibrational and fast relaxational excitations [*29*].

CKN and CRN. In dipolar systems as glycerol, the dielectric susceptibility couples mainly to the reorientational motions of the molecules and its coupling to the density fluctuations, considered by most theoretical approaches, is not well understood. Therefore it seemed interesting to extend these high-frequency dielectric experiments to molten salts which are characterized by mobile ions. Here a good coupling of the ionic motion, tested by dielectric spectroscopy, and the density fluctuations can intuitively be expected. In addition, the investigation of more fragile glass formers is highly desirable as it can be assumed that in these glass formers the additional vibrational excitations giving rise to the so-called Boson peak are less pronounced [*20*].

We have investigated the high-frequency dielectric response of the ionic conductors CKN (T_g = 333 K, m = 94) and CRN (T_g = 333 K, m $\approx$ 80). Binary mixtures of potassium nitrate and calcium nitrate are well studied ion-conducting glass formers (see, e.g., [*30,31*]). The ionic mixture $[Ca(NO_3)_2]_{0.4}[KNO_3]_{0.6}$ was one of the first glass formers on which the high-frequency behavior has been examined using neutron [*7*] and light scattering techniques [*8*]. In addition, we investigated another nitrate mixture, that also can easily be vitrified: $[Ca(NO_3)_2]_{0.4}[RbNO_3]_{0.6}$. In a recent study up to 300 MHz [*32*] in this compound dielectric properties similar to that of CKN have been found. Results from our group on $\varepsilon''(\nu)$ of CKN up to 40 GHz have already been published elsewhere [*33*]. Very recently we were able to extend these measurements up to a frequency of 380 GHz [*13*].

Figure 3 shows ε'' for CKN as a function of frequency ν in a double-logarithmic plot. $\varepsilon''(\nu)$ reveals a minimum similar to that observed with neutron [*7*] and light scattering techniques [*8*]. Indeed, when scaled appropriately, χ'' obtained from the

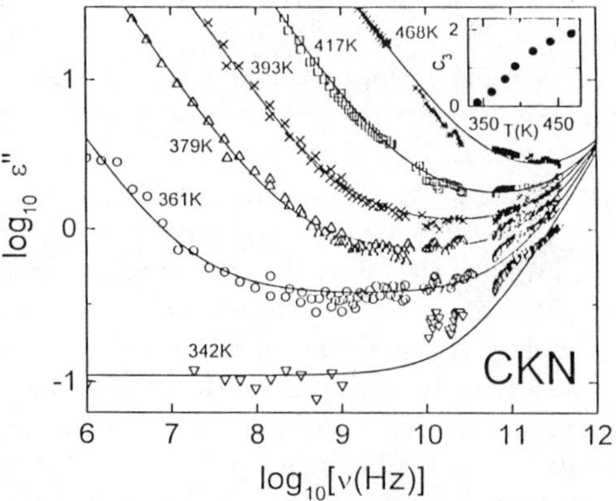

Figure 3. Frequency dependence of the dielectric loss of CKN for various temperatures. The solid lines are fits using $\varepsilon'' = c_1\nu^{-b} + c_2\nu + c_3$. The inset shows the resulting temperature dependence of the constant loss c_3.

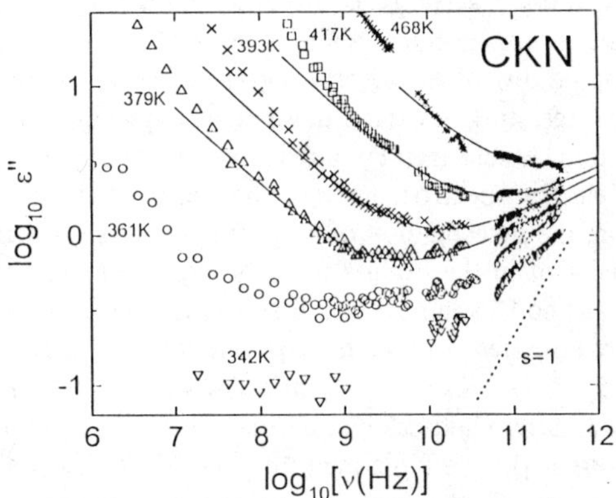

Figure 4. Same data as in Figure 4 but with fits (solid lines) using the MCT expression, Eq. (1), with $\lambda = 0.76$ ($a = 0.3$, $b = 0.54$). The dotted line indicates a linear behavior.

scattering experiments agrees well with our results for $\varepsilon''(\nu)$. Our data also agree reasonably well with the most recent data by the group of Funke [*31*] which, however, are too incomplete to enable an unambiguous evaluation of the susceptibility minimum. The $\varepsilon''(\nu)$-minimum (Figure 3) is relatively broad and, most important, exhibits a high-frequency wing which increases significantly sublinear. In CRN the data resemble the results obtained for CKN with a relatively broad minimum, and a power law exponent at its high-frequency wing which clearly is smaller than unity [*13*]. In Figure 3, at 342 K, $\varepsilon''(\nu)$ seems to become constant for $\nu < 1$ GHz. This may be due to the appearance of a constant loss region as predicted by Wong and Angell [*5*] but at these low values of ε'' the uncertainty of the data is rather high. The lines in Figure 3 are fits using the same formula as in Figure 2(b) for glycerol, i.e. $\varepsilon'' = c_1\nu^{-b} + c_2\nu + c_3$. For temperatures $T \geq 379$ K, data and fit agree reasonably well. For lower temperatures the clearly sublinear increase of $\varepsilon''(\nu)$ at the highest frequencies cannot be described by the fits. Similar to glycerol, the constant loss obtained from the fits is temperature dependent as indicated in the inset of Figure 3.

Figure 4 shows the same data as Figure 3 together with fits using the MCT interpolation formula, Eq. (1) (solid lines). We obtain a consistent description of the $\varepsilon''(\nu)$-minima using $\lambda = 0.76$ (corresponding to a=0.3, b=0.54) for $T \geq 379$ K. For CRN we found $\lambda = 0.91$ (a=0.2, b=0.35) [*13*]. At low frequencies, the model does not fit the data well. This may be ascribed to additional conductivity contributions which lead to a divergence of $\varepsilon''(\nu)$ for decreasing frequencies. At temperatures below 379 K the high-frequency increase of $\varepsilon''(\nu)$ becomes steeper than ν^a. The parameters for CKN agree reasonably well with those obtained by Li *et al.* [*8*] (a=0.273, b=0.458, λ = 0.811) from fits of Eq. (1) to their light scattering results. When comparing Figs. 3 and 4 one has to state that at $T \geq 379$ K both approaches are able to describe the minimum region with comparable quality. However, Eq. (1) results from microscopic theoretical considerations while up to now there is no theoretical foundation of the constant loss phenomenon. At lower temperatures both approaches are not able to take account of the high frequency increase of $\varepsilon''(\nu)$. However, for temperatures below or near the critical temperature T_c (see below) the idealized MCT which leads to Eq. (1) cannot be expected to give a correct description of the data [*1*]. Here hopping processes are assumed to be important and the extended MCT should be used [*1*].

The critical temperature T_c should manifest itself in the temperature dependence of the $\varepsilon''(\nu)$-minimum [*1*]. MCT predicts for $T > T_c$: $\varepsilon''_{min} \sim (T-T_c)^{1/2}$ and $\nu_{min} \sim (T-T_c)^{1/(2a)}$. The MCT also predicts a critical temperature dependence of the time scale of the α-process: $\nu_{max} \sim (T-T_c)^{\gamma}$ with an exponent γ which is determined by the slopes of the susceptibility minimum: $\gamma = 1/(2a) + 1/(2b)$. Since the dielectric loss maxima are obscured by the dc-conduction process, we take ν_{max} from the fits to the imaginary part of the dielectric modulus, $M'' = \mathrm{Im}(1/\varepsilon)$, published elsewhere [*33*]. We are aware of the ongoing controversy about the use of the electrical modulus formalism but we think that at least in CKN this procedure is justified by the fact that the relaxation times evaluated from $M''(\nu)$ at high temperatures closely follow those obtained from mechanical experiments [*33*]. Figure 5 shows the results for ε''_{min}, ν_{min}, and ν_{max} for CKN (closed symbols) and CRN [*13*] (open symbols). Here representations have been chosen that lead to straight lines that extrapolate to T_c if the above critical temperature

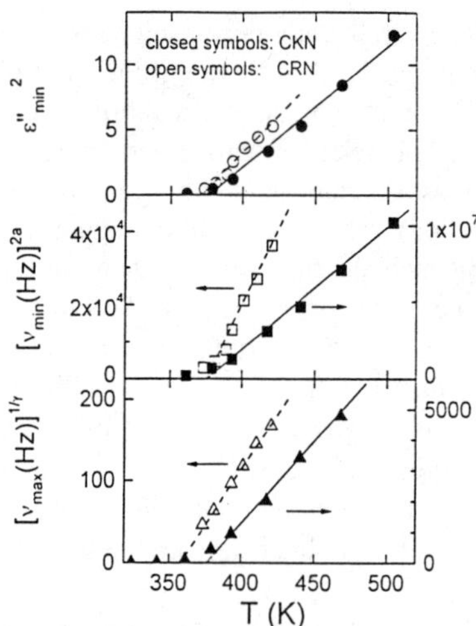

Figure 5. Temperature dependence of the height and position of the dielectric loss minimum and of the α-peak position of CKN (closed symbols) and CRN (open symbols). Representations have been chosen that should result in straight lines according to the predictions of the MCT. For CKN the solid lines extrapolate to a T_c of 375 K for all three quantities. For CRN T_c varies between 360 and 375 K.

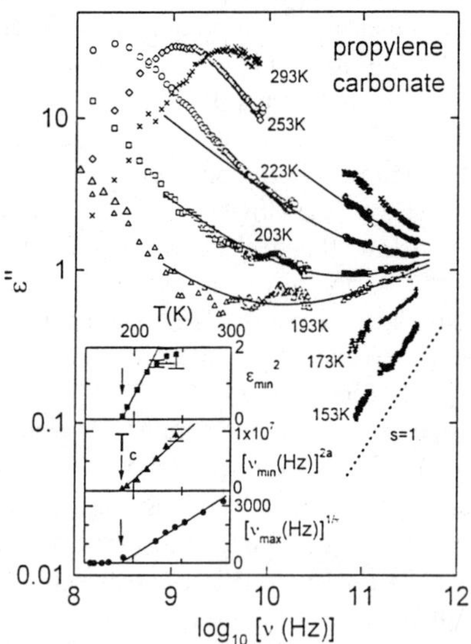

Figure 6. Frequency dependence of the dielectric loss of PC for various temperatures. The solid lines are fits using Eq. (1) with $\lambda = 0.78$ ($a = 0.29$ and $b = 0.5$). The dotted line indicates a linear behavior. The inset shows height and position of the loss minimum and the α-peak position of PC. Representations have been chosen that should result in straight lines according to the predictions of the MCT. The solid lines extrapolate to a T_c of 187 K for all three quantities.

dependencies are obeyed. Indeed for CKN all three data sets can consistently be described with a critical temperature T_c = 375 K as indicated by the solid lines. This compares well to the T_c = 378 K obtained by Li *et al.* from light scattering experiments [*8*]. For temperatures near T_c the data deviate from the predicted behavior, possibly due to hopping processes which are considered in extended versions of MCT [*1*] only. For CRN the three quantities lead to values for T_c around 365 K which, however, differ by approximately 15 K (dashed lines in Figure 5). In this context it is interesting that for λ close to unity the simple scaling relations mentioned above are expected to fail [*34*].

Propylene Carbonate. In this section we want to show first results on PC. Similar to glycerol, PC ($T_g \approx 160$ K) is a dipolar system. However, in contrast to the relatively strong, hydrogen-bonded glycerol, PC can be characterized as a fragile (m = 104) van der Waals liquid. Figure 6 shows the dielectric loss of PC for frequencies above 100 MHz. Data from 10^{-5} Hz up to 1 GHz have already been published in [*35*]. At low frequencies a well developed α-peak is observed which agrees with the results from other groups [*36*]. At high frequencies a minimum shows up. The solid lines are fits using the MCT interpolation formula, Eq. (1) with λ = 0.78 (a=0.29 and b=0.5). A good agreement of data and fits is found. The obtained λ is identical to that determined from light scattering results [*37*] and also consistent with a recent analysis of solvation dynamics experiments [*38*].

The MCT predicts a significant change in the behavior of $\varepsilon''(\nu)$ at T_c: For $T < T_c$ the dielectric loss should exhibit a so-called "knee" at a frequency ν_k, i.e. a change of power law from $\varepsilon'' \sim \nu^a$ at $\nu > \nu_k$ to $\varepsilon'' \sim \nu$ at $\nu < \nu_k$. As seen in Figure 6, at temperatures $T \leq 193$ K the slope of $\varepsilon''(\nu)$ increases significantly and approaches a linear behavior (dotted line). This behavior could be indicative of the "knee" with ν_k located above the investigated frequency range. However, these data show significant error bars; thus experiments up to higher frequencies and with higher sensitivity are in progress to check for the appearance of the "knee" at low temperatures.

In the inset of Figure 6 the critical behavior is examined. ε_{min}, ν_{min}, and ν_{max} are plotted in a way that according to MCT should result in straight lines extrapolating to T_c. The three quantities follow this prediction and a T_c of 187 K is obtained. This lies in the same range as the T_c obtained from light scattering (179 K) [*37*], neutron scattering (180-200K) [*39*] and solvation dynamics experiments (176 K) [*38*].

Salol. Salol (phenyl-salicylate, $T_g \approx 218$ K) can be characterized as a van der Waals liquid with an intermediate fragility of m $\approx$ 63. Up to now we have performed measurements using one BWO and near room temperature only. However, even this limited data set is sufficient to give clear evidence of a minimum in $\varepsilon''(\nu)$ if plotted together with the dielectric data of Hofmann et al. [*23*], obtained at frequencies below 10 GHz (Figure 7). Again the $\varepsilon''(\nu)$ obtained by our high-frequency measurements increases weaker than linear (dotted line).

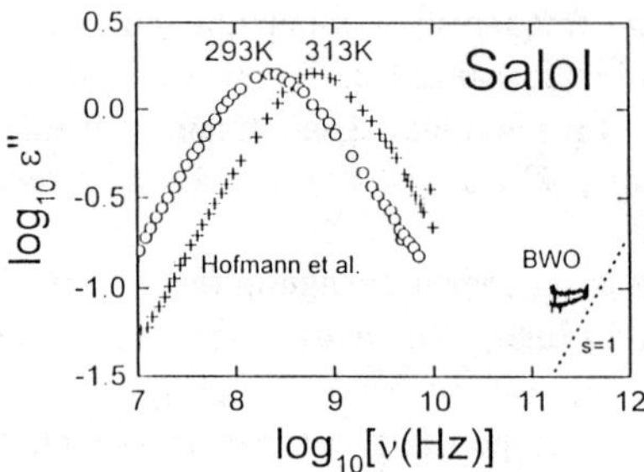

Figure 7. Frequency dependence of the dielectric loss of Salol for two temperatures. The data below 10 GHz have been taken from Hofmann *et al.* [*23*]. The dotted line indicates a linear behavior.

Conclusions

In conclusion, we have presented high-frequency dielectric loss spectra of five materials which belong to very different classes of glass formers. The ε''-spectra of all materials investigated reveal a susceptibility minimum at high frequencies indicating a quite general behavior. In all cases the observed minimum is relatively broad and cannot be ascribed to a pure crossover from the α-relaxational process to the far-infrared response. However, introducing an additional temperature dependent constant loss, the data in glycerol and to a lesser extend also in CKN can be described rather well. A microscopic explanation of such a constant loss contribution is still missing. For CKN, CRN, glycerol, and PC the frequency and temperature dependencies of the measured dielectric response have been compared to the predictions of the simplest version of MCT. For CKN, CRN, and PC the data are in rather good accord with the model predictions. For the relatively strong, hydrogen bonded glass former glycerol the increase of $\varepsilon''(\nu)$ above the minimum is steeper than expected from the simple model. A possible explanation for this discrepancy is the appearance of additional vibrational excitations which contribute mainly to the neutron and light scattering results and give rise to the Boson peak. Remainders of these contributions seem to be found in our dielectric results and a more sophisticated evaluation [*21*] is necessary to test the agreement with the theory. Apart from any theoretical interpretation our dielectric data show unambiguously that there are non-trivial additional processes contributing to the susceptibility of glass forming liquids at high frequencies.

Acknowledgments

We gratefully acknowledge stimulating discussions with C.A. Angell, R. Böhmer, W.Götze, K.L. Ngai, W. Petry, L. Sjögren, and A.P. Sokolov. We thank Yu.G. Goncharov, R. Kohlhaas, and H. Rall for performing some of the measurements and A. Maiazza for the preparation of the CKN and CRN samples.

References

1. Götze, W.; Sjögren, L. *Rep. Progr. Phys.* **1992**, *55*, 241.
2. Ngai, K. L. *Comments Solid State Phys.* **1979**, *9*, 121. Ngai, K. L. in *Disorder Effects on Relaxational Processes*; edited by Richert, R. and Blumen. A.; Springer: Berlin, 1994; p.89.
3. Dixon, P. K.; Wu, L.; Nagel, S. R.; Williams, B. D.; Carini, J. P. *Phys. Rev. Lett.* **1990**, *65*, 1108.
4. Menon, N.; Nagel, S. R. *Phys. Rev. Lett.* **1995**, *74*, 1230.
5. Wong, J.; Angell, C. A. in *Glass: Structure by Spectroscopy;* M. Dekker: New York, 1974; p. 750.
6. Petry, W.; Wuttke, J. *Transp. Theory Statist. Phys.* **1995**, *24*, 1075.
7. Knaak, W.; Mezei, F.; Farago, B. *Europhys. Lett.* **1988**, *7*, 529.
8. Li, G.; Du, W. M.; Chen, X. K.; Cummins, H. Z.; Tao, N. Z. *Phys. Rev. A* **1992**, *45*, 3867.
9. Rössler, E.; Sokolov, A. P.; Kisliuk, A.; Quitmann, A. P. *Phys. Rev. E* **1994**, *49*, 14967.

10. Wuttke, J.; Hernandez, J.; Li, G.; Coddens, G.; Cummins, H. Z.; Fujara, F.; Petry, W.; Sillescu, H. *Phys. Rev. Lett.* **1994**, *72*, 3052.
11. Wuttke, J.; Petry, W.; Coddens, G.; Fujara, F. *Phys. Rev. E* **1995,** *52*, 4026.
12. Lunkenheimer, P.; Pimenov, A.; Dressel, M.; Goncharov, Yu. G.; Böhmer, R.; Loidl, A. *Phys. Rev. Lett.* **1977**, *77*, 318 .
13. Lunkenheimer, P.; Pimenov, A.; Loidl, A. *Phys. Rev. Lett.* **1997**, in press.
14. Volkov, A. A.; Goncharov, Yu. G.; Kozlov, G. V.; Lebedev, S. P.; Prokhorov, A. M. *Infrared Phys.* **1985**, *25*, 369. Volkov, A. A.; Kozlov, G. V.; Prokhorov, A. M. *Infrared Phys.* **1989**, *29*, 747.
15. Mopsik, F. I. *Rev. Sci. Instrum.* **1984,** *55*, 79.
16. Böhmer, R.; Maglione, M.; Lunkenheimer, P.; Loidl, A. *J. Appl. Phys.* **1989**, *65*, 901.
17. Born, M.; Wolf, E. *Principles of Optics*, 6th edition; Pergamon Press: Oxford, 1980.
18. Angell, C. A. *J. Chem. Phys. Solids* **1988**, *49*, 863.
19. Böhmer, R.; Angell, C. A. *Phys. Rev. B* **1992**, *45*, 10091.
20. Sokolov, A. P.; Steffen, W.; Rössler, E. *Phys. Rev. E* **1995**, *52*, 5105.
21. Franosch, T.; Götze, W.; Mayr, M.; Singh, A.P. *Phys. Rev. E* **1997**, in press.
22. Schönhals, A.; Kremer, F.; Hofmann, A.; Fischer, E. W.; Schlosser, E. *Phys. Rev. Lett.* **1993**, *70*, 3459.
23. Hofmann, A.; Kremer, F.; Fischer, E. W.; Schönhals, A. in *Disorder Effects on Relaxational Processes*; edited by Richert, R. and Blumen. A.; Springer: Berlin, 1994; p. 309
24. Lunkenheimer, P.; Pimenov, A.; Schiener, B.; Böhmer, R.; Loidl, A. *Europhys. Lett.* **1996**, *33*, 611.
25. Cummins, H. .; Li, G.; Du, W.; Pick, R. M.; Dreyfus, C. *Phys. Rev. E* **1996**, *53*, 896.
26. Strom, U.; Hendrickson, J. R.; Wagner, R. J.; Taylor, P. C. *Solid State Commun.* **1977**, *15*, 1871. Strom, U.; Taylor, P. C. *Phys. Rev. B* **1977**, *16*, 5512. Liu C.; Angell, C. A. *J. Chem. Phys.* **1990**, *93*, 7378.
27. Roland, C. M.; Ngai, K. L.; Lewis, L. J. *J. Chem. Phys.* **1995**, *103*, 4632.
28. Roland, C. M.; Ngai, K. L. *J. Chem. Phys.* **1995**, *103*, 1152.
29. Sokolov, A. P.; Kisliuk, A.; Quitmann, D.; Kudlik, A.; Rössler, E. *J. Non-Cryst. Solids* **1994**, *172-174*, 138. Sokolov, A.P. (private communication)
30. Howell, F. S.; Bose, R. A.; Macedo, P. B.; Moynihan, C. *J. Phys. Chem.*, **1974**, *78*, 639. Angell, C. A.; *Chem. Rev.* **1990**, *90*, 523.
31. Ngai, K. L.; Cramer, C.; Saatkamp, T.; Funke, K. in *Proceedings of the Workshop on Non-Equilibrium Phenomena in Supercooled Fluids, Glasses, and Amorphous Materials, Pisa, Italy, 1995*, World Sientific: Singapore, 1995; p. 1.
32. Pimenov, A; Lunkenheimer, P.; Nicklas, M; Böhmer, R.; Loidl, A.; Angell, C. A. (submitted to *J. Non-Cryst. Solids*)
33. Pimenov, A.; Lunkenheimer, P.; Rall, H.; Kohlhaas, R.; Loidl, A. *Phys. Rev. E* **1996**, *54*, 676.
34. Sjögren,L. (private communication).
35. Böhmer, R.; Schiener, B.; Hemberger, J.; Chamberlin, R. V. *Z. Phys. B* **1995**, *99*, 91.
36. for a compilation of dielectric results on PC, see: Angell, C.A .; Boehm, L.; Oguni, M.; Smith, D. L. *J. Molecular Liquids* **1993**, *56*, 275.
37. Du, W. M.; Li, G.; Cummins, H.Z.; Fuchs, M.; Toulouse, J.; Knauss, L.A. *Phys. Rev. E* **1994**, *49*, 14967.
38. Ma, J.; Vanden Bout, D.; Berg, M. *Phys. Rev. E* **1996**, *54*, 2786.
39. Börjesson, L.; Elmroth, M.; Torell, L.M. *Chem. Phys.* **1990**, *149*, 209. Börjesson, L.; Howells, W.S. *J. Non-Cryst. Solids* **1991**, *131-133*, 53.

Chapter 14

Structural Relaxation of Supercooled Liquids from Impulsive Stimulated Light Scattering

Yongwu Yang and Keith A. Nelson[1]

Department of Chemistry, Massachusetts Institute of Technology, Cambridge, MA 02139

A time-domain light scattering technique was applied to the study of structural relaxation strength and dynamics in the fragile supercooled liquids salol and 60/40 calcium/potassium nitrate. Acoustic, thermal diffusion, and relaxation modes were observed. In both glass formers, the relaxation mode dynamics on ns-ms time scales could be described well by the stretched exponential function and the temperature dependence of the relaxation mode strength or nonergodicity parameter showed a square-root cusp-like feature at a characteristic crossover temperature, consistent with mode-coupling theory predictions. The acoustic velocity and damping rate of salol were measured in the MHz-GHz range, from which the acoustic modulus spectra were constructed and compared to the susceptibility spectra obtained from dynamic light scattering.

When a liquid is slowly cooled below its melting temperature T_m, it usually undergoes a first-order phase transition into an ordered crystalline phase. However, upon sufficiently fast cooling through T_m, many liquids can remain in supercooled amorphous states indefinitely and can form amorphous solid states, or glasses, at or below the glass transition temperature T_g. Although the static structure of the system appears to change little upon cooling from the high-temperature liquid to the low-temperature glass, the dynamics of structural evolution undergo dramatic change. The simplest evidence is provided by the viscosity η, which increases by more than thirteen orders of magnitude upon cooling from T_m (at which typically $\eta \approx 0.1$ Poise) to T_g (often defined as the temperature at which $\eta \approx 10^{13}$ Poise). The corresponding time scales for structural relaxation are roughly 10 picoseconds (10^{-11} sec) at T_m and 10^3 seconds at T_g. Structural evolution of a liquid continuum can be described in terms of volume (or density) and shear responses, expressed as complex bulk and shear moduli, or alternatively in terms of longitudinal (compressional) and shear responses expressed as complex elastic constants. Structural relaxation effects on longitudinal acoustic waves can be observed on various time (i.e. acoustic frequency) scales through inelastic neutron and light scattering, and ultrasonics. Quasi-elastic neutron and light scattering measurements, conducted in the frequency and time domains, are dominated

[1]Corresponding author.

by structural relaxation contributions in many cases and therefore also can provide quite direct information about structural relaxation dynamics. There are a number of indirect, macroscopic probes of structural relaxation, including measurements of dielectric relaxation (important in its own right and measurable over a frequency range as wide as 10^{-3}-10^{10} Hz in the most favorable cases), enthalpy relaxation (measured as a dynamical "specific heat spectroscopy", in which the temperature response to ac heating is determined), and thermal and mass transport. In addition, a host of microscopic responses coupled to structural relaxation including molecular orientation, nuclear spin relaxation, fluorescence depolarization, and many others have been measured and the results used to infer structural relaxation dynamics. The quantitative relations between structural relaxation dynamics and other macroscopic and microscopic responses are generally not rigorously established, however, and in many cases there are clear differences among the different dynamical responses, so inferences based on indirect measurements require considerable care. A wide range of measurement methods used to extract structural relaxation dynamics from glass-forming liquids has been reviewed recently (1).

The study of structural relaxation in supercooled liquids has intensified in recent years, largely as a result of advances in both experimental and theoretical approaches to the problem. Experimental methods newly developed or tailored for the study of glass-forming materials have permitted study of macroscopic and microscopic response functions, heterogeneity, and other properties over previously inaccessible time and distance scales, providing increasingly detailed information to test and guide theoretical models (1). On the theoretical side, the recently developed mode coupling theory (MCT) of supercooled liquids (2-3) has offered key conceptual features, most notably a hidden singularity at a "crossover" temperature T_c predicted to lie between T_m and T_g, and several experimentally testable predictions concerning these features, thereby stimulating many of the recent measurements. Improved models of the liquid-glass transition and progress in the difficult effort of numerical calculations relevant to glass-forming liquids have also had significant impact (1).

In this paper recent measurements on glass-forming liquids using a time-domain light scattering technique called impulsive stimulated scattering (ISS) are discussed. As discussed below, ISS offers a dynamic range of more than six decades, covering roughly 10^{-10}-10^{-3} sec time scales, and therefore permits characterization of structural relaxation dynamics in supercooled liquids throughout much of the temperature range between T_m and T_g (4). Acoustic, structural relaxation, and thermal diffusion modes all contribute to ISS data. In many cases the contributions are well separated temporally and the first two contributions can be used to deduce structural relaxation dynamics. In addition, the total relaxation strength, expressed as the Debye-Waller factor or, in terms of mode-coupling theory, the nonergodicity parameter, can be determined directly from ISS data. Both the structural relaxation dynamics and strength, measured as functions of temperature, can be compared to key MCT predictions. In this paper, we report ISS studies on two model fragile glass formers, the organic molecular liquid salol and the ionic salt mixture calcium/potassium nitrate (CKN) in a 60:40 mole ratio $Ca(NO_3)_2$:$K(NO_3)$. Experimental results on the structural relaxation modes reported earlier (5-8) are reviewed and new results on the acoustic modes are presented. All the results are compared to MCT predictions.

Background

Mode Coupling Theory. Mode coupling theory of the liquid-glass transition deals with a closed set of generalized oscillator equations of motion for the normalized density autocorrelation functions $\phi_q(t) = <\rho(q,t)\rho(q,0)>$ (the wavevector q will be treated as a scalar for isotropic fluids), in which the relaxation kernel is expressed in terms of the nonlinear interactions between the density fluctuations. It provides detailed quantitative predictions for complex structural relaxation. [For more information on MCT and its experimental tests, see a comprehensive review article (9) and a theme issue in Transport Theory and Statistical Physics (10).]

In its idealized version with thermally assisted hopping processes ignored, MCT describes a transition at a critical temperature T_C from a weak-coupling ergodic or liquid-like state to a strong-coupling nonergodic or ideal glass state. For $T > T_C$, fluctuations away from the equilibrium density relax back to equilibrium via a two-step process: a fast (picosecond) partial relaxation of $\phi_q(t)$, labeled the "β" relaxation, down to a plateau value, followed by a low-frequency, highly nonexponential "α" relaxation of $\phi_q(t)$ all the way down to zero, i.e. back to the equilibrium density. See Figure 1. While the β relaxation dynamics show a weak temperature dependence, the α relaxation exhibits dramatic, critical slowing down with decreasing temperature. At T_C, α relaxation never occurs, resulting in structural arrest: fluctuations of the density never fully relax back to the equilibrium state, and so the system no longer has access to all of phase space, i.e. no longer displays ergodic behavior. The density-density correlation function $\phi_q(t)$ does not relax to zero, but (through the β relaxation process which persists below T_c) only to the asymptotic "plateau" value, f_q^c, whose magnitude measures how far from equilibrium the system remains. The long-time limiting value f_q^c is the nonergodicity parameter, or Debye-Waller factor, at the crossover temperature T_C. At still lower temperatures the system is increasingly far from equilibrium, i.e. for $T < T_c$, $\phi_q(t \rightarrow \infty) = f_q$ increases with decreasing T.

In the extended MCT with thermally assisted hopping processes included, both α relaxation and ergodicity are restored at and below T_C. Thus, the ergodic-to-nonergodic transition is smeared out by thermally assisted hopping. However, an "effective" nonergodicity parameter can be defined, corresponding to the value of $\phi_q(t)$ not at $t \rightarrow \infty$, but at the end of the β relaxation, at the "plateau" level which in the extended model for temperatures below T_c does not persist for all times but, like at temperatures above T_c, only until the beginning of the α relaxation. The effective nonergodicity parameter is predicted to show a square-root cusp, i.e.

$$f_q(\mathrm{T}) = \begin{cases} f_q^c + O(\sigma) & (\mathrm{T} > \mathrm{T_c}), \\ f_q^c + \mathrm{h_q}(\sigma)^{1/2} + O(\sigma) & (\mathrm{T} < \mathrm{T_c}), \end{cases} \tag{1}$$

with dimensionless parameter $\sigma = (T_C - T)/T_C$ and amplitude h_q. (See Refs. 9 and 11 for detailed discussion of this result.) Here the temperature T_C indicates a change in the nature of the α relaxation. The α relaxation dynamics are controlled primarily by anharmonic processes above T_C and by activated hopping processes below T_C. Since the α relaxation appears in neutron scattering as a quasi-elastic peak, the integrated area of the α peak is a measure of the total amount of relaxation necessary to reach

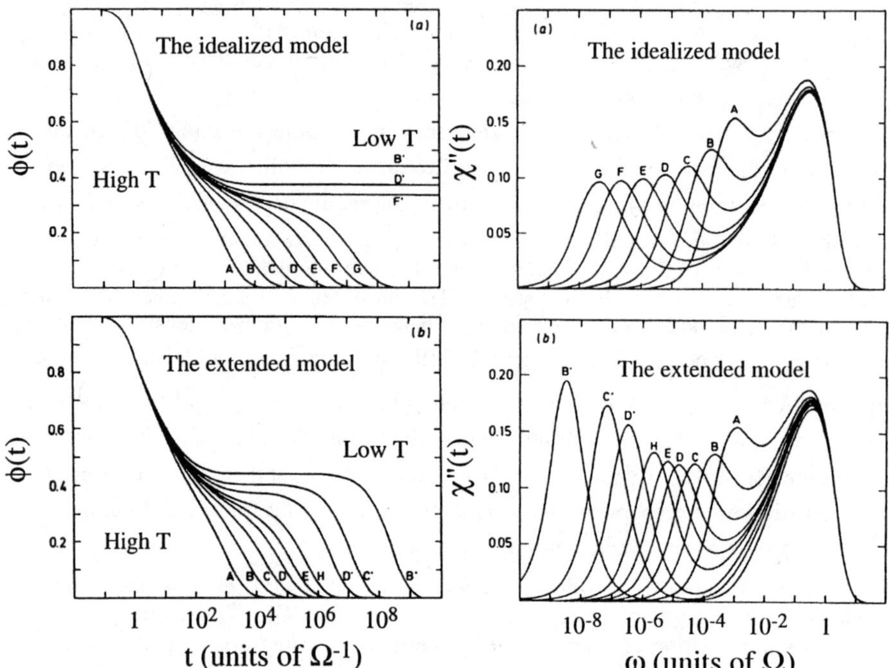

Figure 1. Density auto-correlation function $\phi(t)$ and its corresponding susceptibility spectrum $\chi''(\omega)$ for the idealized and extended mode-coupling theory. Ω is a microscopic frequency corresponding to near-neighbor vibrations. The figure shows fast and slow decay features in $\phi(t)$ and $\chi''(\omega)$, with the slower feature reaching progressively longer time scales (or lower frequency scales) as T is reduced. The curves labeled X' represent temperatures at or below T_c. In the idealized model, nonergodic behavior is seen at these T values, indicated by the disappearance of the slow relaxation feature and the decay of $\phi(t)$ to a nonzero level. In the extended model, the slower relaxation feature is restored. Adapted from Ref. (9).

equilibrium following the β relaxation, i.e. a measure of the total α relaxation strength or the effective nonergodicity parameter.

Concerning α and β relaxation dynamics, MCT shows that they both exhibit non-Debye behavior. On long time scales, there is no exact analytical solution to the MCT equations, but α relaxation can be described reasonably well by the stretched exponential or Kohlrausch-Williams-Watts (KWW) function $e^{-(\Gamma_R t)^\beta}$, where $\Gamma_R=1/\tau_R$ is the characteristic structural relaxation rate (τ_R is the corresponding structural relaxation time) and β is the stretching exponent, with low values of β (whose range is given by $0 < \beta \leq 1$) describing highly nonexponential relaxation dynamics. On intermediate time scales surrounding the plateau region, both α and β relaxations show power-law time-dependent decays. The imaginary part of the susceptibility spectrum for the density, i.e. of the compressibility spectrum, $\chi''(\omega) \sim \omega\phi''(\omega)$, where $\phi(\omega)$ is the complex Fourier transform of $\phi(t)$, exhibits low-frequency α and higher-frequency β relaxation features corresponding to the two processes as shown in Fig. 1. The minimum between them occurs at some frequency ω_{min}, above which the β relaxation wing follows the power law $\chi''(\omega) \sim (\omega/\omega_{min})^a$, and below which the α relaxation wing follows the von Schweidler law $\chi''(\omega) \sim (\omega/\omega_{min})^{-b}$. In this regime, the relaxation dynamics show universal behavior for all those quantities which have a nonzero projection onto density fluctuations or products of density fluctuations (9).

Impulsive stimulated light scattering. In the ISS experiments described here, two parallel-polarized, picosecond excitation pulses are overlapped spatially and temporally to form an optical interference or "grating" pattern inside the sample. The grating spacing Λ and wavevector q are given by $\Lambda=2\pi/q=\lambda_E/2\sin(\theta_E/2)$, where λ_E and θ_E are the wavelength of and angle between the excitation pulses respectively. There are two excitation mechanisms of present interest. In impulsive stimulated Brillouin scattering (ISBS), the spatially varying electric field exerts a sudden ("impulsive") electrostrictive stress at the grating wavevector, launching two counterpropagating acoustic waves with wavevectors $\pm q$. The time-dependent strain response to the applied stress is described by the density-density response function, $G_{\rho\rho}(q,t>0) \propto e^{-\Gamma_A t}\sin[\omega_A(q)t]$ where $\omega_A(q)$ is the acoustic frequency and $\Gamma_A(q)$ the acoustic damping rate. The response is monitored through time-dependent diffraction of probe light which is incident on the grating pattern at the phase-matching or "Bragg" angle. The diffracted signal, $I(q,t) \propto |G_{\rho\rho}(q,t)|^2$, shows time-dependent oscillations and decay from which the acoustic frequency $\omega_A(q)$, velocity $v(q) = \omega_A/q$, and damping rate $\Gamma_A(q)$ can be determined and the complex elastic modulus $M(s) = M'(s)+iM''(s) = -\rho_0(s/q)^2$, where $s = \omega_A - i\Gamma_A$, can be calculated.

The relaxational part of the acoustic modulus spectrum, M_R, is related to the acoustic frequency and damping rate through the relations (12-14)

$$\frac{M'_R}{\rho_0} = \frac{\omega_A^2 - \Gamma_R^2}{q^2} - c_0^2, \qquad (2)$$
$$\frac{M''_R}{\rho_0} = \frac{2\omega_A \Gamma_R}{q^2}.$$

Here ρ_0 is the equilibrium density, c_0 is the acoustic velocity in the low-frequency limit and Γ_R is the acoustic damping rate due to the structural relaxation contribution, determined by subtracting a "baseline" contribution from the damping rate measured. Note that, consistent with treatments of polarized light scattering from simple and viscous liquids, we write the ISBS response as $G_{\rho\rho}$ when in fact it is one-dimensional longitudinal strain, not isotropic density modulation which includes a shear as well as longitudinal contribution, which is observed. For the samples under discussion, the shear modulus is negligibly small compared to the longitudinal modulus even at the lowest temperatures and highest acoustic frequencies examined.

For absorbing materials, there is an additional excitation mechanism, referred to as impulsive stimulated thermal scattering (ISTS). In ISTS, optical absorption at the interference maxima (the grating "peaks") and subsequent rapid thermalization lead to sudden, spatially periodic heating which images the interference pattern. Thermal expansion at the peaks launches counterpropagating acoustic waves with wavevectors $\pm q$, as in the response to the ISBS excitation mechanism. ISTS also gives rise to steady-state, nonoscillatory thermal expansion at the heated peaks and compression at the unheated nulls which persists until thermal diffusion from peaks to nulls returns the sample to a uniform temperature and density. The time-dependent strain response to sudden, spatially periodic heating is given by the density-temperature (strictly speaking, density-entropy) correlation function $G_{\rho T}(q,t)$, and diffracted signal intensity can often be described in the nanosecond-milllisecond temporal range by the functional form (12, 13)

$$I_{TS}(q,t) \propto \left|G_{\rho T}(q,t)\right|^2$$
$$= \left\{A\left[e^{-\Gamma_H t} - e^{-\Gamma_A t}\cos(\omega_A t)\right] + B\left[e^{-\Gamma_H t} - e^{-(\Gamma_R t)^\beta}\right]\right\}^2 \qquad (3)$$

In this expression the first ("A") term, in which $\Gamma_H(q)$ is the thermal diffusion rate, describes the complete density response for simple liquids, including transient acoustic oscillations and steady-state thermal expansion which decays due to heat transport. The second ("B") term describes an additional component to the density response observed in glass-forming liquids at some temperatures between T_m and T_g. In such fluids the density response to sudden heating includes slow components, due to the α structural relaxation process, which do not contribute to the transient overshoot of the steady-state thermal expansion level and the resultant acoustic oscillations. Instead the slow components yield a gradual approach of the density to its steady-state value. This nonoscillatory rise is often described well on nanosecond-millisecond time scales by the stretched exponential function. On faster time scales it may be possible to observe β

relaxation dynamics in the time-dependent thermal expansion response, but such observations have not been reported to date.

In summary, the complete time-dependent density response observed through ISTS can be described in terms of acoustic, structural relaxation, and thermal diffusion modes. Structural relaxation dynamics can be determined either through their influence on the acoustic mode, as in ISBS, or through observation of the structural relaxation mode. Concerning the latter, note that strictly speaking, $G_{\rho T}(q,t)$ measured in ISTS includes not only the density response to stress, i.e. $G_{\rho\rho}(q,t)$, but also the temperature response to heating, i.e. the time-dependent specific heat, and the stress response to temperature (15). That is, following impulsive heating the temperature needs to rise in the heated regions and thermal stress needs to develop before there can be any density response. In the discussion that follows these are assumed to have no significant effects on the dynamical response observed. This assumption may be justified if the temperature and stress responses, which are nonhydrodynamic, are fast compared to the density response at the wavevectors and sample temperatures examined.

In addition to the dynamics of these three modes, the relaxation strength or the effective nonergodicity parameter f_q in the low-q limit can be obtained from the amplitude ratio of the relaxation mode and the total steady-state response (12,13), i.e.,

$$f_{q\to 0} = 1 - \frac{c_0^2}{c_\infty^2} = \frac{B}{A+B}, \qquad (4)$$

when the acoustic, relaxational, and thermal diffusion modes are well separated from each other, i.e. when $\omega_A >> \Gamma_R >> \Gamma_H$. This equation relates ISTS data to the low-q Debye-Waller factor whose temperature-dependence is given by Eq. (1) in MCT. The result is understandable in simple physical terms. In ISTS, sudden heating causes local density relaxation events which collectively give rise to macroscopic thermal expansion. The relaxation that can occur on short time scales contributes to the transient acoustic response (of magnitude $\propto$ A) as well as the total steady-state response [of magnitude $\propto$ (A + B)]. Relaxation that occurs on slow time scales (compared to the acoustic period) does not contribute to the magnitude of the acoustic response, but does contribute an amount (of magnitude $\propto$ B) to the total steady-state thermal expansion. In terms of Fig. 1, the rapid relaxation that contributes to the acoustic response is just the β-relaxation process through which $\phi(t)$ decays to a plateau level. The remaining slow relaxation is the α relaxation whose strength is being measured. The nonergodicity parameter is the extent of slow relaxation, measurable through ISTS data as the amplitude B of the slow relaxation component compared to the total amplitude A + B.

The conditions $\omega_A(q) >> \Gamma_R >> \Gamma_H(q)$ pose limitations on the temperature range over which the relaxation mode and the nonergodicity parameter can be characterized with a single wavevector since the characteristic structural relaxation rate Γ_R varies sharply with temperature, overlapping the acoustic frequency ω_A at higher T and the thermal diffusion rate Γ_H at lower T. However, the temperature range can be enlarged at high and low T through use of large and small wavevectors respectively, since Γ_R is q-independent at the low-q range accessible to ISTS while $\omega_A \propto q$ and $\Gamma_H \propto q^2$ in this range. In practice, ISTS excitation angles ranging from less than 0.5° to nearly 90° can be used to permit access to a wide wavevector range.

For even weakly absorbing materials, the signal contribution due to ISBS is generally negligible compared to that due to ISTS, especially at small wavevectors since the magnitude of the ISBS contribution is proportional to the square of wavevector. At large wavevectors or for quite transparent liquids, the ISBS contribution must be considered and may become dominant.

Experimental

Experimental implementation of the ISS technique has been described earlier (6,13,16) and is summarized briefly here. The excitation pulses are derived from the output of a Q-switched, mode-locked, and cavity-dumped Nd:YAG laser which yields a 500-μJ, 1.064-μm pulse of 100 ps duration at a repetition rate of up to 1 kHz. The pulse is split with a 50% beam-splitter into the two excitation pulses that are cylindricially focused and crossed inside the sample. The resulting material responses are monitored through measurement of time-resolved diffraction of probe light incident at the phase-matching ("Bragg") angle, using either of two alternate probe systems.

For the long-time dynamics and for acoustic modes with frequencies less than 500 MHz, the electro-optically gated output of a single-mode argon laser is used as a quasi-cw probe beam. Diffracted signal is directed into an amplified fast photodiode and temporally resolved with a digitizing oscilloscope. The digitized signal is transferred to a computer for storage and subsequent analysis. In this setup, the entire time-resolved material response is recorded in a single laser shot, and averaging over a few hundred shots produces excellent signal/noise ratios in several seconds.

To resolve acoustic dynamics with frequencies higher than 500 MHz, a variably delayed probe pulse is employed so that the time resolution is only dependent on excitation and probe pulse durations, not the detection electronics. At a given delay time the diffracted signal intensity provides information about the sample response at only that time following excitation. By varying the delay time, the data are collected "point by point" on the time axis by variably delaying a picosecond probe pulse. This probe pulse is derived from a second mode-locked and Q-switched Nd:YAG laser. Its output consists of a train of 100 ps, 1.064 μm pulses. The largest pulse is selected electro-optically by a Pockels' cell and then frequency-doubled by a β-barium borate (BBO) crystal to yield 532-nm probe light. The timing delay between excitation and probe pulses is controlled electronically by shifting the phase of a common mode-locker radio-frequency source to the probe laser and electronically delaying the timing of the Q-switches and the single pulse selector concurrently (16). The diffracted signal is directed into a low-bandwidth amplified photodiode. The output of the photodiode is sent to a lock-in amplifier whose reference frequency is synchronized with a mechanical chopper running at an arbitrary frequency less than the laser repetition rate. The mechanical chopper is placed in the probe beam in order to reduce scattered light due to the excitation beams. The lock-in is connected through a GPIB interface to a personal computer, which also controls the time delay of the probe pulse. For each delay time between the excitation and probe pulses, the computer records the diffracted signal intensity from the lock-in amplifier. Typically, several hundred laser shots are averaged at each delay time.

Data were collected at multiple excitation angles between 0.5° to 45° (6, 7, 13). At each excitation angle, data were recorded upon cooling of the sample from the liquid state at high temperature through the supercooled region to temperatures close to T_g in intervals of 5K in high and low-temperature regions and 2.5K at intermediate temperatures Before recording data, the temperature of the sample was allowed to stabilize for about 10 minutes to within ±0.05K of the specified temperature.

Results

ISTS Data. Typical ISS data are shown in Fig. 2 for salol at various temperatures at $q=0.7433\ \mu m^{-1}$. At this small q the ISBS contribution to the signal is negligible. ISTS occurs through weak O-H vibrational overtone absorption of the 1.064 μm excitation light. The data at all temperatures exhibit damped acoustic oscillations at short times and a steady-state signal level which decays due to thermal diffusion on long time scales. At intermediate temperatures, the signal level rises slowly toward its steady-state level, reflecting slow, nonexponential structural relaxation dynamics. The features of the acoustic, thermal diffusion, and relaxation modes change as the sample temperature is reduced and the α relaxation dynamics move toward longer time scales. The acoustic frequency ω_A increases monotonically, reflecting the gradual stiffening of the material, and the acoustic damping rate Γ_A reaches a maximum at about 270.9K at which the relaxation dynamics occur on the time scale of the acoustic oscillation period at this wavevector. At lower temperatures, as the dynamics shift to time scales longer than the oscillation period, they are observable directly in the data in the form of the relaxation mode. At still lower temperatures, the relaxation mode is not observable because thermal expansion to the steady-state level at the grating peaks is slow compared to thermal diffusion from the peaks to the nulls. The thermal diffusion rate increases monotonically with lowering temperature and shows a rapid increase around 241K at this wavevector. This corresponds to the temperature at which the specific heat response (i.e. the time scale on which the temperature can respond to heat input) becomes slow compared to the thermal diffusion dynamics at the experimental wavevector.

For all wavevectors, the data exhibit the same T-dependent trends. Since $\omega_A \propto q$ and $\Gamma_H \propto q^2$, the relaxation dynamics overlap the acoustic oscillation period and the thermal diffusion time at temperature ranges which both increase with wavevector. For the acoustic mode, this results in a maximum in attenuation rate and in the rate of dispersion at temperatures that increase with wavevector. The thermal diffusion rate also undergoes its sharpest T-dependent changes at temperatures that increase with wavevector.

Raw data such as that displayed in Fig. 2 were fitted to Eq. (3). Fits to the data yielded the dynamical parameters describing the acoustic, thermal diffusion, and α relaxation modes as well as the relative amplitude B/A. In what follows we first show the temperature dependence of the α relaxation strength or Debye-Waller factor of two glass formers, salol and CKN, and compare the results with MCT predictions. Then we present the analysis of the acoustic mode dynamics in salol.

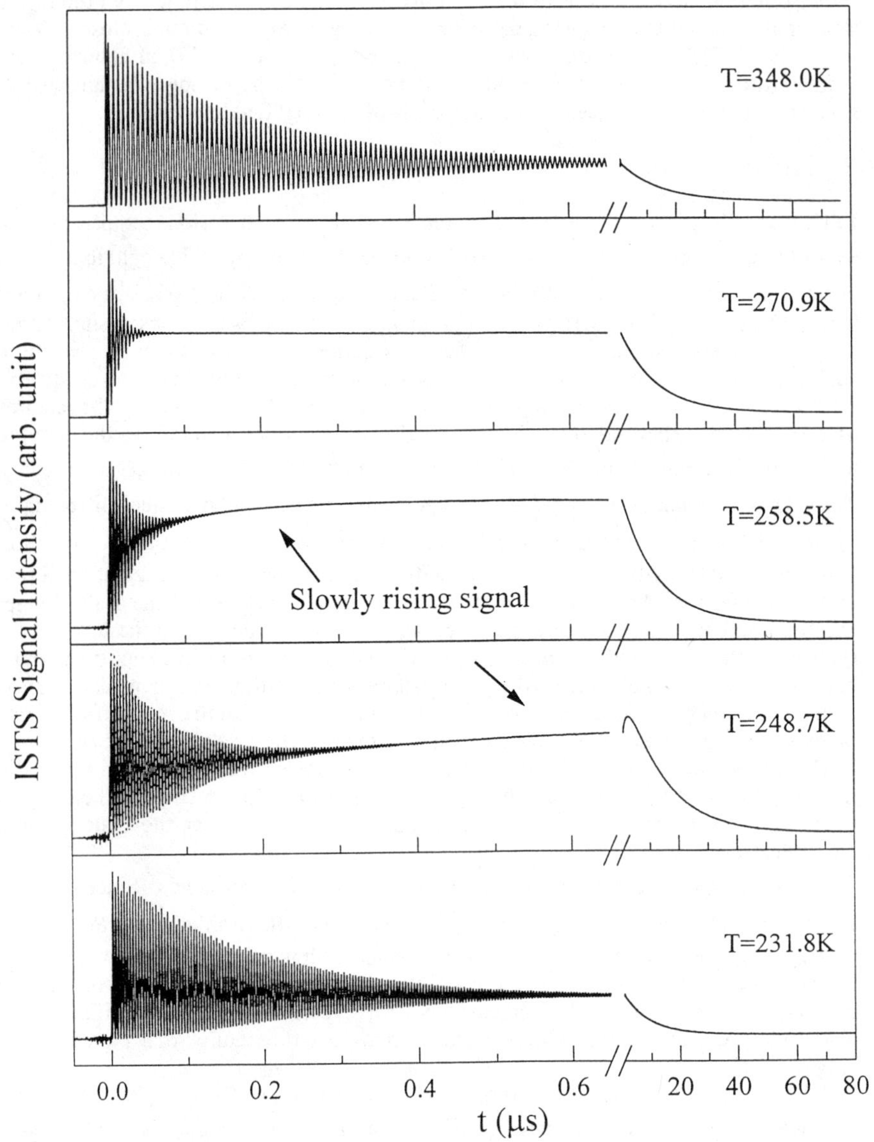

Figure 2. ISTS data (solid curves) and fits (dotted curves) to Eq. (3) of salol at wavevector $q = 0.7433\ \mu m^{-1}$ at several temperatures recorded with the quasi-cw probe system. The data show short time acoustic oscillations and the long time thermal diffusion dynamics. In the intermediate temperature range, ISTS signal slowly reaches the steady-state level, revealing nonexponential structural relaxation dynamics which become slower as T is reduced.

Nonergodicity parameters. Figure 3 shows the T-dependent Debye-Waller factor $f_{q\to 0}$ (symbols) of salol. Its behavior can be fit well by Eq. (1), yielding a square-root cusp at a crossover temperature of 266K±1K (solid curve) (5,6). This T_c value is in good agreement the results of neutron scattering (17) and depolarized light scattering (DLS) (18), confirming a *q*-independent value of T_c as expected.

ISTS measurements have been made to the "CKN" mixture of ionic salts $[Ca(NO_3)_2]$ and $[KNO_3]$ with molar ratio of 2:3 (7). The Debye-Waller factor values $f_{q\to 0}(T)$ (symbols) obtained with different wave vectors are plotted in Fig. 4. It is evident that $f_{q\to 0}(T)$ shows a weak cusp-like anomaly as predicted by MCT. The solid curve in Fig. 4 represents the best fit to Eq. (1) which gives a crossover temperature $T_c = 378K \pm 2K$. This T_c value is also in good agreement with the results from neutron scattering (19) and DLS (20).

Acoustic dynamics. Data were collected from salol at a wide range of temperatures with 20 excitation angles from 0.5° to 45° (13,14). For each excitation angle, the acoustic frequency ω_A and damping rate Γ_A were obtained by fitting the data to Eq. 3. The results, shown in Fig. 5, show the T-dependent and *q*-dependent trends discussed qualitatively above.

The structural relaxation contributions Γ_S to the total acoustic damping rates Γ_A were derived by subtracting the "background" contributions whose values were assumed to be constant or weakly and linearly dependent on T, based on the values of damping rates measured away from the relaxation peaks. The values of Γ_S were used to obtain the structural relaxation part of the acoustic modulus $M_R(\omega_A)/\rho_0$ through Eq. (2). Since data were recorded at many sample temperatures and excitation wavevectors, interpolation between measured values was possible to construct acoustic modulus spectra at constant T. Figures 6 and 7 show the real and imaginary parts of $M_R(\omega_A)/\rho_0$ at various temperatures. The modulus spectrum moves through the middle of the ISTS frequency range at around 270K, as shown by Fig. 6a which indicates α peak in M''_R/ρ_0. At higher temperatures, the peak of the modulus spectrum is at higher frequencies than those within our "window" of acoustic frequencies, and we observe the low-frequency wing of the α relaxation feature in M''_R/ρ_0. Above 290K, the relaxation spectra move out of our experimental frequency window to much higher frequencies, and characterization of the relaxation dynamics is not possible. At lower temperatures (see Fig. 7a), the center of the α relaxation feature in M''_R/ρ_0 moves to lower frequencies than those within our window, and we see the high-frequency wing of this feature. At the lowest temperatures, the minimum in the relaxation spectrum and the low-frequency tail of the β relaxation feature in M''_R/ρ_0 appear to be barely visible at the high-frequency edge of our acoustic frequency window.

Fits generated by assuming the stretched exponential (KWW) structural relaxation function are shown by the solid curves in Figures. 6 and 7. Within experimental uncertainty, the stretching parameter β shows no T-dependence with $\beta \approx 0.50$ in the temperature range of 263K to 290K. Below 263K, M'_R/ρ_0 (Fig. 7) is almost constant across the acoustic frequency window probed, and at the lowest temperatures both the

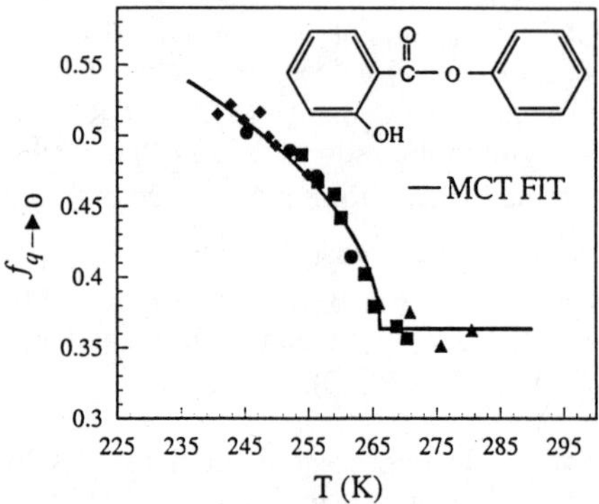

Figure 3. Temperature dependence of the Debye-Waller factor $f_{q \to 0}$ of salol determined from ISTS data at several wavevectors (◆ q = 0.1352 μm^{-1}; ● q = 0.2235 μm^{-1}; ■ q = 1.250 μm^{-1}; ▲ q = 4.532 μm^{-1}) via Eq. (4) . The results show a square-root cusp at about 266K. The solid line shows a fit to the MCT Eq. (1).

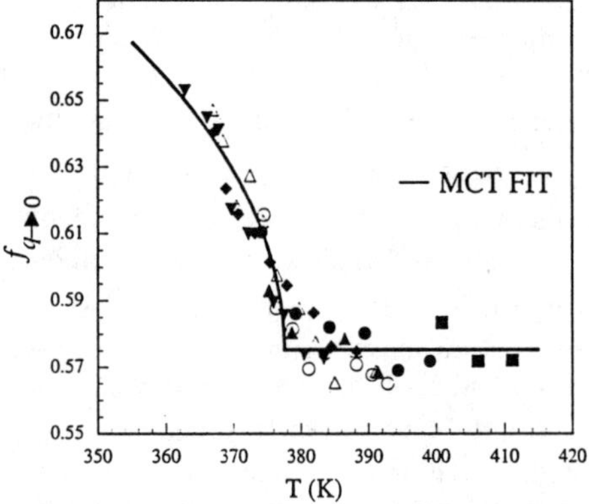

Figure 4. Debye-Waller factor $f_{q \to 0}$ at several wavevectors (▼ q = 0.227 μm^{-1}; O q = 0.232 μm^{-1}; Δ q = 0.235 μm^{-1}; ◆ q = 0.336 μm^{-1}; ▲ q = 0423 μm^{-1}; ● q = 0.623 μm^{-1}; ■ q = 0.896 μm^{-1}) of CKN versus T. The best fit (solid line) to MCT Eq. (1) yields a crossover a temperature T_c = 378K.

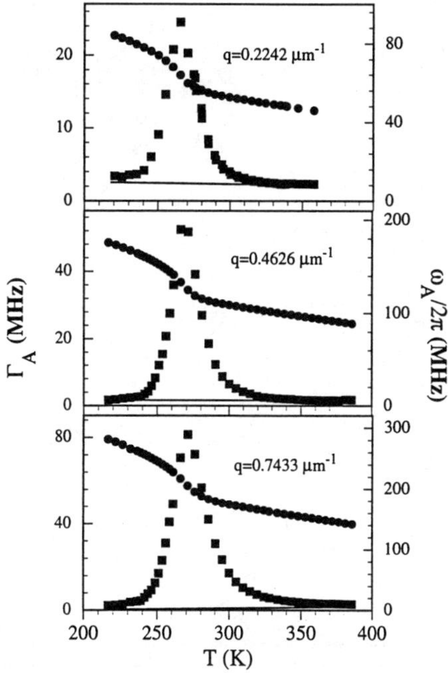

Figure 5. Temperature dependences of longitudinal acoustic frequencies $\omega_A/2\pi$ (●) and damping rates Γ_A (■), from fits to ISTS data at q=0.7433 μm^{-1} shown in Figure 2 and two lower wavevectors. The damping rate shows a maximum and the acoustic frequency shows greatest dispersion when the structural relaxation spectrum overlaps with the acoustic frequency, i.e. at 260-270K for the three wavevector values.

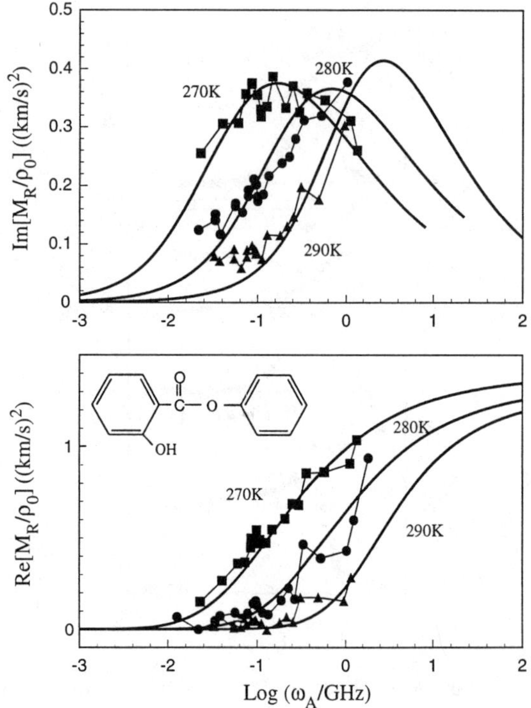

Figure 6. The structural relaxational part of the reduced acoustic modulus above 270K. The imaginary parts are shown in the top and the real parts in the bottom. The solid curves represent fits with the KWW relaxation function. Within the experimental uncertainty, β was found to be constant from 263K to 290K.

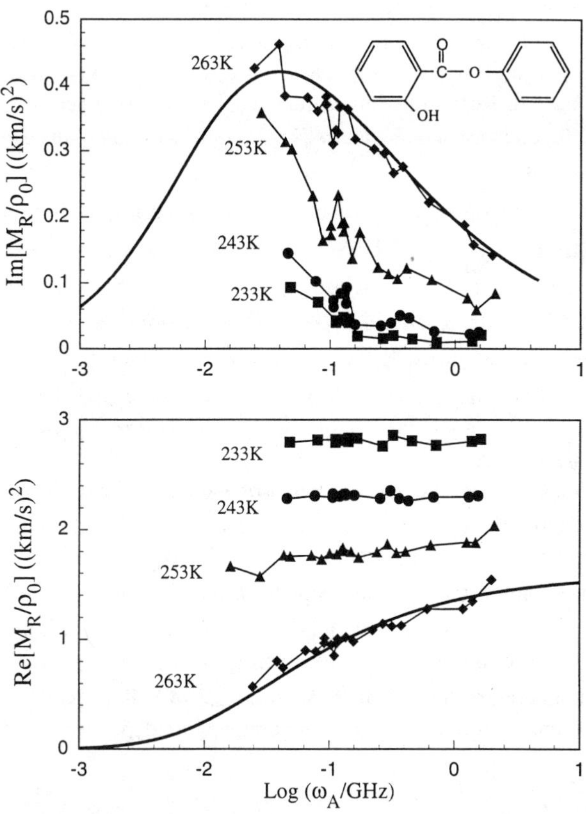

Figure 7. The frequency dependence of the structural relaxational part of the reduced acoustic modulus at and below 263K. The real part M'_R/ρ_0 is nearly frequency-independent in the acoustic frequency window probed except at 263K. We therefore fit the data only at 263K with α-relaxation KWW function (solid curves).

α and β relaxation features contribute. Therefore we have not tried to fit the spectra to the KWW function.

As predicted by the MCT factorization property, which indicates that all the time-dependent responses coupled to density should have the same functional form, the acoustic modulus spectrum and the susceptibility spectrum from DLS can be described by a single β-susceptibility spectrum $\chi''(\omega)$ up to a multiplying factor. In the $\log\chi''$ versus $\log\omega$ plot, all the spectra should be made to overlap by shifting the spectra along the ordinate. The β-susceptibility spectra $\chi''(\omega;T)$ have been obtained from the depolarized light scattering spectra in the range sub-GHz to 5THz (21). We therefore plot in Fig. 8 the DLS spectra (thin solid lines) and the reduced acoustic modulus spectra M''_R/ρ_0 (open symbols) shown in Fig. 7, along with the β-susceptibility spectra $\chi''(\omega)$ (thick solid lines) at several temperatures on a log-log scale. As can be seen in Fig. 8, the acoustic modulus spectra M''_R/ρ_0 at the three lower temperatures overlap with the low-frequency part of DLS spectra and coincide with the predicted β-susceptibility spectra $\chi''(\omega)$. At 263K only part of data overlaps with χ'', presumably due to the proximity of the α relaxation peak. Although only part of the β-susceptibility spectra could be observed in M''_R/ρ_0, which do not by themselves permit determination of the minimum frequency, nevertheless the results show that the β-susceptibility spectra $\chi''(\omega)$ derived from DLS and application of MCT provide a consistent description of acoustic data.

To account for differences in the units of the two curves, we have shifted the M''_R/ρ_0 curves through scaling along the ordinate to overlap the DLS spectra. The temperature dependences of the amount shifted, $\log(h_{DLS}/h_M)$ (the ratio of critical amplitudes), are shown in the inset. Since we are dealing with M''_R/ρ_0, not M''_R directly, we have not accounted for T-dependence in the density ρ_0. Also as discussed in (22), the critical amplitude h_{DLS} from fitting of the DLS spectra with the MCT predictions is temperature dependent. T-dependences of both density ρ_0 and amplitude h_{DLS} may therefore contribute to the smooth temperature dependence of $\log(h_{DLS}/h_M)$.

Conclusions

ISTS experiments permit characterization of the α relaxation dynamics and determination of the relaxation strength or Debye-Waller factor $f_{q\to 0}$. The temperature dependent Debye-Waller factors $f_{q\to 0}(T)$ in salol and CKN both show weak square-root cusp-like anomalies at distinct crossover temperatures T_c. Well above T_c, the α relaxation dynamics obey scaling laws and the relaxation times show power-law T-dependences. Our findings are largely consistent with the predictions of the MCT.

Impulsive stimulated light scattering experiments also provide the acoustic dynamics, from which the structural relaxation dynamics can be derived. Longitudinal acoustic modes in salol were characterized over 2 decades of acoustic frequencies from 10 MHz to several GHz. The relaxation dynamics were analyzed in terms of the frequency-dependent structural relaxational contribution to the acoustic modulus and compliance spectra, which were constructed from the measured acoustic frequencies

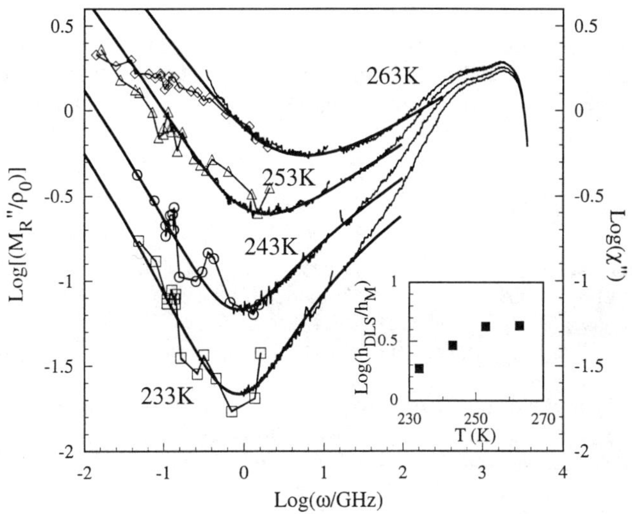

Figure 8. Relaxation spectra of salol from ISS acoustic data (symbols), together with DLS susceptibility spectra (thin lines) and the extended MCT fits (thick lines). The spectra shown in Fig. 7 have been shifted along the ordinate by an amount $\log(h_{DLS}/h_M)$, shown in the inset, to optimize overlap with the DLS susceptibility spectra.

and damping rates (after subtracting background contributions to the latter). The results above 263K were fitted to the relaxational spectrum of the KWW function, and were found to be consistent with expectations for α relaxation dynamics. Within experimental uncertainty, the KWW stretching parameter β was found to be T-independent in the range 263K to 290K. At lower temperatures, the α relaxation spectrum moves to frequencies lower than those within our acoustic frequency window, and the acoustic properties are influenced by the β relaxation dynamics. The results below 263K were found to be consistent with scaled β-susceptibility spectra derived from DLS measurements on salol.

While our results are consistent with many MCT predictions, direct experimental characterization of the acoustic dynamics in the high frequency (1-20 GHz) range remains necessary. In the acoustic dynamics analysis shown above, this major part of the β relaxation spectrum was not mapped out, and the minimum frequencies in the T-dependent spectra were not determined. Our analysis simply shows that the acoustic modulus data in the 10-1000 MHz acoustic frequency range probed can be scaled to the higher-frequency susceptibility spectra determined through DLS. The most compelling test would come from independent measurements of the acoustic modulus spectra, determination of the β relaxation spectra (including the minimum frequencies) in the structural relaxation parts, and comparison of the results to those of DLS.

Acknowledgments

We wish to thank W. Götze for many helpful discussion concerning the data analysis, and H.Z. Cummins and M. Fuchs for providing their depolarized light scattering data and the extended mode coupling theory fits. This work was supported in part by NSF Grant No. DMR-9317198.

References

1. See, for example, *Journal of Non-Crys. Solids* **1994**, 172-174.
2. Bengtzeliu, U.; Götze, W.; and Sjölander, A. *J. Phys. C* **1984**, 17, 5915.
3 Leutheusser, E. *Phys. Rev. A* **1984**, 29, 2765.
4. Halalay, I. C.; Yang, Y.; and Nelson, K.A. *Transport Theory Stat. Phys.* **1995**, 1053.
5. Yang, Y. and Nelson, K.A. *Phys. Rev. Lett.* **1995**, 74, 4883.
6. Yang, Y. and Nelson, K.A. *J. Chem. Phys.* **1995**, 103, 7732.
7. Yang, Y. and Nelson, K.A. *J. Chem. Phys.* **1996**, 104, 5429.
8. Yang, Y.; Muller, L.J.; and Nelson, K.A. *Mat. Res. Soc. Symp. Proc.* **1996**, 407, 145.
9. Götze, W. and Sjögren, L. *Rep. Prog. Phys.* **1992**, 55, 241.
10. Special issure devoted to relaxation kinetics in supercooled liquids--mode coupling theory and its experimental tests, *Transport Theory and Statistical Physics* **1995**, 24.
11. Cummins, H.Z.; Li, G.; Du, W.M.; and Hernandez, J. *Physica A* **1994**, 204, 169.
12. Yang, Y. and Nelson, K. A. *J. Chem. Phys.* **1995**, 103, 7722.
13. Yang, Y. *PhD thesis* **1996**, Massachusetts Institute of Technology.
14. Yang, Y., Silence, S.M., Duggal, A.R., and Nelson, K.A. *in preparation.*
15. Yan, Y.-X.; Cheng, L.-T.; and Nelson, K.A. J. Chem. Phys. **1988**, 88, 6477.
16. Duggal, A. R. *PhD thesis* **1992**, Massachusetts Institute of Technology.
17. Toulouse, J.; Boddens, G.; and Pattnaik, R. *Phys. A* **1993**, 201, 305.
18. Li, G.; Du, W.M.; Sakai, A.; and Cummins, H.Z. *Phys. Rev. A* **1992**, 46, 3343.
19. Mezei, F.; Knaak, W.; and Farago, B. *Phys. Rev. Lett.* **1987**, 58, 571.
20. Li, G.; Du. W.M.; Chen, X.K.; and Cummins, H.Z. *Phys. Rev. A* **1992**, 45, 3867.
21. Cummins, H.Z.; Du, W.M.; Fuchs, M.; Götze, W.; Hildebrand, S.; Latz, A.; Li, G.; and Tao, N.J. *Phys. Rev. E.* **1993**, 47, 4223.
22. Fuchs, M; Cummins, H.Z.; Du, W.M.; Götze, W; Latz, A; Li, G.; and Tao, N.J.; *Phil. Mag. B* **1995**, 71, 771.

Chapter 15

Multiple Time Scales in the Nonpolar Solvation Dynamics of Supercooled Liquids

J. Ma, John T. Fourkas[1], D. A. Vanden Bout[2], and M. Berg[3]

Department of Chemistry and Biochemistry, University of South Carolina, Columbia, SC 29208

Recent experimental and theoretical work on the dynamics of solvation in nonpolar supercooled liquids is reviewed. Transient hole burning experiments have shown that solvation dynamics occur on a wide variety of time scales. A major division into phonon-like and structural relaxation is apparent as the viscosity of the solvent increases in the supercooled region. The structural component is strongly nonexponential and extends over many decades in time. Mode-coupling theory provides a consistent explanation of the structural relaxation behavior versus temperature. A continuum model links nonpolar solvation to shear relaxation of the solvent following a size change of the solute upon electronic excitation. The relationship between phonon-like and structural dynamics is examined within the continuum model, which distinguishes them as different dynamical regimes of a single coordinate, and is contrasted with "spectroscopic" models, which treat the different dynamics as arising from distinct solvent coordinates.

The dynamics of supercooled liquids can be measured by a variety of different experiments, each of which measures a differently weighted average of the many coordinates present in the liquid. The majority of detailed results on short time dynamics has come from just three experiments: dielectric relaxation, light scattering and neutron scattering (*1*). Solvation dynamics, i.e. the response of a solvent to changes in the electronic state of a solute, is a new and distinctly different type of experiment for examining liquid dynamics. Because it focuses on the short time and length scales which must underlie effects seen on all scales, it offers a unique and valuable perspective on liquid dynamics. Furthermore, solvation dynamics specifically weights the liquid coordinates by their ability to affect the electronic surfaces which control chemical reactions. It thereby creates a bridge connecting the general understanding of liquid dynamics to specific problems of solvent effects in chemistry and biology.

[1]Current address: Department of Chemistry, Boston College, Chestnut Hill, MA 02167

[2]Current address: Department of Chemistry, University of Minnesota, Minneapolis, MN 55418

[3]Corresponding author.

We have been active in developing the transient hole burning experiment as a method of measuring solvation dynamics and in applying it to nonpolar solvents in the supercooled region (*2-10*). A large amount of information on solvation in low viscosity, polar solvents is also available (*11-14*), but interest in the distinctly different solvation mechanisms in nonpolar solvents has been growing (*15-19*). Even in polar solvents, the amount of information on solvation in supercooled liquids is limited (*20-24*).

A recurring theme in our results is the wide range of time scales which affect solvation. A primary division of the dynamics into a viscosity dependent and a viscosity independent component is made. The viscosity independent component is assigned to phonon-like dynamics within a fixed structure, and the viscosity dependent component is assigned to relaxation of that structure. The structural relaxation is found to be nonexponential and covers a wide range of time scales. In some cases, standard nonexponential treatments are strained to fit the observed range of time scales.

Mode-coupling theory (MCT) (*25-27*) provides a consistent explanation of the temperature dependence of the structural relaxation. The α- and β-relaxation regions overlap significantly, and a global fitting to complete response functions is essential. MCT also predicts that certain parameters should have the same value in different experiments. Comparison of the solvation based fits with MCT analyses of light scattering data shows that some, but not all, of these parameters are in agreement with this prediction.

Although MCT makes *a priori* and general predictions about the form of structural relaxation data, it does not predict specific relaxation times, does not make statements about the interaction mechanism in nonpolar solvents, and does not provide any detail about the phonon-like relaxation component. We have developed a continuum model of nonpolar solvation which postulates an interaction resulting from solute size changes and successfully relates solvation data with viscosity and ultrasound measurements. The continuum theory also provides a simple model system in which the interaction of phonon and structural relaxation can be examined. In particular, it illustrates clearly the differences between theories which treat phonon and structural dynamics as different time regimes of a single dynamical process and theories which treat phonon and structural dynamics as distinct and separable processes.

Transient Hole Burning Measurements of Solvation Dynamics

The basic principles of the transient hole burning (THB) measurement are summarized in Figure 1. The free energy of the solute ground state and excited state vary with the arrangement of local solvent molecules, indicated in Figure 1 by a one-dimensional solvation coordinate. In equilibrium in the ground state, the solute molecules occupy a distribution of solvation configurations. Each configuration has a different transition energy to the excited state, resulting in a broad absorption spectrum (Figure 1a). The sample is irradiated with a short pulse of light which is resonant with only a subset of solvent configurations. A local depletion or "hole" in the ground state distribution is created (Figure 1b). The resulting excited state molecules are created in a distribution which is narrower than and displaced from the equilibrium distribution. Following the excitation, the solvent begins to relax to reestablish equilibrium in both the ground and excited states. Both the hole and the excited state distribution increase in width $\sigma(t)$ and develop a Stokes' shift between their peak frequencies $\Sigma(t)$ (Figure 1c).

Solvation dynamics can be monitored through either the time-dependent widths or time-dependent Stokes' shifts (*8*). An important advantage of this technique is that the measurements are absolute rather than relative. In the absence of any solvent movement, the Stokes shift and solvent broadening are zero. The equilibrium values of these quantities are obtained independently from steady-state spectroscopy. Thus a

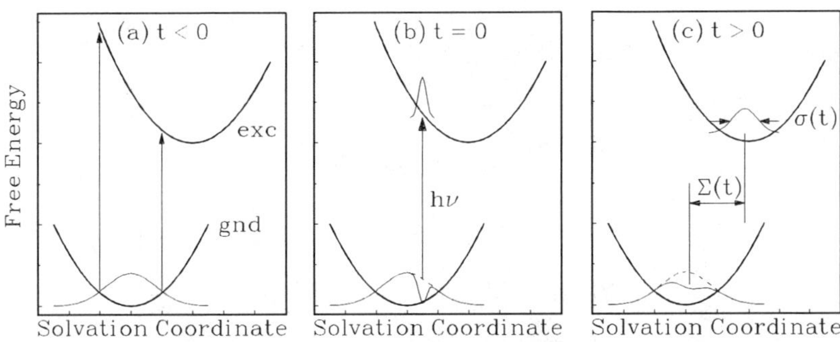

Figure 1. A schematic illustration of the transient hole burning experiment. (Reproduced with permission from ref. 2. Copyright 1996 American Physical Society.)

THB measurement, at even a single time, gives the fraction of the total relaxation to that point,

$$R_\Sigma(t) = 1 - \frac{\Sigma(t)}{\Sigma_{eq}}, \qquad R_\sigma^2(t) = 1 - \frac{\sigma^2(t)}{\sigma_{eq}^2} \tag{1}$$

There is no need to extrapolate to or fit either long or short time values.

Phonon Versus Structural Dynamics

When transient hole burning was first performed over a temperature range spanning the low viscosity liquid, the supercooled region and the glass, it was immediately apparent that there were two distinct components to the relaxation (*4-7*). This result is illustrated in Figure 2. At room temperature, the 1.5-ps THB width is almost identical to the equilibrium absorption width. The relaxation is almost entirely complete by this time. As the temperature is lowered, the THB width becomes narrower than the absorption width, indicating that a portion of the relaxation occurs after 1.5 ps. However, as the supercooled region is traversed, a significant THB width remains at 1.5 ps. A significant fraction of the relaxation remains subpicosecond, even as the viscosity of the solution diverges. Thus, at least two relaxation components exist: one which slows down with increasing viscosity, and one which does not.

The viscosity independent component can be associated with the phonon dynamics of a solid by continuing the hole burning experiments below the glass transition temperature T_g. In the low temperature solid, permanent hole burning (PHB) is more convenient than THB. In PHB, photochemical destruction of the excited molecules allows the hole width to be measured several minutes after the initial bleaching. Below T_g, PHB and THB give the same result, indicating that there is no significant relaxation between 1.5 ps and several minutes.

The widths in the solid glass fit a simple model for phonon-induced line broadening in solids (*4*). The curve in Figure 2 assumes a single phonon band centered at 30 cm^{-1}. Extending the curve into the supercooled liquid region shows that the fast component seen in THB is simply the extension of these phonon dynamics into the liquid. At short times, the liquid appears to have an effectively static structure, about which rapid vibrational motion occurs. At higher temperatures, this simple model is not accurate, both because of the changing density, and because the other relaxation component begins to affect the THB measurements.

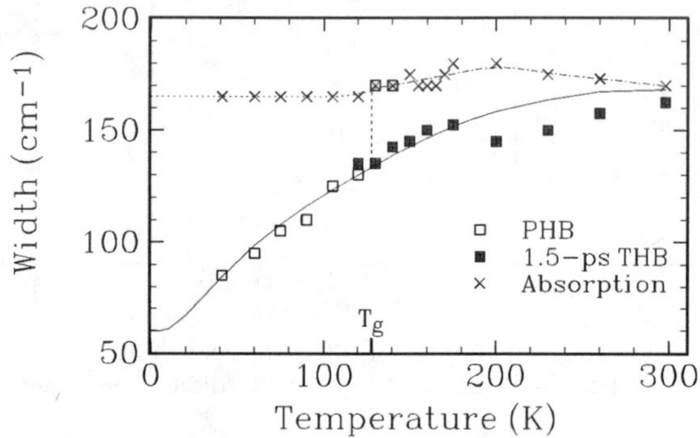

Figure 2. Solvent-induced widths of the electronic transition of dimethyl-*s*-tetrazine in *n*-butylbenzene as measured by transient hole burning (THB), permanent hole burning (PHB) and absorption spectroscopy. (Adapted from ref. 4)

The second relaxation component is due to the relaxation of the temporary structure which supports the phonon dynamics. As the temperature is raised above T_g, the PHB width suddenly jumps to the full absorption width. At this temperature, the second component becomes dynamic on the laboratory time scale instead of frozen. Near T_g, this component has little effect on the THB results. Near room temperature however, this component becomes fast enough to broaden the line at 1.5 ps, causing the THB and absorption widths to merge.

In the intermediate temperature regime, the THB width was measured at various delay times in the ps to ns region, and the average relaxation time of the second component was determined. The results are plotted in Figure 3. The relaxation times are proportional to the viscosity over a range of four decades. This result contrasts

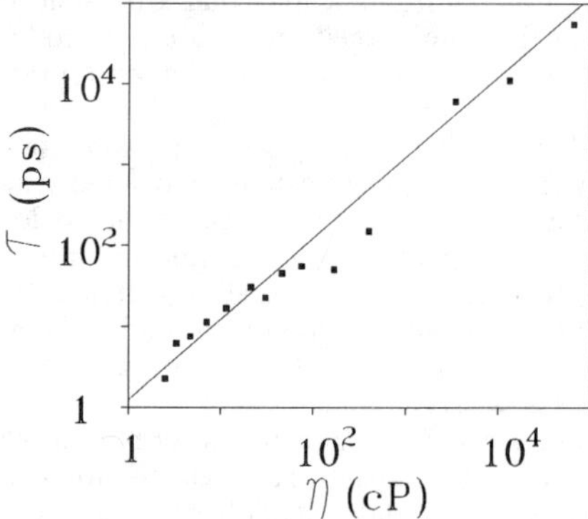

Figure 3. Average relaxation time of the structural component of the solvent relaxation versus viscosity. Measured for dimethyl-*s*-tetrazine in *n*-butylbenzene. (Reproduced with permission from ref. 5. Copyright 1993 American Physical Society.)

strongly with the early, phonon-induced relaxation, which shows no evidence of slowing at high viscosity. Because the viscosity is the most common measure of the overall structural relaxation time of a liquid, the second relaxation component can be linked with structural relaxation. This conclusion is reinforced by the quantitative modeling discussed later in this paper.

The same pattern illustrated here for *n*-butylbenzene (*4, 5*) has also been seen in several other liquids (*2, 3, 6, 7*). Presumably the phonon-induced component seen in solvation is essentially the same as the "microscopic" peak seen in scattering experiments, and the structural component is essentially the inelastic peak seen in low resolution scattering. The important point is that both components contribute significantly to solvation. Many theories of chemical dynamics in solution assume a single time scale for the dynamics. These theories may be plausible at sufficiently low viscosities, where the time scales of phonon and structural dynamics are comparable. However, such theories must fail at a qualitative level if the viscosity is even moderately high, because the phonon and structural dynamics will occur on different time scales.

Comparison to Mode-Coupling Theory

To date, THB experiments have focused on the structural relaxation component. Even within this single component, a wide variety of time scales are important. The most salient feature of the relaxation curves is that they are nonexponential, and the shape of the curves is temperature dependent. As the temperature is lowered and the primary structural relaxation time increases, some relaxation persists even at very early times.

One approach to explaining these features is mode-coupling theory (MCT) (*25-27*). This theory does not attempt to model details of the particular interactions involved in solvation or to predict numerical values of observed quantities. Instead it uses general features of liquid dynamics to make *a priori* predictions about the shapes of relaxation functions and the form of the relaxation time temperature dependence. The major predictions which can be tested with our data are:

1) The shape of the relaxation function is approximated by

$$R(t) \approx \begin{cases} R_\alpha(t/\tau_\alpha) \approx f_c \exp\left[-(t/\tau_\alpha)^\beta\right], & t \sim \tau_\alpha, \quad R \le f_c \\ R_\beta = f_c + \dfrac{d}{\tau_\beta^a} g_\lambda(t/\tau_\beta), & t \sim \tau_\beta, \quad R \sim f_c \end{cases} \qquad (2)$$

Different approximations apply to two different time regions: the α- and β-regions. The α-region, which covers the intermediate-to-late portion of the relaxation, obeys a simple scaling law equivalent to time-temperature superposition. This region can be approximated with a stretched exponential. The β-region, which covers the early-to-intermediate portion of the structural relaxation, obeys a more complicated form based on the temperature independent function g_λ. This function can be calculated for various values of λ, the exponent parameter. The value of the exponent a is also determined by the value of λ. Thus MCT predicts a range of relaxation times even greater than given by a simple stretched exponential. Two characteristic times, τ_α and τ_β, are needed to describe the relaxation function.

2) The temperature dependence of the relaxation times is given by:

$$\tau_a = c_\alpha (T - T_c)^{-\gamma}, \quad T > T_c \qquad (3)$$

$$\tau_\beta = c_\beta |T - T_c|^{-1/2a} \tag{4}$$

The exponents a and γ are defined by the value of λ in equation 2 and do not represent new adjustable parameters. Because τ_α and τ_β change differently with temperature, equations 2-4 predict a relaxation shape which changes with temperature.
3) MCT does not make predictions of the various constants appearing in equations 2-4; these must come from a detailed treatment of the specific interactions probed by a specific experiment. However, three of the parameters, T_c, λ and c_β, are predicted to be independent of the specific experiment and should be transferable between experiments.

The solvation of *s*-tetrazine in propylene carbonate is a good example of a system which is difficult to fit into a standard analysis. The response function spreads over an increasing range of times as the temperature is lowered, strongly violating the time-temperature superposition principle. Figure 4 shows a MCT explanation of the shape. Figure 4a is an α-scaling plot generated by overlapping the lower portions of the response functions. For intermediate-to-long times, the temperature scaling does work, and a stretched exponential fit to this region is shown. At short times, the R_α approximation fails, but this failure is expected from MCT.

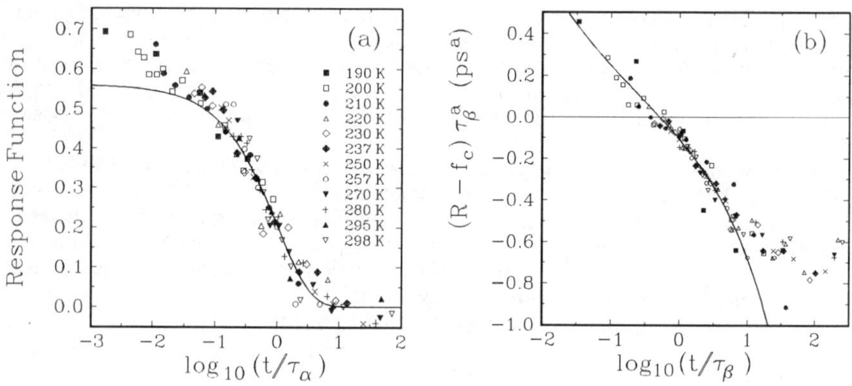

Figure 4. The solvation response function in propylene carbonate showing agreement with the α-scaling law at long time (a) and the β-scaling law at short time (b). Fit parameters: f_c = 0.65, λ = 0.78, β = 0.8. (Reproduced with permission from ref. 2. Copyright 1996 American Physical Society.)

For short to intermediate times, R_β is the appropriate approximation. A β-scaling plot is shown in Figure 4b along with a fit to g_λ. The β-approximation gives a good account of the short time data for which the α-approximation fails. Similarly, the β-approximation fails at long times, but this region is accounted for by the α-approximation. Note that both the α- and β-fits are constrained to share the same value of f_c. Thus MCT gives good fits to the changing shape of the response function as the temperature changes.

Each of the scaling plots generates a set of scaling times, τ_α and τ_β. The temperature dependence of these times is predicted to show a power law divergence at a critical temperature T_c (equations 3 and 4). Figure 5a tests this relationship. The exponents are predetermined by the value of λ used in fitting g_λ (Figure 4b). A common value of T_c is required for both τ_α and τ_β. In general, the MCT fit is good. The poorer agreement at high temperatures may be due to the anticipated failure of MCT far from T_c or an increase in experimental error with short relaxation times. In

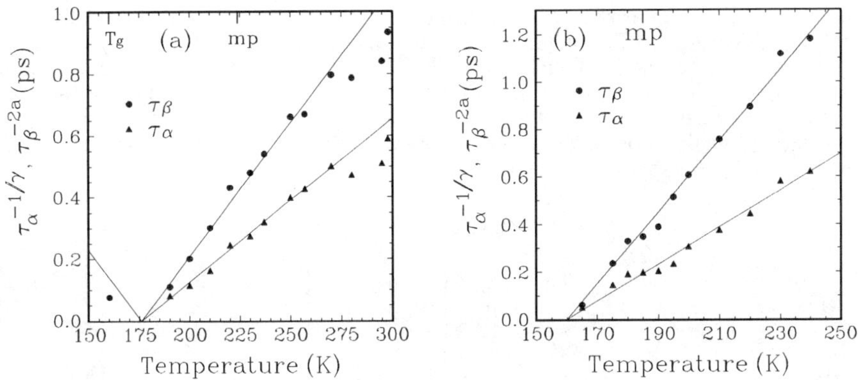

Figure 5. The temperature dependence of the times obtained from α- and β-scaling plots (Figures 4 and 7) for (a) propylene carbonate and (b) *n*-butylbenzene. A power law divergence to a common value of T_c is predicted by MCT. (Reproduced with permission from ref. 2. Copyright 1996 American Physical Society.)

either case, good agreement is found down to a viscosity of at least 3.4 cP, a value typical of a "normal" liquid.

In addition to the solvation data presented here, Du et al. have published an extensive MCT analysis of light scattering data in propylene carbonate (*28*). In the solvation analysis, we have adopted their value of λ = 0.78 and independently determined T_c. Our value of T_c = 176 K is very close to their value of 179 K. These results are in agreement with the MCT prediction that λ and T_c should be transferable between experiments. However, our value of $c_\beta = 3.6\text{x}10^3$ ps $K^{1/2a}$ is substantially smaller than the light scattering value of $1.0\text{x}10^3$ ps $K^{1/2a}$, in disagreement with MCT.

Unlike propylene carbonate, solvation data in *n*-butylbenzene do not show a strong change in shape with temperature. Figure 6 shows a master plot of these data. Within the experimental error, a single stretched exponential can be used to account for the

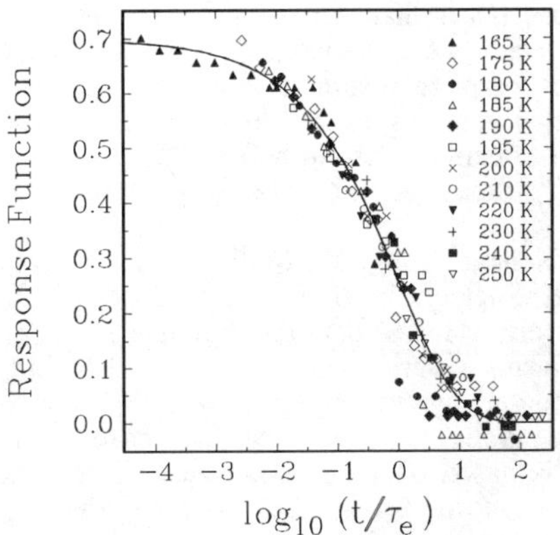

Figure 6. A master plot of the solvation response function of dimethyl-*s*-tetrazine in *n*-butylbenzene along with a stretched exponential fit (β = 0.45). (Reproduced with permission from ref. 2. Copyright 1996 American Physical Society.)

data. This result appears to contradict the temperature dependent shape predicted by MCT. However, a MCT analysis can also be successful within experimental error.

Figure 7 shows such an analysis. In the α-scaling plot (Figure 7a), the overlap of the lower portions of the response functions is optimized. The upper portions are allowed to deviate due to β-relaxation. Figure 7b shows that β-relaxation provides a satisfactory explanation of the early time behavior of the response functions. Again both α- and β-fits share the same value of f_c.

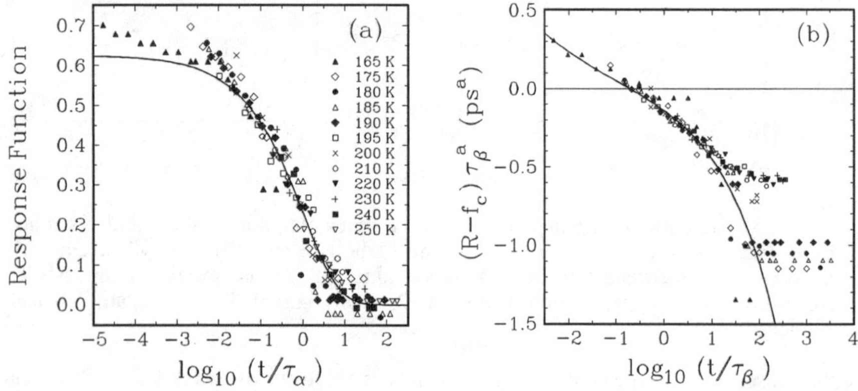

Figure 7. The solvation response function in *n*-butylbenzene showing agreement with α-scaling laws at long time (a) and β-scaling laws at short time (b). Fit parameters: $f_c = 0.625$, $\lambda = 0.86$, $\beta = 0.5$. (Reproduced with permission from ref. 2. Copyright 1996 American Physical Society.)

The temperature dependence of the resulting scaling times is shown in Figure 4b. As in propylene carbonate, the times show the predicted power law behavior with a common divergence at $T_c = 160$ K. Thus MCT provides an entirely consistent picture of the data, which is at least as good as more standard analyses, as exemplified by Figure 6. This result suggest that the wide range of systems which have been analyzed by standard methods may also be consistent with MCT.

A key point in these fits is that the α- and β-regions are not distinct, but often overlap heavily. As a result, a simultaneous fit to both regions as short and long time approximations to the complete response function is needed. For example, in *n*-butylbenzene, the overlap is strong enough that f_c is never seen as a plateau value in the data. β-Relaxation effects are strong before R_α reaches its short time asymptote. An attempt to fit α and β time regions separately will fail, but a simultaneous fit can be successful.

Yang, Muller and Nelson (Yang, Y.; Muller, L. J.; Nelson, K. A. Proceedings of the Material Research Society, in press) have used MCT to analyze impulsive stimulated thermal scattering (ISTS) data in *n*-butylbenzene. They find a value of T_c = 150 K. This is close to the solvation value of T_c = 160 K, but the difference is larger than can be accommodated by the expected errors in the analyses. One possible reason for the difference is that the ISTS analysis is based on data below T_c, where as the solvation analysis is based on data above T_c. A comparison between different temperature regions is more demanding of both theory and experiment than comparisons over a common range.

A Continuum Model of Nonpolar Solvation

Mode-coupling theory provides an *a priori* means of understanding some of the general features and trends of the solvation data. However, it does not provide insight into the physics of the solvation process, nor does it provide quantitative predictions or physical interpretations of the various constants involved. We have developed a continuum model for solvation in nonpolar systems as a complement to the MCT approach (*3, 17*). The continuum model is phenomenological in the sense that it requires assumptions about or independent experimental data on the mechanical moduli of the solvent. However, it provides quantitative fits to the data and a simple physical interpretation of the processes involved in solvation.

Continuum models have often been used to describe molecular scale dynamics in solution, often with surprising success. However, it has not generally been recognized that the existence of distinct phonon and structural processes is predicted by these models. After describing the model and showing its success in describing experimental data, we will use it to explore the relationship of these two processes.

The continuum model postulates that the coupling of the solvent to the solute arises from a change in the effective size of the solute when it changes electronic state. This may represent a change in bond lengths or in the size of the electron cloud of the solute. It may also represent a change in polarizability, which alters the nearest neighbor attractive forces. In any of these cases, the forces are between the centers-of-mass of the solute and solvent molecules. In contrast, solvation in polar systems generally involves torques between the molecules from dipole-dipole and other angle dependent forces (*11*).

The solute is modeled as a spherical cavity within a viscoelastic continuum, which represents the solvent. The size of the cavity in the ground state is determined by a balance of the solvent pressure and solute-solvent attractions tending to collapse the cavity and the solute-solvent repulsion acting against collapse. In the excited state, the solvent-solute forces are altered, and a slightly different cavity size is attained at equilibrium. The restoring force is assumed to be the same in the ground and excited states and to be linear over the size change involved.

The solvation process is modeled by taking the system at equilibrium in the ground state and suddenly changing to the excited state forces at $t = 0$. Initially the cavity is at the ground state size and is out of equilibrium. The solvent responds to the new forces as a viscoelastic continuum, i.e. it has a time-dependent shear and bulk modulus. As the solvent responds, the solute cavity expands toward its excited state value, and the transition energy drops. Within this simple model, the resulting time-dependent Stokes' shift can be calculated analytically given knowledge of the mechanical moduli of the solvent (*17*).

Even if the time-dependent moduli have a single relaxation time, the solutions have two distinct time scales. The fastest component represents a purely elastic response of the solvent and is due to mechanical waves propagating away from the solute. For typical parameters, this time scale is ~100 fs. Clearly this portion of the solution represents the phonon-induced component of solvation, which is subpicosecond at all viscosities and is shared by both solid and liquid systems.

The second portion of the solution represents a viscous response of the solvent. Its time scale is closely linked to the shear relaxation time of the solvent and thus it scales directly with the viscosity. At low viscosities, this response can be subpicosecond and difficult to distinguish from the phonon-induced component, but as the viscosity increases, the solvation time becomes longer, eventually becoming static when the viscosity diverges at the glass transition. This portion of the solution corresponds to structural solvation.

Examples of these solutions are shown in Figure 8 along with the corresponding experimental data (*3, 17*). In each case, a stretched exponential form has been assumed for the time-dependent shear modulus. Although the data cannot be compared directly with the phonon-induced component, the amplitude of the response function following the phonon-induced component can be compared to the amplitude of the early data. Because the THB measurements give absolute values of the response function, there is no arbitrary normalization of the data, and the model must accurately predict the amplitude of the data. In fitting the model to the data, two parameters are important, the high frequency shear modulus G_∞ and the shear relaxation time τ_S. Although these are initially arbitrary fitting parameters, the product of these two parameters should give the bulk viscosity. In fact, they do reproduce the bulk viscosity well (*3, 17*). The fit value of G_∞ is also close to values obtained from ultrasound measurements (*17*). Because of this agreement with independent experimental measurements, we are confident that the theoretical parameters are physically reasonable, and the basic physical picture of the continuum model is correct.

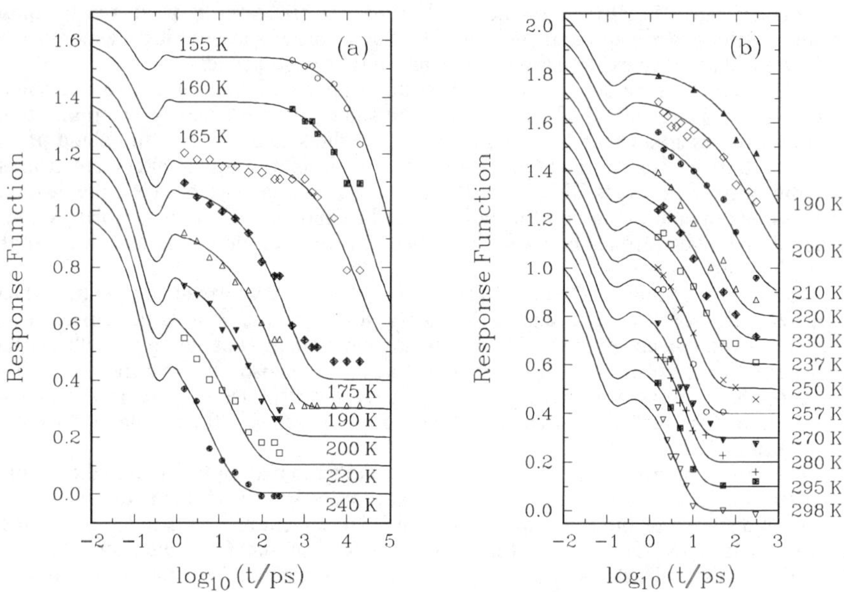

Figure 8. Fits of the continuum model of nonpolar solvation to data from (a) *n*-butylbenzene and (b) propylene carbonate. Each temperature is vertically offset for clarity. (Adapted from Refs. 17 and 3.)

Given that the continuum model is at least a reasonable approximation to real systems, and given that it clearly predicts the existence of phonon-induced and structural components of the relaxation, it is interesting to examine the predictions of this model concerning the relationship of these two components as the viscosity becomes lower, and the two time scales merge. The results suggest some important questions regarding the interpretation of experimental data.

Experimental data are often presented either as a response function in the time domain $R(t)$, or as a spectral density in the frequency domain,

$$S(\omega) = \mathrm{Im}\int_0^\infty e^{-i\omega t}\,\dot{R}(t)\,dt\,. \tag{5}$$

In either case, the effects of phonon and structural relaxation are often treated as distinct, separable contributions

$$R(t) = R_{ph}(t) + R_{st}(t) \qquad S(\omega) = S_{ph}(\omega) + S_{st}(\omega) \tag{6}$$

We refer to this as the "spectroscopic" approach, because in the frequency domain, it is analogous to the decomposition of a spectrum into independent bands corresponding to different transitions. It is valid if phonon and structural motion result from the dynamics of two different coordinates corresponding to quasi-independent terms in a Hamiltonian.

Both the continuum model and MCT take a different, "unified" approach. Phonon and structural components are defined as different time regimes in the dynamics of a single quantity. They represent short and long time asymptotic approximations to a single exact function. If we attempt to represent the exact function with the sum of the long and short time limiting functions, an "interference term" will appear at intermediate times

$$R(t) = R_{ph}(t) + R_{ph-st}(t) + R_{st}(t) \qquad S(\omega) = S_{ph}(\omega) + S_{ph-st}(\omega) + S_{st}(\omega) \tag{7}$$

containing the errors in this procedure.

Figure 9 shows solutions for an exponentially decaying shear modulus and typical values of the model parameters [$G_\infty = 1.2\text{x}10^{10}$ dyne/cm^2, $K_\infty = K_s = (5/3)G_\infty$, $r = 3$ Å, $\rho = 1$ g/cm^3]. The solution is decomposed into a phonon component, a structural component and the interference term. At moderately high viscosity, there is a wide time scale separation between the phonon and structural processes, and the

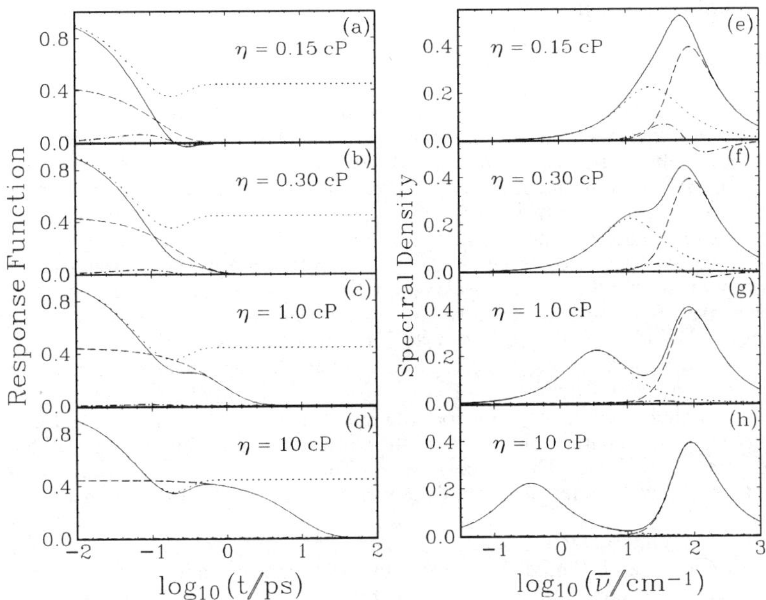

Figure 9. Time (a-d) and frequency (e-h) domain solutions of the continuum model at various viscosities. The exact solution (solid) as well as the phonon (dashed), structural (dotted) and "cross terms" (dash-dotted) are shown.

interference term is negligible (Figure 9c and 9d). In this case, the spectroscopic and continuum models are not experimentally distinguishable.

At low viscosities, the time scale separation between phonon and structural processes breaks down, and the interference term becomes significant (Figure 9a and 9b). It is no longer accurate to discuss phonon and structural processes independently, and a spectroscopic approach to the results is demonstrably incorrect.

Experimental data on liquid dynamics is often analyzed by fitting a function to the long time / low frequency data, subtracting this function from the data, and treating the remainder with theories appropriate to short time dynamics. This example points out a potential problem with this scheme. The structural component is a perfect fit to the long time / low frequency data by construction. Subtracting it from the full results leaves a good representation of the phonon dynamics if the viscosity is sufficiently high. For low viscosities, the phonon dynamics will be contaminated with the interference term.

At all viscosities, it is tempting to decompose the spectral density into at least two bands. These bands can be treated separately with, for example, a Brownian oscillator model. However, in the current case it is clear that such oscillators have no physical significance. There are not two quasi-independent coordinates present within the continuum model. The two bands represent two time regimes within a single dynamical system.

It is sometimes suggested that spectral densities should be viewed as a density of some type of modes intrinsic to the solvent weighted by coupling strengths to the solute. In the case of the continuum model, the intrinsic modes of the solvent are known exactly; they are the Debye modes and follow a Debye density of states. The solvation spectral density does not bear a close resemblance to the Debye density of states. Although a coupling function could be defined, it would be such a complex function of frequency, that it is questionable if the decomposition of the response leads to greater physical insight.

Conclusions

We have demonstrated that solvent dynamics on a variety of time scales can influence the electronic energy levels of a chemical system in solution. Within the class of structural dynamics, stretched exponential decays, which incorporate more than one time scale, have often been seen in supercooled liquids. Our solvation experiments on structural relaxation support the MCT idea that there is an even greater range of time scales involved. This idea is embodied as a set of β-relaxation dynamics preceding the stretched exponential, α-dynamics. As a result, very rapid dynamics remain active at viscosities at which the average relaxation time is quite long. Because chemical dynamics are often most sensitive to the fastest dynamics, this result has important implications in understanding the viscosity dependence of chemical processes.

On an even larger scale, we have shown that phonon dynamics can be just as important as structural dynamics. These two components have a clear experimental distinction in their contrasting behavior with increasing viscosity. We have also shown that the origin of separate phonon and structural dynamics can be understood with simple continuum ideas. The success of this approach obviates any need to invoke concepts such as "single particle" dynamics to understand the origin of phonon dynamics.

The continuum model also serves to contrast two different approaches to treating multiple time scales in liquid dynamics. In the "spectroscopic" approach, multiple quasi-independent coordinates are postulated, each coordinate responsible for a different time scale. The continuum model demonstrates that this is not the only approach possible. It is also possible that multiple and distinct time scales arise from the inherently complex dynamics of a single dynamical quantity. One experimentally distinguishable difference between these approaches is in the predicted behavior as the structural and phonon time scales merge at low viscosity.

In the near future, we expect investigations into this viscosity regime to be increasingly important. The glass transition temperature, where observable properties diverge, has long been recognized as an essential point in understanding disordered materials. More recently, interest has focused on a point at moderately high viscosities, where there may be a crossover from transport by diffusion to transport by hopping. We suggest that the low viscosity region where structural and phonon dynamics separate into distinct time scales is an equally important point in understanding the fundamental dynamics of liquids. We hope the current work will spur further more sophisticated investigations into this important region.

Acknowledgments

This paper is based on work supported by the National Science Foundation. MB received support from a NSF Presidential Young Investigator Award and a Sloan Research Fellowship. JTF received support from a NSF Postdoctoral Fellowship.

Literature Cited

1. *Disorder Effects on Relaxational Processes: Glasses, Polymers, Proteins*; Richert, R.; Blumen, A., Eds.; Springer-Verlag: New York, NY, 1994.
2. Ma, J.; Vanden Bout, D.; Berg, M. *Phys. Rev. E* **1996**, *54*, pp. 2786-96.
3. Ma, J.; Vanden Bout, D.; Berg, M. *J. Chem. Phys.* **1995**, *103*, pp. 9146-60.
4. Fourkas, J. T.; Berg, M. *J. Chem. Phys.* **1993**, *98*, pp. 7773-85.
5. Fourkas, J. T.; Benigno, A.; Berg, M. *J. Chem. Phys.* **1993**, *99*, pp. 8552-58.
6. Yu, J.; Berg, M. *J. Phys. Chem.* **1993**, *97*, pp. 1758-64.
7. Yu, J.; Berg, M.; *J. Chem. Phys.* **1992**, *96*, pp. 8741-49.
8. Yu, J.; Earvolino, P.; Berg, M. *J. Chem. Phys.* **1992**, *96*, pp. 8750-56.
9. Kang, T. J.; Yu, J.; Berg, M. *J. Chem. Phys.* **1991**, *94*, 1991, pp. 2413-23.
10. Kang, T. J.; Yu, J.; Berg, M. *Chem. Phys. Lett.* **1990**, *174*, pp. 476-80.
11. Maroncelli, M. *J. Mol. Liq.* **1993**, *57*, pp. 1-37.
12. Reynolds, L.; Gardecki, J. A.; Frankland, S. J. V.; Horng, M. L.; Maroncelli, M. *J. Phys. Chem.*, **1996**, *100*, pp. 10337.
13. Horng, M. L.; Gardecki, J.; Papazyan, A.; Maroncelli, M. *J. Phys. Chem.* **1995**, *99*, pp. 17311.
14. Jimenez, R. Fleming, G. R.; Kumar, P. V.; Maroncelli, M. *Nature* **1994**, *369*, pp. 471.
15. Saven, J. G.; Skinner, J. L. *J. Chem. Phys.* **1993**, *99*, pp. 4391.
16. Bagchi, B. *J. Chem. Phys.* **1994**, *100*, pp. 6658.
17. Berg, M. *Chem. Phys. Lett.* **1994**, *228*, pp. 317-22.
18. Stratt, R. M.; Cho, M. *J. Chem. Phys.* **1994**, *100*, pp. 6700.
19. Ladanyi B. M.; Stratt, R. M. *J. Phys. Chem.* **1996**, *100*, pp. 1266.
20. Richert, R.; Stickel, F.; Fee, R. S.; Maroncelli, M. *Chem. Phys. Lett.* **1994**, *229*, pp. 302.
21. Richert, R. *Chem. Phys. Lett.* **1993**, *216*, pp. 223.
22. Richert, R. *Chem. Phys. Lett.* **1992**, *199*, pp. 355.
23. Richert, R.; Wagener, A. *J. Phys. Chem.* **1991**, *95*, pp. 10115.
24. Richert, R. *Chem. Phys. Lett.* **1990**, *171*, pp. 222.
25. Götze, W.; Sjögren, L. *Trans. Theory Stat. Phys.* **1995**, *24*, pp. 801.
26. Götze, W.; Sjögren, L. *Rep. Prog. Phys.* **1992**, *55*, pp. 241.
27. Götze, W. In *Liquids, Freezing Glass Transition*; Hansen, J. P., Levesque, D., Zinn-Justin, J., Eds.; Elsevier Science B.V.: Amsterdam, 1991.
28. Du, W. M.; Li, G.; Cummins, H. Z.; Fuchs, M.; Toulouse, J.; Knauss, L. A. *Phys. Rev. E* **1994**, *49*, pp. 2192.

Polyamorphism and Supercooled Water

Chapter 16

Polyamorphic Transitions in Network-Forming Liquids and Glasses

J. L. Yarger[1], C. A. Angell[2], S. S. Borick[2], and G. H. Wolf[2]

[1]Department of Chemistry, University of California, Berkeley, CA 94720
[2]Department of Chemistry, Arizona State University, Tempe, AZ 85287–1604

The occurrence of polyamorphism in stable liquids, supercooled liquids, and solid amorphous materials is reviewed. Tetrahedrally bonded network systems have been the primary focus of many experimental and computer simulation studies of amorphous phase transitions. Of these network-forming systems, H_2O and Si have been shown to undergo a first-order phase transition between two distinct amorphous phases. Nevertheless, many other tetrahedral networks, such as SiO_2 and GeO_2, show a much more gradual structural change. This brings up questions to the generality of polyamorphic behavior in network glasses. A proposed explanation for the variations in behavior among tetrahedral glasses is given in terms of the effective yield strength, percent volume collapse, and cooperativity. Comparison with crystal-to-amorphous phase transformations in compositionally similar systems yields additional insight.

It has long been known that the loss of internal (structural) equilibrium during cooling is responsible for the dependence of many physical properties on the temperature and stress history of the sample. Early glass workers attributed these property variations to residual stresses within the glass, and consistent properties were obtained through the stress-relaxing process of annealing. Further work, however, suggested that internal stresses alone could not account for all property variations in glasses with differing thermal or stress histories. In a 1921 paper entitled "Polymorphism and Annealing of Glass," Lebedeff (*1*) proposed that structural changes during annealing occur analogously to the polymorphic transitions between quartz, cristobalite, coesite, and other crystalline forms of silica. Lebedeff's proposal inspired much of the early work and discussions on structural relaxation in glasses, but the extension of his idea to the existence of actual macroscopic polyamorphic changes between glassy phases has only recently been contemplated.

The possibility of actual macroscopic polymorphism in the amorphous state, called "polyamorphism" (2), arose recently as a result of the experimental observation of large hysteretically reversible changes in the density of amorphous solid water (ASW) obtained on samples of pressure amorphized Ih ice (*3*). The quasi-thermodynamic character of these transitions, which occur between low-density (LDA) and high-density (HDA) amorphous phases below the glass transition temperature of water, have been connected to the existence of a real (though metastable) line of first order phase transitions in the supercooled liquid state of water (*4*). Computer simulations of glass-glass transitions (*5, 6, 7, 8, 9*) have been used to model the experimentally observed polyamorphism in ASW and further extend the transition into the supercooled state.

Mishima has reported the existence of a phase boundary between the high density and low density amorphs of H_2O (*10*), which would seem to be a requirement for accepting that polyamorphism is a well-defined phenomenon. Although many reports of sudden density changes in glasses under compression have since appeared, a phase boundary between isochemical (same composition) phases of different densities has only been observed in one other case (*11*). The extent to which distinct structural "polyamorphs" (in a true or quasi-thermodynamic sense) occur in a broader class of glassy materials is an important question. Answering this question requires an understanding of the mechanistic and physical factors that determine the manifestation of polyamorphic behavior.

In most cases, the physical and structural properties of a glass change quite gradually with changes in temperature, pressure or any other external field variable. For this reason, most of the early work on glasses describes their behavior within the context of structural relaxation and kinetics. This includes the temperature (or pressure) induced glass transition which is understood as a kinetic consequence of the crossing of the fixed experimental measurement time scale and the intrinsic liquid relaxation time scale, the latter being strongly temperature (or pressure) dependent. While the changes in thermodynamic properties (e.g. heat capacity) at the glass transition can be described semi-quantitatively within a phenomenological non-linear relaxation kinetic framework, it is becoming clear from recent studies that the loss of internal equilibrium ("breaking of ergodicity") is somewhat more complicated than the usual description would suggest (*12, 13*). It now seems that there exists a range of dynamic processes, with differing time scales, that cause structural organization and clustering in the supercooled liquid near T_g (*14, 15, 16*). The subject of the structural and dynamic heterogeneity near T_g is the topic of many of the papers in these proceedings and will not be discussed here. Instead, we will focus on recent experiments which suggest that, even in the glassy state well below T_g, cooperative phenomena can occur and sometimes appear as a "quasi-thermodynamic" phase transformation between isochemical structural polymorphs or "polyamorphs."

Review of Polyamorphism

We use the term "polyamorphism" to describe the existence, in a condensed amorphous system of fixed composition, of two or more structurally distinct phases, the transition between which occurs across a defined interface. In the cases of metastable liquid phases, the transition will be first order in the normal phase transition sense, and will require conventional nucleation and growth of the second phase. A closely related version of the same phenomonon can occur within the glassy state though, due to the absence of normal diffusion, the transition is displaced in pressure and the mechanism must be different (*17*). Consequently, the transition is less sharply defined, and the transition character is less obvious. In the case of silicon, it has been suggested that the polyamorphic transition from metallic liquid to tetrahedral semiconductor is also a liquid-to-amorphous solid transition. The reverse process is a "melting", because of the enormous differences in atomic mobility assumed for the two phases (*18, 19, 20*). In the broadest sense, even the liquid-vapor transition constitutes a polyamorphic phase transition. As in the latter case, there may only be limited temperature-pressure range in which the polyamorphic transition between condensed phases appears as a first-order thermodynamic discontinuity. The line of first-order transition may terminate at a critical point (T_c) beyond which only continuous changes between different structural states can occur. Polyamorphism should not be confused with first-order liquid-liquid or glass-glass phase separation in multi-component systems. These more common transitions are driven principally by chemical segregation, and are well understood in terms of competing enthalpic and entropic thermodynamic influences associated with chemical mixing. Polyamorphic transitions, however, are manifested solely as isochemical transitions between two different structural organizations. On the other hand, the occurrence of polyamorphism in one component of a binary mixture can be responsible for chemical segregation in the binary system even if the high temperature mixing is ideal (*21*).

While reports of polyamorphism are rare, and established cases even rarer, the possibility of liquid-liquid transitions in isotropic systems has been recognized for several decades (22). The existence of separate liquid species was implicit in the "two fluid" models proposed to explain the maxima in the melting curves of Cs (*23*) and Ba (*24*), and was an explicit prediction of analytical theories, using softened core repulsion potentials, proposed by Stell and coworkers (*25*). Liquid-liquid transitions of undetermined order have also been proposed for C (*26, 27*), Te (*28, 29*), Se (*30, 31*), S (*32*), I (*33*), and Sb (*34*) on the basis of conductivity and thermobaric studies. The case of Y_2O_3-Al_2O_3 liquid is unique as a transition which can be "caught in the act" since it occurs close to the glass transition temperatures of the coexisting phases (*11*). The transition between amorphous solid and liquid silicon referred to earlier is seen clearly in computer simulations (*35*) where the reverse transition from over-coordinated liquid to immobile tetrahedral phase is seen as a first order transition occurring somewhat below a density maximum, closely paralleling the behavior proposed for water at high pressures (*8, 9*).

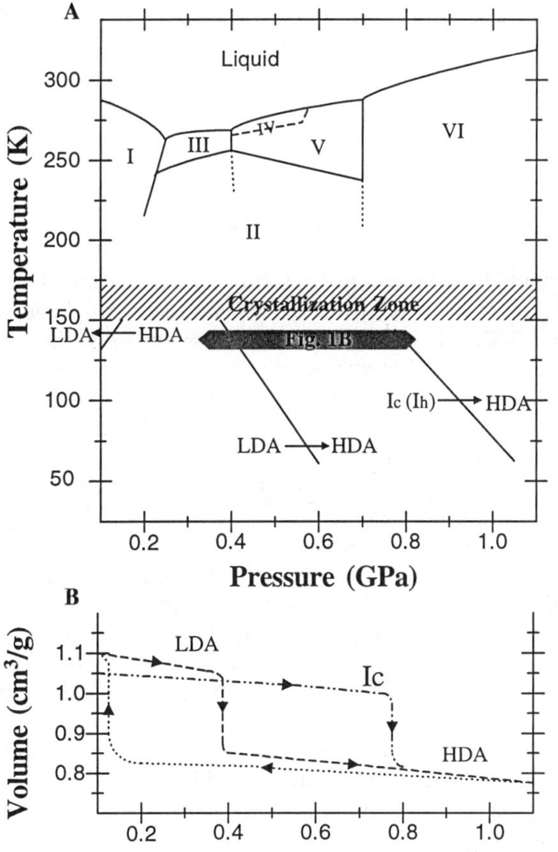

Figure 1. (A) A qualitative pressure-temperature diagram of H_2O showing both polymorphs and polyamorphs. (B) The non-equilibrium amorphous-to-amorphous, or polyamorphic, LDA ⇒ HDA and HDA ⇒ LDA transitions as well as the crystal-to-amorphous Ic ⇒ HDA pressure-volume plots. Data is extracted from O. Mishima et al (*4*).

The best characterized example of polyamorphism between two phases in the glassy state is the transition between LDA and HDA in amorphous solid water (*4*). Figure 1 shows the phase diagram, together with the measured density-temperature relations for ice and amorphous phases at 130 K (data taken from Mishima et. al. (*4*)). Sharp volume changes and hysteretic effects are signatures that the transition is first-order. It remains first-order until about 140 K, above which crystallization prevents further measurement, so the critical point suggested by the MD studies cannot be observed experimentally. Computer simulations of ST2 (*36*) water showed evidence of a critical point in the supercooled liquid just above 235 K, at a pressure of about 200 MPa, which were supported by follow-up calculations on TIP4P and SPC water (*37*). Tanaka (*38*), also performing TIP4P calculations, argued that the critical point should be located at -200 MPa but otherwise agreed on the polyamorphic nature of low temperature water.

While water provides the best studied, and least ambiguous case of solid polyamorphism, there are a number of other cases which show comparable but less sharply defined behavior. Almost all of these are systems with low coordination number open network as their low pressure ground state structures. An interesting exception would be the case of the organic liquid triphenyl phosphite, though the true polyamorphic character of the "glacial" phase remains to be properly established (*39*).

Polyamorphism vs. hysteretic relaxation in tetrahedral glasses

Since the discovery of polyamorphism in amorphous solid water, much attention has been given to related structural transitions in other tetrahedral systems. The two classic examples are SiO_2 and GeO_2. Here we consider the observations on these two systems and two other recently studied materials, BeF_2 and GaSb, to see how generally the description "polyamorphism" can be applied. Alternative descriptions will be discussed along with observable characteristics that might be helpful in characterizing the nature of the various transitions.

In the liquid phases, both SiO_2 and GeO_2 exhibit the anomalous pressure effect on viscosity for which water is considered remarkable, and SiO_2 also exhibits a (weak) density maximum. As suggested by these similarities, both SiO_2 and GeO_2 glasses exhibit rather abrupt increases of density on sufficient compression, which reverses hysteretically on decompression. However, the transition in these cases observed at room temperature is much more gradual than the analogous transition observed in amorphous ice at low temperatures. The recent measurements on BeF_2 (*40*) and GaSb (*41*, *42*, *43*) also show gradual behavior similar to that found in silica and germania.

To illustrate these differences, we compare in Fig 2. the different pressure-volume relations using a scaled variables representation in which the reference state is the volume at the pressure where densification is completed and the compressibility returns to a normal solid state value. Data are shown for H_2O, GeO_2, and BeF_2 glasses. Silica is not included because the equation of state (*44*) has not been determined in the region of pressure where Brillouin (*45*), Raman (*46*), IR (*47*), and x-ray (*48*) studies all show the structural changes to occur. However, qualitative models based on these spectroscopic studies indicates a more gradual transition that is observed in other tetrahedral systems. Of these three systems shown, GeO_2 at 298 K and H_2O at 130 K show a spontaneous reversion to the low density form on decompression. BeF_2 at 298 K does not, though this may be a function of the temperature of study, since HDA water does not revert to LDA water when the temperature of the cycle is 77 K.

The differences in abruptness of conversion from low density to high density forms and vice versa, which distinguish H_2O from its four structural homologues, raises an important question: are the more gradual transitions in SiO_2, BeF_2, GeO_2 and GaSb sufficient to warrant description as polyamorphic changes, or should a continuous relaxation (e.g. a two-state exchange mechanism (*49*)) description be used? Is the cooperativity of the process in the latter cases, unlike the H_2O case, insufficient to generate a first-order transition? It is notable that in the case of H_2O the coordination remains tetrahedral throughout the transformation from LDA to HDA. In GeO_2 and SiO_2 cases, on the other hand, a transformation involves a change

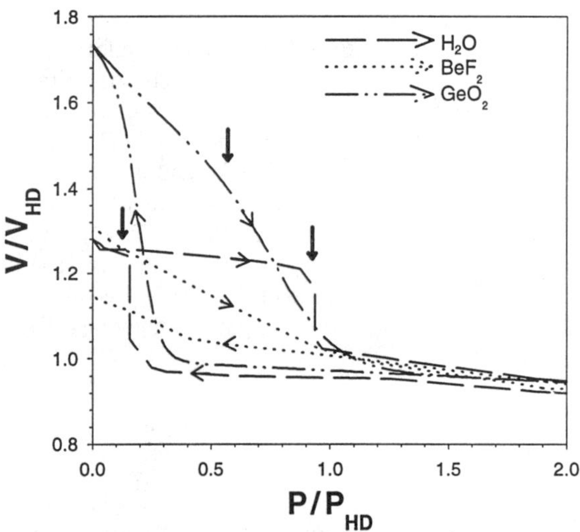

Figure 2. High density (HD) scaled pressure-volume equation of state for compression and decompression of H_2O (*4*), GeO_2 (*62*), and BeF_2 (*40*). The arrows (↓) indicate the onset for transformation during compression.

in cation coordination number from four to six. In BeF_2, the nature of the cation coordination through the transition is unknown, though computer simulation studies would suggest it is similar to that in SiO_2 (*50*). Solid state NMR work is currently in progress to elucidate the structural nature of this transition (*51*).

A relatively large volume change is associated with the collapse of tetrahedral network glasses. In order for these volume changes to be first-order "like", the transition must have a high degree of cooperativity. Therefore, it may be reasoned that the degree to which the structure is disordered may be an important consideration in cooperativity among solid state transitions. In tetrahedral glasses, the first cation coordination sphere is typically regular tetrahedra with small distributions in local bond distances and angles. The disorder becomes pronounced, however, in the second coordination sphere or the linking together of tetrahedra. Inter-tetrahedral angle distributions are shown in Fig 3 for amorphous GeO_2 (*2*), SiO_2 (*52*), BeF_2 (*53, 54, 55*), and H_2O (*56*). Amorphous ice is shown to have a narrow distribution and exhibits a cooperative first-order transition when compressed isothermally. GeO_2, on the other hand, has a wider distribution of intertetrahedral angles and thus behaves less cooperatively, showing a gradual structural change upon compression. This trend is followed in SiO_2 and BeF_2, which show even more gradual transition regions. The limit to cooperativity can be seen in the crystal-to-amorphous transition of analogous compounds. In hexagonal (Ih) and cubic (Ic) ice, the crystal-amorphous transition is first-order (*57, 58*), however, in other tetrahedral systems the transition is often considerably broadened (*59*). The lack of first-order behavior in all crystal-amorphous transformations indicates that other factors besides cooperativity must play a role in the transition. The enormous volume collapse, and hence stresses, that accompany the structural change likely inhibit a coherent change at these relatively low temperatures, where atomic mobilities are low and the shear strength is high. The change in cation coordination of GeO_2 and SiO_2 (quartz crystal and glass) at high pressure results in a collapse of individual tetrahedra. However, if the yield strength of the material is high, the surrounding tetrahedra may not coherently collapse. Rather, a localized volume

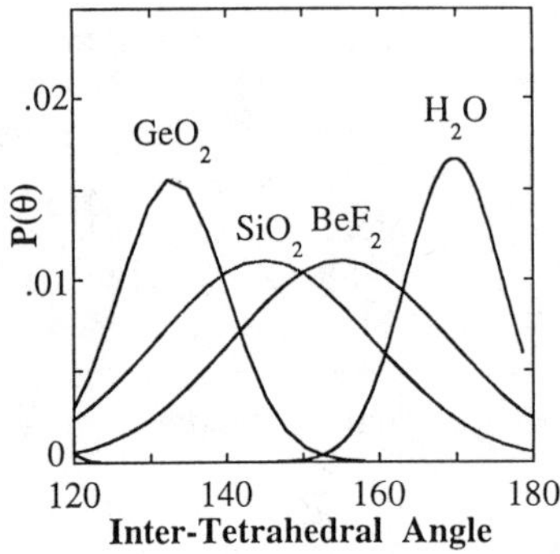

Figure 3. The population distribution of inter-tetrahedral angles for GeO_2, SiO_2, BeF_2, and H_2O is shown. For GeO_2, SiO_2, and BeF_2 information is extracted from x-ray (*48, 55*), NMR (*52*), Raman (*2, 46*), and neutron (*54*) experiments. The distribution for H_2O was taken from computer simulations (*56*).

collapse of a tetrahedral unit in the amorphous network would temporarily cause a pressure gradient with the high density material being at a lower pressure than the surrounding low density material. If this hypothetical scenario were true, it could easily be tested by increasing the temperature to a region where the material has a minimum shear strength. Evidence for heterogeneous pressure comes from the gradual nature of volume changes in higher strength materials. For example, high coordinate phases have been predicted and experimentally shown to have lower yield strengths than the corresponding tetrahedral phase (*60*). In GeO_2, the compression of the tetrahedral phase has a more gradual transition (^{iv}Ge - ^{vi}Ge) than that observed in the reversible decompression (^{vi}Ge-^{iv}Ge) (*61, 62*). This could be attributed to the lower yield strength and a smaller inter-tetrahedral distribution of the high density phase.

The notion of pressure-induced heterogeneity in the amorphous state is certainly not novel. At low temperatures, especially below the glass transition temperature, a quasi-thermodynamic transition can be suppressed to allow metastable pressurization of the low-density polymorph into the stability field of the high-density polymorph. The ultimate limit of this metastability is at the spinodal P-T line where no thermodynamic barrier to the transformation exists. If a mechanical instability were related to a critical inter-tetrahedral angle, then the distribution displayed in glasses could cause varying regions of the sample to transform at different pressures with the high yield strength enabling the support of heterogeneous domains. The large volume change would cause a reduced pressure, therefore, providing heterogeneity in the structure and pressure throughout the sample. This indicates that pressure may no longer be a good thermodynamic variable ($P_{int} \neq P_{ext}$) in glasses with high yield strengths, even under hydrostatic compression.

Despite the qualitative evidence and explanations we have presented, it is not possible, given the current experimental data, to discern the mechanisms of structural transitions in SiO_2, GeO_2, BeF_2 and other tetrahedral network glasses.

Unlike amorphous ice, none of these systems displays a visually observable phase separation through the transition region (*2*). The only evidence for any structural heterogeneity is the increase in parasitic (Mia) visible light scattering during Brillouin and Raman experiments in the diamond anvil cell throughout the pressure region of the structural change. Such an increase in light scattering is consistent with the nucleation and limited growth of a large number of small domains. It is interesting to note that Kivelson (*39*) observed a similar increase in turbidity in phase-separated triphenyl phosphite. Taken together, the experimental data certainly does not rule out polyamorphism in tetrahedral glasses. If yield strength is an important material consideration, as we suppose it is, then it is not unreasonable to suggest that a new phase may be nucleating, but is prevented from growing at an appreciable rate because of a high strength network surrounding it. At the same time, one cannot rule out the possibility of a continuous structural relaxation process. Computer simulations in the liquid state of tetrahedral networks, however, indicate evidence for liquid-liquid phase separation. Complementary experiments are underway in our laboratory that will help resolve the underlying nature of these glassy state transitions and provide a test for the validity of the above arguments for inhibition in the glassy state polyamorphism. A most conclusive experiment would be to measure the equation of state of one of these materials at temperatures above T_g and below T_c. Experimentally, this is most feasible in the GeO_2 and BeF_2 systems because of the lower T_g and moderate pressure range for the transition.

Conclusions

A review of polyamorphic behavior in known systems is given. Possible reasons for the large variation in transition behavior in similar network glasses are discussed. Cooperativity, material yield strength and the percent volume collapse are proposed to contribute to the inability of network glasses to exhibit true first order phase transitions.

Acknowledgments

The authors are indebted to Kathy Culbreath and Dr. Susan De Paul for assistance with this manuscript. This work was supported by the National Science Foundation under Grant DMR 9108028-002.

Literature Cited

(1) Lebedeff, A. A. *Trans. Opt. Inst. Petrograd.* **1921**, 2, 1.
(2) Wolf, G. H.; Wang, S.; Herbst, C. A.; Durben, D. J.; Oliver, W. F.; Kang, Z.C.; Halvorson, K. In *High-Pressure Research: Applications to Earth and Planetary Sciences*; Syono, Y., Ed.; American Geophysical Union; Terra Scientific Publishing Company: Washington, D. C., **1992**; 503-517.
(3) Mishima, O.; Calvert, L. D.; Whalley, E. *Nature* **1985**, 314, 76.

(4) Mishima, O. *J. Chem. Phys.* **1994**, 100, 5910.
(5) Poole, P. H.; Sciortino, F.; Essmann, U.; Stanley, H. E. *Nature*, **1992**, 360, 324.
(6) Harrington, S.; Zhang, R.; Poole, P. H.; Sciortino, F.; Stanley, H. E. *Phys. Rev. Lett.* **1997**, 78, 2409.
(7) Roberts, C. J.; Debenedetti, P. G. *J. Chem. Phys.* **1996**, 105, 658.
(8) Poole, P. H.; Grande, T.; Sciortino, F.; Stanley, H. E.; Angell, C. A. *Comp. Mat. Sci.,* **1995**, 4, 373.
(9) Roberts, C. J.; Panagiotopoulos, A. Z.; Debenedetti, P. G. *Phys. Rev. Lett.*, **1996**, 77, 4386.
(10) Mishima, O.; Takemura, K., Aoki, K. *Science* **1991**, 254, 406.
(11) Aasland, S.; McMillan, P. F. *Nature* **1994**, 369, 633.
(12) Angell, C. A. *Science* **1995**, 267, 1924.
(13) Ediger, M. D.; Angell, C. A.; Nagel, S.R. *J. Phys. Chem.* **1996**, 100, 13200.
(14) Schmidt-Rohr, K.; Spiess, H.W. *Phys. Rev. Lett.* **1991**, 66, 3020.
(15) Cicerone, M. T.; Ediger, M. D. *J. Chem. Phys.* **1995**, 103, 5684.
(16) Schiener, B.; Bohmer, R.; Loidl, A.; Chamberlin, R. V. *Science* **1996**, 274, 752.
(17) Mishima, O. *Nature* **1996**, 384, 546.
(18) Angell, C. A.; Borick, S.; Grabow, M. Proc. 17th Conf. Liquid and Amorphous Semiconductors (in press) 1997.
(19) Thompson, M. O.; Galvin, G. J.; Mayer, J. W.; Peercy, P. S.; Poate, J. M.; Jacobson, D. C.; Cullis, A. G.; Chew, N.G. *Phys. Rev. Lett.* **1984,** 52, 2360.
(20) Donovan, E. P.; Spaepen, F.; Turnbull, D.; Poate, J. M.; Jacobson, D. C. *J. Appl. Phys.* **1985**, 57, 1795.
(21) de Neufville, J. P.; Turnbull, D. *Faraday Society Discussions.* **1970**, 50, 182.
(22) Of course liquid-liquid transition are very well known in non-isotropic liquids, this is the field of liquid crystals.
(23) Kennedy, G. C.; Jayaraman, A.; Newton, R. C. *Phys. Rev.* **1962**, 126, 1363.
(24) Jayaraman, A.; Klement, W.; Kennedy, G. C. *Phys. Rev. Lett.* **1963,** 10, 387.
(25) Stell, G.; Hemmer, P. C *J. Chem. Phys.* **1973**, 56, 4274.
(26) Ferraz, A.; March, N. H. *Phys. Chem. Liq.* **1979,** 8, 289.
(27) Thiel, M.; Ree, F. H. *Phys. Rev. B.* **1993**, 48, 3591.
(28) Brazhkin, V. V.; Voloshin, R. N.; Popova, S. V.; Umnov, A. G. *J. Phys.: Condens. Matter* **1992**, 4, 1419.
(29) Voloshin, R. N.; Brazhkin, V. V.; Popova, S. V. *High Press. Res.* **1994**, 13, 51.
(30) Brazhkin, V. V.; Voloshin, R. N.; Popova, S. V. *JETP Lett.* **1989**, 50, 424.
(31) Brazhkin, V. V.; Popova, S. V.; Voloshin, R. N. *Phys. Lett. A* **1992**, 166, 383.
(32) Brazhkin, V. V.; Voloshin, R. N.; Popova, S. V.; Umnov, A. G. *Phys. Lett. A* **1991**, 154, 413.
(33) Brazhkin, V. V.; Popova, S. V.; Voloshin, R. N.; Umnov, A. G. *High Press. Res.* **1992**, 6, 363.
(34) Umnov, A. G.; Brazhkin, V.V. *High Press. Res.* **1995**, 13, 233.
(35) Angell, C. A.; Borick, S.; Grabow, M. *J. Non-Cryst. Solids* **1996**, 205, 463.
(36) Madura, J.; Impey, R. W.; Klein, M. *J. Chem. Phys.* **1983**, 79, 926.
(37) Poole, P. H.; Sciortino, F.; Essmann, U.; Stanley, E. *Nature* **1992**, 360, 324.
(38) Tanaka, H. *J. Chem. Phys.* **1996**, 105, 5099.

(39) Ha, A.; Cohen, I.; Zhao, X.; Lee, M.; Kivelson, D. *J. Phys. Chem.* **1996**, 100, 1.
(40) Yarger, J. L., Ph.D. Dissertation (Arizona State University, 1996).
(41) Sidorov, V. A.; Brazhkin, V. V.; Khvostantsev, L. G.; Lyapin, A. G.; Sapelkin, A. V.; Tsiok, O. B. *Phys. Rev. Lett.* **1994**, 73, 3262.
(42) Brazhkin, V. V.; Lyapin, A. G.; Khvostantsev, L. G.; Sidorov, V. A.; Tsiok, O. B.; Bayliss, S. C.; Sapelkin, A. V.; Clark, S. M. *Phys. Rev. B* **1996**, 54, 1808.
(43) Brazhkin, V. V.; Lyapin, A. G.; Popova, S. V.; Sapelkin, A. V. *JEPT Lett.* **1994**, 59, 131.
(44) Meade, C.; Jeanloz, R. *Phys. Rev. B* **1987**, 35, 236.
(45) Grimsditch, M. *Phys. Rev. Lett.* **1984**, 52, 2379.
(46) Hemley, R. J.; Mao, H. K.; Bell, P. M.; Mysen, B. O. *Phys. Rev. Lett.* **1986**, 57, 747.
(47) Williams, Q.; Jeanloz, R. *Science* **1988**, 239, 902.
(48) Meade, C.; Hemley, R. J.; Mao, H. K. *Phys. Rev. Lett.* **1992**, 69, 187.
(49) Frauenfelder, H.; Sligar, S. G.; Wolynes, P. G. Science **1991**, *254*, 1598.
(50) Brawer, S. A. *J. Chem. Phys.* **1981**, 75, 3522.
(51) Yarger, J. L.; De Paul, S. M.; Pines, A. University of California, Berkeley, unpublished data.
(52) Dupree, E.; Pettifer, R. F. *Nature* **1984**, 308, 523.
(53) Brawer, S. *J. Chem. Phys.* **1980**, 72, 4264.
(54) Leadbetter, A. J.; Wright, A. C. *J. Non-Cryst. Solids* **1972**, 7, 156.
(55) Narten, A. H. *J. Chem. Phys.* **1972**, 56, 1905.
(56) data extracted from computer simulations using both TIP4P and ST2.
(57) Mishima, O.; Calvert, L. D.; Whalley, E. *Nature* **1984**, 310, 393.
(58) Floriano, M. A.; Handa, Y. P.; Klug, D. D.; Whalley, E. *J. Chem. Phys.* **1989**, 91, 7187.
(59) Ponyatovsky, E. G.; Barkalov, O. I. *Mat. Sci. Rep.* **1992**, 8, 147.
(60) Meade, C.; Jeanloz, R. *Science* **1988**, 241, 1072.
(61) Itie, J. P.; Polian, A.; Calas, G.; Petiau, J.; Fontaine, A.; Tolentino, H. *Phys. Rev. Lett.* **1989**, 63, 398.
(62) Smith, K. H.; Shero, E.; Chizmeshya, A.; Wolf, G. H. *J. Chem. Phys.* **1995**, 102, 6851.

Chapter 17

Facts and Speculation Concerning Low-Temperature Polymorphism in Glass Formers

Daniel Kivelson[1], J.-C. Pereda[1], K. Luu[1], M. Lee[1], H. Sakai[1], A. Ha[1], I. Cohen[1], and Gilles Tarjus[2]

[1]Department of Chemistry and Biochemistry, University of California, Los Angeles, CA 90095
[2]Laboratoire de Physique Theorique des Liquides,Université Pierre et Marie Curie, 4 Place Jussieu, 75252 Paris Cedex 05, France

We have studied a low-T, "apparently-amorphous," rigid phase which is distinct from the glass and forms via a first order transition from the supercooled liquid at T's above T_g in the fragile glass-former triphenyl phosphite. We speculate that this phase may not be truly amorphous (but perhaps a mesoscopically modulated defect-ordered phase), that it may occur in all glass-formers (although not readily recognizable if the transition temperature lies below T_g), and that it might be the source of the giant "Fischer clusters" observed in orthoterphenyl and of the HDA phase of H_2O.

Discussion. In common with the work reported by Stanley*(1)* and by Yarger*(2)* at this symposium, we have been probing the metastable, "apparently" amorphous phases (distinct from the glass) that have been found in glass-forming systems at temperatures below the melting point. Whereas the other two talks are focused on liquids such as H_2O and SiO_2 that form tetrahedral networks, we focus on "fragile" liquids which do not form such networks.*(3-5)* This general phenomenon has been called*(6-11)* "polyamorphism" or simply "polymorphism;" although we have previously made use of the first of these terms, we have difficulty accepting it because we retain some doubt, as explained below, that these low-T phases are all truly amorphous, and for this reason we denote them as "apparently-amorphous."*(10,11)* This ambiguity arises because, although in large-angle x-ray studies the low-T phase appears to be amorphous, inconclusive low-angle x-ray studies leave the impression, as does the frustration-limited domain (FLD) theory mentioned below, that the low-T phase may not be truly amorphous.

Our interest in these low-T phases arises from the very fact that they have been discovered(*4-26*) and are not yet well-understood, as well as from the fact that they may be the phases that have been predicted by the recently introduced FLD theory as a *necessary consequence* of the behavior of supercooled liquids.*(27,28)* The FLD theory suggests that the low-T phases are defect-ordered with supermolecular (*i.e.*, larger than molecular or mesoscopic) characteristic lengths.*(29,30)* Whether these theoretically predicted phases are the same as those we have denoted as the observed "apparently-amorphous" phases has not as yet been determined. We shall say more about this below.

This low-T phase phenomenon involves a transition from supercooled liquid to an apparently-amorphous (rigid/solid) phase which we have called*(10)*, perhaps unfortunately, the "glacial phase." The glacial phase is to be distinguished (as explained below) from the "glass" which is distinctly amorphous, and of course, from the normal crystal which has clear long-range order. A glass is a supercooled liquid that has been cooled below the temperature known as the "glass-transition temperature," T_g; below T_g some relaxation processes (α-relaxations) are too slow to allow the liquid to equilibrate in the experimentally available time. Formation of a glass at T_g does not involve a true phase transition because if the experimental time is increased, T_g is decreased.*(3-5)* We, therefore, expect the glass (below T_g) to have much the same amorphous structure as the supercooled liquid just above T_g. The liquid to glacial phase transition, however, seems to be a true first order transition, and at least for systems such as that we have studied, takes place at temperatures well above T_g where the liquid, though quite viscous, still flows and relaxes readily in ordinary experiments.*(10)* An interesting aspect of this is that the transition temperature, T_D, at which the supercooled liquid transforms to the glacial phase, may lie above T_g in some systems and below in others. Indeed, we have found*(10,11)* $T_D > T_g$ in triphenylphosphite (TPP), and in this article we present evidence for why we believe T_D may possibly also lie above T_g in H_2O, but below T_g in orthoterphenyl (OTP). Of course, below T_g equilibration occurs so slowly that if $T_D < T_g$, the supercooled-to-glacial-phase transition is not likely to be found; however, as we discuss below, ubiquitous impurities which dissolve preferentially in the glacial phase could broaden the transition range sufficiently that a small degree of two-phase coexistence might be observed above T_D (and T_g) with the amount of glacial phase decreasing with increasing T. With this in mind, we hypothesize that the large clusters*(31)* reported in supercooled OTP (and in several other supercooled liquids as well)*(32-36)* could be examples of what might be called macroscopic "impurity-stabilized glacial inclusions" at temperatures above T_D, *i.e.*, at $T > T_g > T_D$.

There is also the possibility that the low-T phases reported*(1,2,4-9,12-26)* in tetrahedrally networked fluids may be related to the glacial phase, In Fig. 1 we present a speculative schematic phase diagram for TPP and OTP, while in Fig. 2 we draw what is believed to be the corresponding diagram for H_2O. We find it intriguing to associate the low-density amorphous phase (LDA) and the high-density amorphous phase (HDA) of H_2O with the supercooled liquid and glacial phase, respectively, of TPP.*(37-39)* More generally we speculate that the low-T apparently-amorphous glacial phase is actually a defect-ordered (metastable) phase to which (in accordance with the FLD theory) *all* supercooled liquids (barring crystallization) convert at sufficiently low T.*(27-30)* The presumed necessary existence of this low-T phase may, among other things, explain the avoidance of the Kauzmann catastrophe, *i.e.*, the apparent violation of the third law found by extrapolating heat capacity data of supercooled liquids to T's below T_g.*(40,41)*

Here we feel justified in presenting a rich blend of speculation and experimental results because the phenomena discussed have as yet been only marginally studied, and models, even though tentatively presented, are required in order to give direction to future studies.

Experimental Overview. Our experimental studies have been focused primarily on the fragile glass-former triphenyl phosphite (TPP).*(10,11)* This liquid is readily supercooled below its melting point ($T_m \approx 295$ K), but at about 238 K it usually crystallizes. However, if the sample is quick-quenched to T's below about 230 K, crystallization can be delayed for long periods of time, the delay increasing markedly

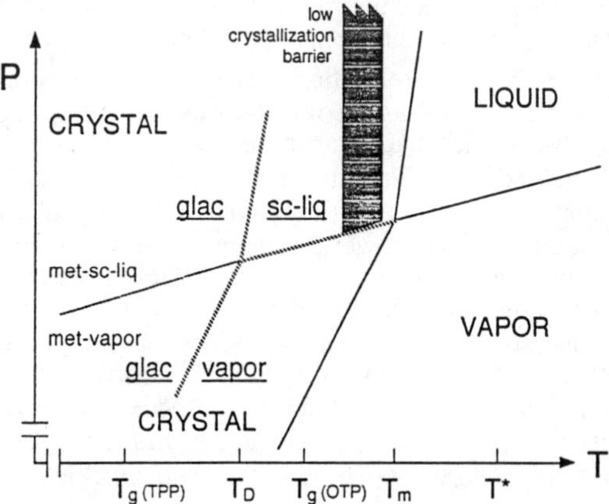

Fig. 1. Hypothesized Schematic Phase Diagram for Metastable TPP and OTP. Capitals indicate stable phases, underlined labels indicate metastable phases (sc-liquid being the supercooled liquid), and met-phases are phases that are metastable not only relative to the stable but to an underlined metastable state as well. Note that the apparent quadruple points connect phases of different relative stability.

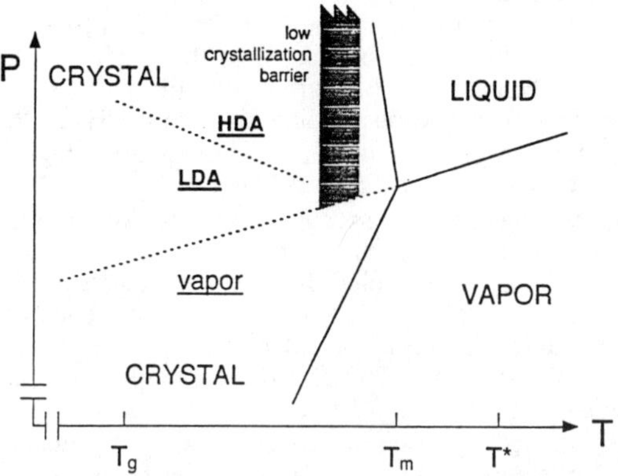

Fig. 2. Hypothesized Schematic Phase Diagram for Metastable H_2O. Capitals indicate stable phases and underlined labels indicate metastable phases. Note that the apparent quadruple point connects phases of different relative stability.

for decreasing T. In fact, below 230 K crystallization, when it occurs, appears to be heterogeneous, starting on the container walls, at vapor-liquid interfaces, and on various immersed objects. At temperatures between about 221 K and 213 K the liquid TPP appears to undergo a first order transition to a solid phase which we have denoted the glacial phase. The transition has been observed in several ways which have been reported in detail elsewhere;*(10,11)* here we merely summarize some of our results, both those already published and those based on our more recent, continuing studies.

Turbidity. One can determine the turbidity of a sample by measuring the intensity of the transmitted light, and one finds that with time liquid TPP at about 213 K turns more and more turbid; after about 4 hr the sample turns nearly opaque. Astonishingly, during the next 4 hr the sample "clears up," but the system is now rigid and sometimes exhibits cracks. We believe that the increasing turbidity is due to the formation of increasing numbers and size (diameters of the order of the wavelength of visible light) of glacial phase "clusters" and that the subsequent clearing occurs because the clusters coalesce into a homogeneous solid glacial-phase mass. At slightly higher T, at about 219 K, the entire process is accelerated and the clearing is less complete, *i.e.*, the system remains slightly turbid. The acceleration is presumably due to the fact that in the less viscous liquid dynamical processes are speeded up; the lack of complete clearing is attributed to incomplete conversion of the liquid to the glacial phase, *i.e.*, to less conversion at higher T. We dispel the possibility that the remnant turbidity (due to small dispersed liquid regions) that is found in the glacial phase when formed at higher T might be a consequence of poor structural ordering due to rapid phase transformation (and not to lower glacial/liquid ratios) by showing that the very clear glacial sample (formed at 213 K) became slightly turbid when warmed to about 220 K. Thus we interpret these turbidity data in terms of a liquid-glacial equilibrium, not just at a single phase transition temperature T_D, but one that has been spread over a temperature range from about 213 K to 222 K, with higher amounts of glacial phase at lower T. The turbidity experiments indicate that the two coexisting phases are interdispersed, but that the regions of uniform structure must be quite large, many of them being comparable in diameter to the wavelength of visible light. These uniform regions can be denoted as clusters, giant clusters since they consist of billions of molecules. Particularly when the amount of glacial phase is small, it is useful to think of the separate regions of glacial phase as "clusters," small ones in the sense that they are dispersed and give rise to turbidity, but gigantic in the sense that they are actually macroscopic one-phase regions. What is of particular interest to us here is that the coexistence of glacial and liquid phases found between about 213 K and 222 K seems to be an equilibrium (or very long-lived) phenomenon.

Most of our turbidity experiments were not sensitive to very small amounts of glacial clusters that might be present, and so it is possible that the temperature range in which small long-lived concentrations of clusters may exist is actually much broader than indicated above. Although here we are primarily concerned with apparent equilibrium properties, and not the dynamical ones, the rate of crystallization relative to that of glaciation can have great impact on our results, in particular, it might mask the presence of low concentrations of glacial clusters because such low concentrations of glacial clusters appear to catalyze heterogeneous crystallization. We reach this conclusion by noting that glacial samples that have been heated very slowly crystallize rather efficiently above 225 K, whereas liquid samples that have been quick-quenched to 233 K and then slowly cooled do not crystallize at all above 221 K, and they crystallize below 221 K only after exhibiting initial glacial phase growth. Interestingly, in some glaciated samples (but in samples of questionable purity) that have been rapidly heated above melting, we have observed (by means of light scattering) very low, but finite concentrations of long-lived giant clusters .*(11)*

These turbidity studies leave us with the question of why the apparently first order liquid-glacial phase transition should be spread over a temperature range, thereby behaving like a chemical equilibrium between solid-like giant clusters and molecularly dispersed liquid. It is difficult to envisage such a chemical-like equilibrium, involving such huge clusters, as evolving from the rather weak intermolecular forces unless these forces coordinate into a collective role such as in a phase transition. But if indeed the cluster formation is driven by a phase transition, then except at the transition temperature T_D, one should find only one phase or the other, not both, since the pressure is relatively unchanged in our experiments. Of course, if there is an impurity that dissolves differentially in the two phases, then the transition may be spread over a temperature range, and small concentrations of one phase (glacial, but now impure) may appear as dispersed, small, but macroscopic clusters. We chose to call these "impurity-stabilized macrscopic clusters or inclusions." Relatively low dopant levels could account for low cluster concentrations at temperatures quite far from T_D. If a dopant were to dissolve preferentially in the glacial phase, small concentrations of this dopant would give rise to low concentrations of impurity-stabilized macroscopic clusters at temperatures well above T_D.

Calorimetry. We carried out a number of studies to confirm that the supercooled-liquid-to-glacial transition was indeed first order. A liquid TPP sample was first brought to 213 K, allowed to glaciate, and then heated at 1 K per min, while another sample was brought to 213 K, merely allowed to equilibrate thermally for a few minutes, and then heated at the same rate.*(11)* The glaciated sample underwent a clear calorimetric and visual transition to the crystal at about 230 K, while the liquid sample turned to crystal at about 225 K. Similar results were obtained by von Miltenburg and Bok in their more precise adiabatic calorimetric studies.*(42)* We have described our calorimetric studies elsewhere,*(10,11)* but since the observations depend at least as much on dynamic as on thermodynamic features, here we focus on the fact that there are indeed differentiating thermodynamic signatures for the three phases: supercooled liquid (and glass), glacial, and crystal.

Despite our focus on thermodynamics, we wish to comment further on the dynamical behavior of crystallization. Above we proposed that the presence of glacial clusters might catalyze crystallization, and yet the calorimetric studies indicate that crystallization of unglaciated liquid occurs at a lower T than that of glacial phase. We believe that on the time scale of our calorimetric experiments the liquid has probably glaciated slightly and that the glacial clusters readily nucleate crystallization, whereas the glacial samples do not crystallize directly and must first melt appreciably before helping the liquid to crystallize. In this connection it is interesting that a liquid TPP sample equilibrated for about 20 min at 218 K and then brought to 223 K, crystallized in a few hours, whereas a similar sample equilibrated at 218 K for only 3 min did not crystallize at 223 K. Note also that when plunged into a bath at a T above the crystal melting point, a glacial sample originally at 213 K first crystallizes and then melts, whereas a liquid sample prepared at 213 K and not given time to glaciate before being plunged into the bath, does not crystallize at all.

Density. We find that the glacial phase is denser than the crystal phase.*(11)* This conclusion is based upon the observation that test tubes filled with glacial phase break when heated slowly, presumably because of the conversion of one rigid phase (glacial) to a less dense, rigid one (crystal).

NMR. To make sure that at the molecular level the three phases are truly different, we examined their NMR T_1 relaxation times and their x-ray scattering.*(11)* We prepared glacial and liquid samples, as described above, but instead of heating them at a constant rate, we quenched them to T's below $T_g \approx 180$ K; the supercooled liquid then became, by definition, a glass. We also took the crystal down to this low T. Each sample had a different phosphorus spin-lattice relaxation time (T_1), a fact that demonstrates that the three phases (crystal, vitreous and glacial) are distinct; this is especially interesting because it indicates that the glass (vitreous phase) and the glacial phase are distinct.

x-ray. The x-ray data were, at first glance,*(11)* less revealing than the NMR data. The x-ray pattern of the crystal (powder) revealed a full panoply of sharp rings (Bragg peaks), but the glacial phase was similar (though not identical) to the glass which was itself quite similar to the liquid at higher T's. These x-ray data justify the use of the term "amorphous" for the glacial phase, as well as for the glass and liquid, but they give little motivation for our addition of the modifier "apparently" in the term "apparently-amorphous" introduced above. A theoretical reason, mentioned above, to keep this modifier is that one might expect a low-T defect-ordered phase to have scattering patterns that could be difficult to interpret and which, under low-resolution and low sensitivity, might appear amorphous. As an example, these defect-ordered phases could be composed of one-dimensional mesoscopically-modulated structures, such as striped phases or density waves with supermolecular wavelengths.*(29,30)* For the present, at least, we believe that the question of whether the TPP glacial phase is "amorphous" or "apparently-amorphous" remains open.

Impurities. We have examined the role of impurities in TPP, in particular that of phosphoric acid, H_3PO_4, which seems to be present in many TPP samples. Instead of pursuing the difficult (and possibly unattainable) goal of obtaining a pure sample, we decided to study the effect of heavy doping with phosphoric acid. To date these studies have been merely exploratory and as yet inconclusive. However, heavily doped samples in which the phosphoric acid was well dissolved, did not glaciate (nor crystallize), but remained liquid even at 218 K; it is possible that the phase transition temperature T_D has been lowered below T_g by preferential distribution of phosphoric acid in the liquid phase, or at least lowered below the point where the dynamics are sufficiently rapid for glaciation to be observed. These experiments are continuing.

Although our understanding of the role of impurities is fragmentary and although impurities do not seem to affect the viscosity appreciably, they do affect the nature of the glaciation transition. As discussed above, impurities could account for the temperature spread over which glacial/liquid equilibrium is observed. In particular, impurities could account for the stabilization of limited macroscopic amounts of glacial phase (clusters) by attributing their formation to the collective effects of phase formation, and attributing the limitation of their growth (as a function of temperature) to the presence of impurities. As a next step, we hypothesize that the giant clusters (Fischer clusters) reported in orthoterphenyl (OTP) by Fisher and coworkers*(31)* may also be impurity-stabilized glacial-phase fragments, but that in OTP the transition temperature $T_D < T_g$; in this case, the transition would have been broadened above T_D, even above T_g. This could occur if the dopant dissolved preferentially in the glacial phase of OTP. Although the samples of OTP used by Fischer and coworkers*(31)* were highly purified and carefully handled, it would not take much impurity (perhaps even a small level that always tends to build up either through decomposition or polymerization or leaching from the container walls) to stabilize glacial-phase clusters at the very low cluster concentrations that were observed. It should be noted that the possibility that these large OTP clusters are large dielectric fluctuations associated with the approach to a low-T critical point (*i.e.*, to a continuous transition point) is negated by the observation that cluster-free OTP samples can be produced, and that it is only after clusters are initially produced via an apparently heterogeneous process that they seem to persist in equilibrium with the liquid;*(32-36)* this is indicative of a high barrier for glaciation, a property associated with first order transitions. It would appear that once the clusters are formed, they serve as catalyst for glacial nucleation thereby enabling the equilibrium at different temperatures to be readily established.

Experimental problems. Our experiments on TPP have a level of irreproducibility, largely associated, we believe, with the not-always-predictable nor controllable onset of heterogeneous crystallization, in part due to impurities.*(10,11)* A more detailed analysis will be published elsewhere. Despite the current level of irreproducibility, emphasized by the imprecision with which actual numbers are presented in this article, we believe the general information reported here to be robust.

Our current low-angle x-ray studies remain the most tantalizing and incomplete part of our study. With temperature control inadequate for phase transition studies and with limited access to the diffractometer, we have been able to obtain only limited scattering data on poorly characterized glacial-liquid TPP mixtures.

We have yet to check for anisotropy.

Theoretical Overview. The frustration-limited domain (FLD) model predicts a defect-ordered phase at low temperatures, a phase that appears to be a modulated low-dimensional phase with mesoscopic (supermolecular) correlation lengths.*(27-30)* We have hypothesized that the glacial phase in TPP is such a phase. Within the model, the liquid phase, when supercooled well below the crystal melting point, consists of random supermolecular domains, and the glacial phase is a phase in which these domains are somehow ordered. The FLD model ties these two phases together intimately. If this model, and its application to TPP, are valid, then the glacial phase is not truly amorphous, and so, in the framework presented, is at most "apparently amorphous;" the x-ray data are, as yet, inconclusive on this point. (A recent computer simulation of the FLD model by P. Viot and G. Tarjus exhibits a low-T striped phase.)

We have hypothesized further, based only on analogy, that the low-T non-vitreous forms of H_2O, *i.e.*, the HDA form, and other tetrahedrally networked liquids are also glacial phases, as described above. See Fig. 2. We have attributed the broadening of the liquid-glacial transition over a range of temperatures to the presence of nearly ubiquitous impurities, and within this picture we can offer a rationalization for the existence of giant Fisher clusters.*(31-36)* Although this latter description has yet to be directly tested experimentally, it provides an explanation for the formation of such large clusters, which are presumably held together by relatively weak intermolecular forces, by attributing the stability of the cluster to the collective effects associated with a phase transition, the phase growth limited by the presence of impurities.

Caution. The picture that we have drawn of the existence and properties of the glacial phase, of the glacial phase's universal connection to supercooled liquids, and of its role (stabilized by small amounts of impurity) in accounting for the macroscopic clusters that have been observed at T's above T_D, is a reasonable one, and currently the only one. But it clearly has a way to go before it is either verified or contradicted. We hope that the picture may serve as motivation for future study.

The FLD theory upon which our picture is based is built upon a model of a critical system subjected to long-range, weak frustration which causes the critical point, T*, (which lies above the crystal melting point) to be avoided; T* becomes the crossover point above which the liquid dynamics are molecular and below which they are collective.*(27,28)* In the only specific model of this kind that has been solved to date, that of the frustrated spherical model,*(29)* the transition to the modulated phase is continuous, and not, as observed, first order. It is possible that models not built on a lattice would yield first-order transitions. It is also possible that if the pressure were added as a field variable it would convert the transition to a first-order one.

Quite different pictures are also possible. If the low-T phase, in particular the glacial phase of TPP, is truly amorphous so that the liquid-glacial transition is truly an example of polyamorphism, then the relevant properties are controlled by low-T critical points and spinodals.*(37-39)* We have avoided reference to these phenomena, so the termination points of the phase curves in Figs. 1 and 2 do not necessarily have any significance.

Acknowledgments. We are grateful for the important comments and suggestions, as well as the cautionary advice, we have received from Steven A. Kivelson. We have benefited from discussions with Peter Poole and Eugene Stanley concerning their somewhat related approaches to the problems of polymorphism. We would also like to thank David Eisenberg, Duilio Cassio, and Bryan Vicento for their help with our recent x-ray studies. We would like to acknowledge the input from W. Steffen and A. Patkowski who kindly provided us with unpublished results on the formation of giant clusters in supercooled liquid systems, including OTP. We would like to thank the Research Corporation and the National Science Foundation for their support.

Literature Cited.

(1) Stanley, H. E.; L. L. Cruz; Harrington, S. T.; Poole, P. H.; Sastry, S.; Starr, F. W.; Zhang, R., These proceedings

(2) Yarger, J. L.; Angell, C.A.; Borick, S. S.; Wolf, G. H., These proceedings

(3) Angell, C. A., *J. Non-Crystalline Solids,* **1991**, *131-133*, 13.

(4) Mohanty, U., *Adv. Chem. Phys.* **1995**, *89*, 89.

(5) Rajagonopal, A. K; Ngai, K. L., in *Relaxation in Complex Systems;* Ngai, K. L.; Wright, G. B., Eds.; Naval Research Laboratory, Washington, DC: 1984; pg. 261.

(6) Angell, C. A., *Science* **1995,** *267,* 1924..

(7) Grimsditch, M., *Phys. Rev. Lett.* **1984,** *52,* 2379.

(8) Smith, K. H.; Shero, E.; Chizmeshya, A.; Wolf, G. E., *J. Chem. Phys.* **1995**, *102*, 6851.

(9) Mishima, O., *J. Chem. Phys.* **1994**, *100*, 5910.

(10) Ha, A.; Cohen, I.; Zhao, X-L., Lee, M.; Kivelson, D., *J. Phys. Chem.* **1996**, *100*, 1.

(11) Cohen, I.; Ha, A.; Zhao, X-L.; Lee, M.; Fischer, T.; Strouse, M. J.; Kivelson, D., *J. Phys. Chem.* **1996**, *100*, 8518.

(12) Aasland S.; McMillan, P. F., *Nature* **1994**, *369*, 633.

(13) Yaoita, K; Tsuji, K.; Katayama, Y.; Koyama, N.; Kikegawa,T.; Shimomura, O., *J. Non-Crystalline Solids* **1993**, *156-158*, 157.

(14) Angell, C. A.; Sare, E. J., *J. Chem. Phys.* **1970**, *52*, 1058.

(15) Mayer, E. J., *J. Phys. Chem.* **1983**, *89*, 3474.

(16) FacFarlane, D. R.; Angell, C. A., *J. Phys. Chem.* **1984**, *88*, 759.

(17) Mishima, D.; Calvert, L. E.; Whalley, E., *Nature* **1984**, *319*, 393.

(18) Johari, G. P.; Hallbrucker, A.; Mayer, E., *Nature* **1987**, *330*, 552.

(19) Hallbrucker, A.; Mayer, E., *J. Phys. Chem.* 1987**,** *91*, 503.

(20) Hallbrucker, A.; Mayer, E.; Johari, G. P., *J. Phys. Chem.* **1989**, *93*, 4986 .

(21) Johari, G. P.; Astl, G.; Mayer, E. J., J. Chem. Phys. **1990, 92,** 809.

(22) Bellissent-Funel, M.; Bosio, L.; Hallbrucher, A.; Mayer, E.; Sridi-Dorbez, R.; *Chem. Phys.* **1992**, *97*, 1282.

(23) Johari, G. P.; Ram, S.; Astl, G.; Mayer, E., *J. Non-Crystalline Solids*, **1990**, *116*, 282.

(24) Speedy, R. J., *J. Phys. Chem.* **1992**, *96*, 2322.

(25) Angell, C. A., *J. Phys. Chem.* **1993**, *97*, 6339.

(26) Johari, G. P.; Fleissner, G.; Hallbrucker,A.; Mayer, E., *J. Phys. Chem.* **1994**, *98*, 4719.
(27) Kivelson, D.; Kivelson, S. A.; Zhao, X-L. Z. Nussinov; Tarjus, G., *Physica A* **1995**, *219*, 27.
(28) Tarjus, G.; Kivelson, D.; Kivelson, S. A., These proceedings
(29) Chayes, L.; Emery, V. J.; Kivelson, S. A.; Nussinov, Z.; Tarjus, G., *Physica A*, **1996**, *225*, 129.
(30) Low, U.; Emery, V. J.; Fabricus, K.; Kivelson, S. A., *Phys. Rev. Lett.* **1994**, *72*, 1918.
(31) Fischer, E.W.; Meier, G.; Rabinau, T.; Patkowski, A.; Steffen, W.; Thonnes, W., *J. Non-Crystalline Solids* **1991**, *131-133,* 134.
(32) Fischer, E. W.; Becker, C.; Hagenah, J-U.; Meier, G., *Prog. Colloid. Polym. Sci.* **1989**, *80*, 198.
(33) Gerhartz, B.; Meier, G.; Fischer, E. W., *J. Chem. Phys.* **1990**, *92*, 7110 .
(34) Fischer, E. W., *Physica A* **1993**, *201*, 183.
(35) Patkowski, A.; Fischer, E. W.; Glaser, H.; Meier, G.; Nilgens, H.; Steffen, W., *Prog. Colloidal Polym.* Sci. **1993**, *91*, 35.
(36) Kanaya, T.; Patkowski, A.; Fischer, E. W.; Seils, J.; Glaser, H.; Kaji, K., *Acta Polymer*, **1994**, *45*, 137.
(37) Poole, P. H.; Sciortino, F.; Essmann, U.; Stanley, H. E., *Nature* **1992**, *360*, 324.
(38) Poole, P. H.; Sciortino, F.; Essmann, U.; Stanley, H. E., *Phys. Rev. E* **1993**, *48*, 3799.
(39) Poole, P. H.; Essmann, U.; Sciortino, F.; Stanley, H. E., *Phys. Rev. E.* **1993**, *48*, 4605.
(40) Kauzmann, W., *Chem. Rev.* **1948**, *43*, 219.
(41) Angell, C. A.; Sichina, W., *Ann. NY Acad. of Sciences* **1976**, *279*, 53.
(42) van Miltenberg, K.; Blok, K., *J. Phys. Chem.* **1996**, *100*, 16457 .

Chapter 18

Phase Diagram for Supercooled Water and Liquid–Liquid Transition

Hideki Tanaka

Department of Polymer Chemistry, Graduate School of Engineering, Kyoto University, Sakyo, Kyoto 606–01, Japan

Molecular dynamics simulations have been carried out at constant pressure and temperature to examine phase behaviors of supercooled water. The anomalies of supercooled water in thermodynamic response functions at atmospheric pressure, the phase transition between low and high density amorphous ices (LDA and HDA) and a fragile-strong transition are accounted for by reconciling an idea introducing a second critical point separating LDA and HDA ices with a conjecture that LDA is a different phase from a normal water, called water II. It is found that there exist large gaps around temperature 213 K in thermodynamic, structural and dynamic properties, suggesting liquid-liquid phase transition. This transition is identified as an extension of the experimentally observed LDA-HDA transition in high pressure to atmospheric pressure. In a new phase diagram, a locus of the second critical point is moved into negative pressure and the divergences are accounted for in terms of the critical point and the spinodal-like instability.

Water exhibits various anomalies in thermodynamic response functions such as heat capacity and isothermal compressibility. The heat capacity at constant pressure in ambient temperature is large compared with other liquid. The isothermal compressibility has a minimum around 319 K. Below this temperature, water becomes more compressible liquid as decreasing temperature (1). Those properties in supercooled state tend to diverge with a power law behavior when approaching to T_s, 228 K (2). Before reaching the divergence temperature, water always nucleates. The experimental limit of supercooling is T_2 (233 K) at atmospheric pressure. Therefore, it is impossible to investigate experimentally what happens at molecular level near T_s.

These anomalies have been accounted for by various ideas and conjectures. Among them, the most notable one was due to Speedy, which is called "stability-limit-conjecture" (3). He explained the divergence of thermodynamic properties in supercooled state in conjunction with the liquid-vapor spinodal. The liquid spinodal line begins at the liquid-gas critical point. In temperature-pressure plane, this line decreases monotonically with decreasing temperature and goes into negative pressure. The liquid spinodal line has a minimum at negative pressure and passes back to positive pressure as the temperature decreases further. The anomalous thermodynamic behavior of liquid water in the low temperature region can be related to such a reentrant spinodal line, where the thermodynamic properties diverge like approaching to a critical point. The metastable supercooled water is necessarily changed to stable ice below this temperature according to the stability limit conjecture.

Poole *et al.* carried out molecular dynamics (MD) simulations over a wide range of stable, metastable and unstable liquid-state points to find a reentrant of the spinodal line (4). But the liquid spinodal line decreases monotonically with decreasing temperature and does not reenter into the positive pressure region. Instead, they found a second critical point. They located the second critical point at around 120 MPa and concluded that the anomalies are related to the second critical point from which an LDA (low density amorphous) - HDA (high density amorphous) phase boundary appears (5). The phase diagram they proposed suggests that if crystallization does not intervene, supercooled water is further cooled to LDA without a phase transition.

LDA at 0.1 MPa is hard to be prepared by an ordinary method but can be made by hyperquenching or compression-decompression of ice. The glass transition temperature was observed at T_g, 136 K (6). LDA becomes liquid above T_g. When heated further, LDA changes to cubic ice above T_1, 150 K (7). Therefore, it is again impossible to examine the origin of the divergence from lower temperature side. At high pressure, LDA-HDA transition was experimentally observed. It exhibits pressure-hysteresis and is therefore first order phase transition (8). MD simulations successfully reproduced this transition (9). This ensures that the intermolecular interaction for water is sufficient to describe the low temperature and high pressure behaviors of water although temperature and pressure must be shifted to some degree when compared with experiment.

Speedy proposed the other conjecture that LDA has no continuous path from normal water at atmospheric pressure (10). LDA and HDA are liquid above T_g; the former is called "water II", which is a different phase from normal water. The free energy difference between LDA and ice at temperature T is assumed simply by $\Delta G\ (T) \leq \Delta H(T_1)$ - $T\ \Delta S(T_1)$ where the entropy difference $\Delta S(T_1)$ is unknown but the enthalpy difference $\Delta H(T_1)$ has been reported (7). The free energy of normal supercooled water above T_2 can be calculated from the experimental heat capacity. If those two lines touch in a narrow region between T_s and T_2, a first order transition occurs, which may not be observed experimentally. However, $\Delta S(T_1)$ must be much smaller than the estimated value from the model calculation. If it is a little larger, two free energy lines have an intersection above T_2 implying first order phase transition, which should be detected in laboratory experiment. But it has actually never been observed and the first order transition was rejected. Thus, LDA has no thermodynamically continuous path to normal water.

In order to examine the phase behavior of liquid water, Poole *et al.* (11) and Borick *et al.* (12) developed a lattice gas-like model setting up a relevant partition function. Those model calculations predict two different phase diagrams depending on the magnitude of hydrogen bond energy as an input parameter. Thus, these calculations contribute significantly to our understanding of the origin of the anomalies of supercooled water. But the problem as to the thermodynamic continuity of supercooled water down to LDA is not resolved by those studies.

Supercooled liquids are classified into strong and fragile liquids in temperature dependence of viscosity. Strong liquid has an Arrhenius type temperature dependence while viscosity of fragile liquid against temperature is described by VTF equation (13). Stillinger accounted for strong-fragile character in terms of a topography of potential energy surfaces (14). Strong liquid has a uniformly rough potential surface. On the other hand, a potential surface for fragile liquid is complicated; there are small potential energy basins inside a large crater. Potential energy surface that the trajectory cover depends on the temperature in fragile liquid. Water in normal supercooled state clearly belongs to fragile liquid. Angell proposed that if water II is a distinct phase, it could be strong liquid (13).

In order to gain insight into potential energy surfaces and strong-fragile classification, "quenching" turns out to be a useful tool (15). The trajectory moves in the configuration space as time elapses. A collection of the coordinates correspond to a certain point in configuration space. Starting from a point at any instant in configuration space, it is possible to find the corresponding local minimum, which is called inherent structure or Q-structure. With the aid of this concept, molecular motions can be described by the vibrational motions inside a potential well and a transition from one minimum to another. The free energy of a system is given as well by sum of the potential energy of the minimum, vibrational free energy inside the basin and the number of the minima.

In the present study, MD simulations are carried out at constant temperature and pressure in order to solve the problem as to whether a continuous path between supercooled water and LDA exists and to reconcile various conjectures with experimental evidence on the origin of the divergences at T_s. The difference from previous simulation study consists mainly in ensemble and simulation time. This difference gives rise to different phase behaviors. To justify the newly proposed phase diagram, various quantities are calculated and a self-consistency check is made in terms of Q-structures and the potential energy surfaces.

Results and analyses

MD simulations are performed at various temperatures. The pressure is fixed to 0.1 MPa using Nose-Andersen's constant temperature-pressure method (16,17). The number of molecules N is set to 216. The intermolecular interaction is described by TIP4P potential (18), which is truncated smoothly at 8.655Å. The long range correction for Lennard-Jones part is made. The simulation time ranges from 1 ns to 20 ns, depending on temperature. Those longer runs together with a different ensemble may predict a different location of the critical point. This in turn leads to a different phase diagram. A new phase diagram is proposed. To confirm this, other two sets of MD simulations are carried out where the pressures are set equal to ±200 MPa.

The potential energies of instantaneous (I-) structures are plotted in (Figure 1 (a)). The individual point is an average over 500 configurations. To remove a trivial temperature dependence, harmonic energy, $3RT$, is subtracted. The potential energy decreases almost linearly as decreasing temperature from room temperature to 233 K. There is a large gap in potential energy between 233 and 213 K. The gap is twice as large as the energy difference extrapolated from the slope. On the other hand, the potential energy change is small below 213 K. A fairly abrupt change at +200 MPa is again seen between 193 and 173 K. The difference is somewhat smaller compared with that at 0.1 MPa. On the other hand, the potential energy at -200 MPa changes continuously with temperature. Those seem to suggest a liquid-liquid transition in high pressure but no transition in negative pressure.

The energy decrease (Figure 1 (a)) arises from decrease in both the anharmonic energy and the energy of ground state (Q-structure). The potential energies for Q-structure are give in (Figure 1 (b)). An interesting feature is the fact that the potential energy significantly decreases with decreasing the temperature in the range from 298 K to 233 K. This leads to a conclusion that water in this temperature range is a fragile liquid. Liquid water at room temperature has a continuous path to a supercooled state down to 233 K. It undergoes a transition around 213 K to another state. The potential energy of the new state is substantially (1.5 kJ mol^{-1}) lower than that at 233 K in Q-structure. The potential energy in Q-structure at 193 K is almost the same as the potential energy in Q-structure at 213 K. This lower energy state below 213 K has the potential energy higher by 1 kJ mol^{-1} than that of proton-disordered cubic ice, -55.9 kJ mol^{-1}. If this lower energy phase is identified as LDA, the energy difference in our calculation is consistent with the measured heat release, 1.3 kJ mol mol^{-1} at T_1, 150 K.

A simple extrapolation of the potential energy for Q-structure above 233 K intersects with the energy of cubic ice in its Q-structure at around 160 K. Unless there is a break in the potential energy plot between 233 and 160 K and the potential energy becomes less sensitive to temperature, the potential energy of liquid state becomes lower than that of ice. In order to avoid this, water may undergo a kind of transition and the potential energy in Q-structure should be almost constant against temperature. This is indeed the case at 0.1 MPa (also +200 MPa). This means that water blow 213 K at 0.1 MPa (173 K at +200 MPa) in TIP4P model is a strong liquid. In contrast, the potential energy decrease at -200 MPa ceases around 170 K without such a transition.

In the present simulation, the density of the system is allowed to fluctuate. A similar plot for the density is given in (Figure 2). An abrupt change is again found in density at 0.1 MPa and +200 MPa while no appreciable temperature dependence is seen in the systems under negative pressure, -200 MPa. The conclusion derived from the results for potential energy plots is confirmed. As seen from the figure, the gap in density is larger at high pressure than at atmospheric pressure. The phase behaviors are well characterized by the density difference as in the case of usual gas-liquid phases. Therefore, the density difference can be the most appropriate order parameter.

A new phase diagram of metastable water is shown in (Figure 3). This figure is similar to the picture of Stanley and coworkers (4). A main difference is a locus of the second critical point. It is located in negative pressure region. At atmospheric pressure, HDA is changed to LDA phase when the temperature goes down to the intersect with the HDA→LDA spinodal line. The divergences in supercooled water are ascribed to the spinodal instability and/or the existence of the second critical point. It is reasonable that normal water is regarded as water at temperature above the phase boundary. Water II is different phase from normal water. When water is cooled below the spinodal line, it changes to LDA. In practice, crystallization to ice at T_2 prevents us from observing the transition. We can observe only the symptom of the spinodal instability and/or the effect of the critical point in thermodynamic response functions. It is well known that the temperature obtained with TIP4P potential should shift upward by 15 K in order to compare with experiment. That the divergences occur experimentally at 228 K is in good correspondence to the discontinuity at about 213 K in our MD simulation.

It is essential to show water below 213 K is neither ice nor is partially crystalline form. We will confirm this by examining structure factors. The structure factor for liquid water at 193 K, is shown together with that for ice at 213 K in (Figure 4 (a)). Cubic ice has sharp peaks characteristic of solid phase. On the

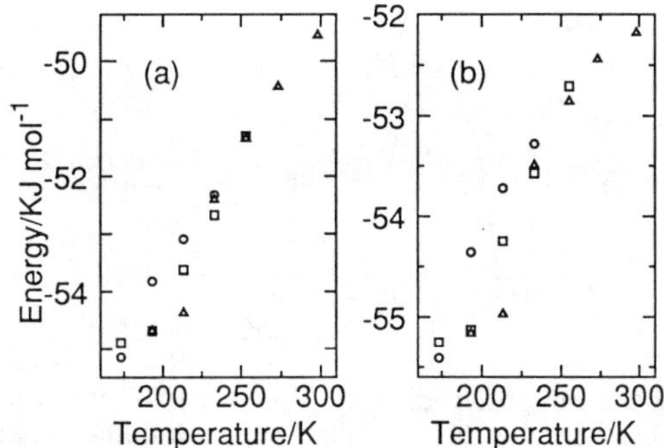

Figure 1. (a) Potential energy of I-structure; (b) Q-structure in kJ mol^{-1}. Triangle; pressure *p*= 0.1 MPa, circle; *p*= + 200 MPa, square; *p*= -200 MPa.

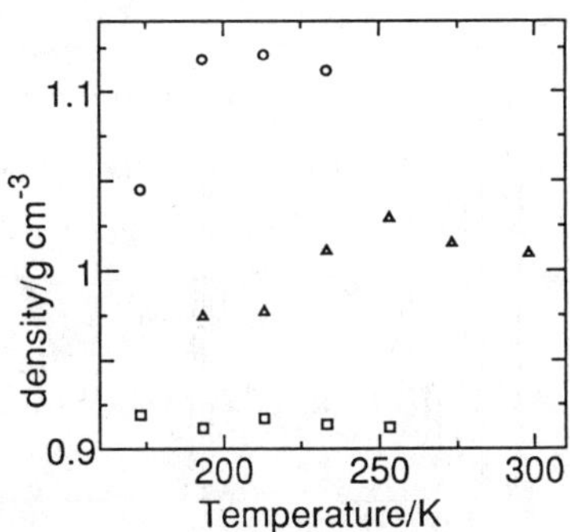

Figure 2. Densities at various temperatures in g cm^{-3} averaged over 500 configurations. Triangle; pressure *p*= 0.1 MPa, circle; *p*= + 200 MPa, square; *p* = -200 MPa.

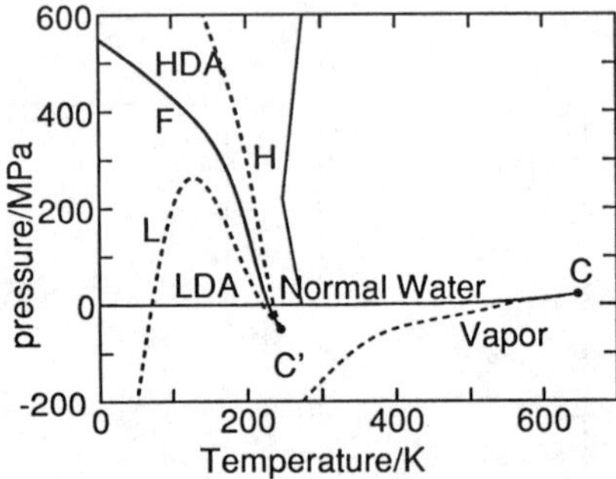

Figure 3. Schematic phase diagram for water. Thin solid lines are coexistence lines of stable phases which separate liquid water from vapor and ice from liquid water. Thin dashed line is liquid spinodal. Boundaries of metastable states are shown by heavy solid line (F) for the coexistence line and by heavy dashed lines (L and H) for LDA spinodal above which LDA becomes unstable and for HDA spinodal below which HDA becomes unstable. The gas-liquid critical point is shown by C. The second critical point is denoted by C'.

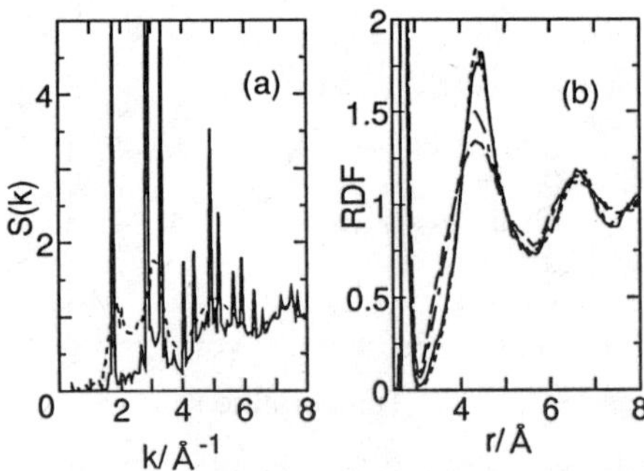

Figure 4. (a) Structure factors of ice at 213 K (solid line) and water at 193 K (dotted line); (b) radial distribution functions for oxygen-oxygen of water at 193 K (solid line), 213 K (dotted line), 233 K (dash-dot line) and 298 K (dashed line).

other hand, the structure factor of water at 193 K has a halo diffraction pattern, typical of liquid or amorphous state. The radial distribution functions of water for Q-structure are given in (Figure 4 (b)). As lowering the temperature from 298 K down to 233 K, only small difference is seen in the second peak. The radial distribution function at 213 K is clearly different from that 233 K, in particular the separation of the second peak from the first one is more distinct. On the other hand, a difference between 213 K and 193 K is negligibly small in Q-structure, which implies that structural change below 213 K is quite small and the system at 213 K is the same phase as that at 193 K.

The distributions of pair interaction energy for Q-structures are depicted in (Figure 5). Only pairs of water molecules separating less than 3.5 Å are taken into consideration. A fairly large number of pairs have interaction energy higher than -10 kJ mol^{-1}, which are regarded as only loosely or not hydrogen-bonded pairs. The decrease in temperature reduces the number of high energy defect pairs while the distribution of strongly hydrogen bonded pairs remains almost unchanged. When the temperature decreases further, the distribution for hydrogen bonded part changes and most of defects disappear. Although water below 213 K does not have a periodic molecular arrangement, local structure and tetrahedral connectivity are much more developed.

In order to examine a structural difference in terms of hydrogen bond connectivity, a simple comparison is made here by calculating hydrogen bond number distributions. Use of hydrogen bonds in Q-structure is preferable, eliminating apparent temperature dependence of hydrogen bond number. A definition of hydrogen bond depends on the potential energy for a pair of molecules. If the interaction energy is below the criterion, -14 kJ mol^{-1}, the pair is regarded as a hydrogen-bonded pair. The distribution of hydrogen bond number per molecule is listed in (Table I). Number of four hydrogen-bonded species increases drastically upon the transition from HDA to LDA phase at pressures 0.1 and 200 MPa. On the other hand, its increase is rather gradual with decreasing temperature at -200 MPa. In (Table II), a difference in hydrogen bond network is examined by calculating numbers of hydrogen bonded rings. If the system containing 216 water molecules is frozen to cubic ice, it is composed of 432 cyclic hexamers only. In liquid state where thermal energy is removed, the ring size distributes in the range from 3 to 7. The most dominant species is still hexagonal ring but the number of cyclic pentamers is comparable to the number of cyclic hexamers, especially above the transition temperature.

The fact that almost all water molecules are tetrahedrally coordinated species in LDA phase is very important to explain why addition of a small amount of salt, alcohol or hydrogen peroxide prevents a system from freezing to ice. Those substances interact strongly with water and take part in the hydrogen bonded network at room temperature. A number of defects in normal water allow the solutes to be dissolved by forming incomplete hydrogen bond network. Those defects are compatible with hydrogen bond network in normal water but incompatible with water II. Those solutes break the almost perfect tetrahedral structure of water II and water is no longer the low entropy LDA structure.

The density of state for intermolecular vibrational motions are compared, which are calculated by performing a normal mode analysis. Those for I- and Q-structures are given in (Figure 6). Difference between those at 298 K and 233 K is insignificant. Difference is large between those at 233 K and 213 K. Lowering the temperature by only 20 K reduces number of modes between 300 to 500 cm^{-1}. Those are mixed modes composed of translations and rotations of individual molecules and lead to a neighboring Q-structure with small excitation energy. In

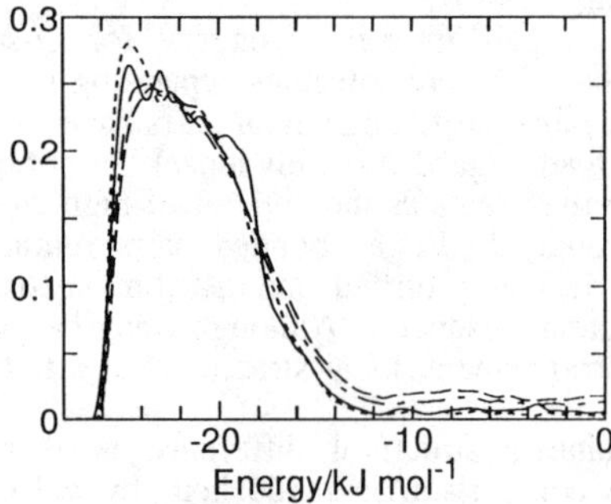

Figure 5. Distributions of pair interaction energy for Q-structures at 193 K (solid line), 213 K (dotted line), 233 K (dash-dot line), 298 K (dashed line).

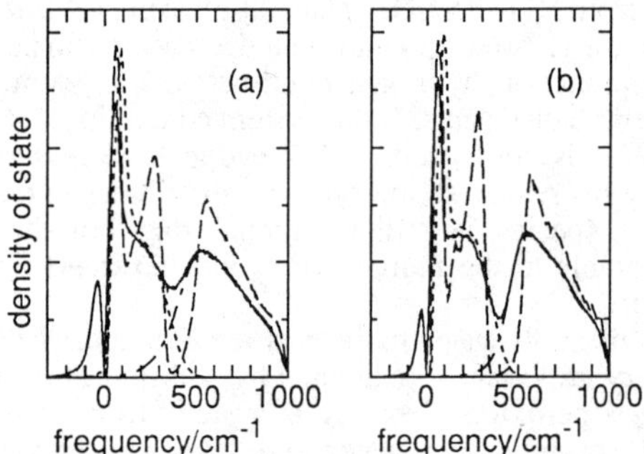

Figure 6. Densities of state for intermolecular vibrational motions (a) at 233 K; (b) at 193 K. I-structure (solid line), Q-structure (dashed line), translational part of I-structure (dotted line) and rotational part of I-structure (dash-dot line).

the case of I-structures, there exist some imaginary frequency modes, which are plotted in negative frequency region of the figure. Those modes are unstable modes and important to the diffusional motions. The percentages of imaginary modes are given in (Table III.) The number of imaginary modes increases with temperature. The Hessian matrices for I-structures are divided into translational and rotational parts and are diagonalized separately. The number of imaginary modes, especially for LDA phase, is extremely reduced by this separation. This indicates that motions in LDA are much more collective and translational and rotational motions are strongly coupled when a transition from a basin to another is induced by unstable modes.

Table I. Hydrogen bond number distribution per molecule at temperature *T*(K) and pressure *p*(MPa)

T	1	2	3	4	5
		p=0.1			
233	0	0.014	0.144	0.802	0.039
213	0.0	0.003	0.060	0.922	0.015
193	0.0	0.004	0.070	0.910	0.016
		p=200			
233	0.001	0.018	0.192	0.741	0.048
213	0.0	0.015	0.175	0.766	0.044
193	0.000	0.010	0.151	0.806	0.033
173	0.000	0.004	0.090	0.883	0.023
		p=-200			
233	0.0	0.007	0.098	0.872	0.022
213	0.0	0.003	0.066	0.918	0.013
193	0.0	0.000	0.022	0.970	0.007

Table II. The numbers hydrogen bonded rings in water at temperature *T* (K) divided by the number of hexagonal rings in cubic ice

T	3	4	5	6	7
193	0.000	0.026	0.1912	0.330	0.059
213	0.000	0.018	0.2033	0.311	0.048
233	0.000	0.041	0.1745	0.230	0.057
298	0.001	0.055	0.1579	0.187	0.053

The mean square displacements (MSDs) from long time MD simulations are linear in time. This ensures the diffusion is normal. The self-diffusion coefficients calculated from the slope of the MSDs are 2×10^{-6}, 9×10^{-8} and 6×10^{-8} cm^2 sec^{-1} at temperature 233, 213 and 193 K, respectively. The transition observed around 213 K is not a glass-transition since molecules even at 193 K still move more than intermolecular distances during our simulation run. The MSDs in lower temperatures (93, 113, 133 and 153 K) are also examined by microcanonical MD simulations whose initial configurations are Q-structures of either LDA or HDA form. HDA has a larger amplitude (Debye-Waller factor) than LDA as shown in (Figure 7). The MSDs both for HDA and LDA are of oscillatory character in short time. LDA structure is more stable than HDA against heating. An overshoot for LDA is more prominent than for HDA. A similar but more pronounced difference for SiO_2 was reported (13).

Table III. Percentage of imaginary modes for I-structures at temperature *T*(K). Full (f), translational part (t) and rotational part (r) Hessian matrices are diagonalized

T	f	t	r
193	3.1	0.01	0.08
233	6.4	0.33	0.53
298	11.3	2.7	2.5

In our phase diagram, water changes to LDA through first order transition, which is in opposition to Speedy's conjecture. It is necessary to check the self-consistency of our phase diagram and show a possibility of low entropy of LDA. The free energy is composed of the potential energy at its minimum, and the vibrational free energy and the configurational entropy. The vibrational free energy is divided into the harmonic, f_h, and anharmonic, f_a, free energy contributions. Our simulation study is based on classical mechanics and the term, f_h, is easily calculated. It is, however, difficult to calculated a reliable free energy arising from the anharmonic vibrations. This term may amount to a few kJ mol^{-1}. However, the difference between LDA and HDA is expected to be small. The configurational entropy, s_c, is dependent on the temperature in the case of fragile liquid. This value changes with the phase transition. No simple way to calculate reliable s_c is known.

If the phase diagram is correct, the free energy of LDA must be equal to that of HDA at T_s. This is because T_s is close to the transition temperature. The transition temperature is tentatively set to 213 K. There is another equilibrium where ice and water exists simultaneously. Neglecting an insignificant pV term at atmospheric pressure, we can write down two equilibrium conditions.

$$u^i + f_h^i + f_a^i - T_m s_c^i = u^h + f_h^h - f_a^h - T_m s_c^h \tag{1}$$

and

$$u^l + f_h^l + f_a^l - T_m s_c^l = u^h + f_h^h - f_a^h - T_m s_c^h \tag{2}$$

where superscripts *l*, *h* and *i* stand for LDA, HDA and ice. The signs of the unknown term f_a are all negative. In general, the anharmonic free energy of liquid state is lower than that of solid at the same temperature. Therefore, the inequality

$$f_a^i - f_a^h \geq 0 \tag{3}$$

must hold at all temperatures. It is reasonable to adopt a similar inequality to the relation between LDA and HDA phases as

$$f_a^l - f_a^h \geq 0 \tag{4}$$

Thus, it becomes possible to evaluate the upper bounds of the configurational entropy changes, $\Delta s_c = s_c^h - s_c^i$ and $\Delta s_c' = s_c^h - s_c^l$. That is,

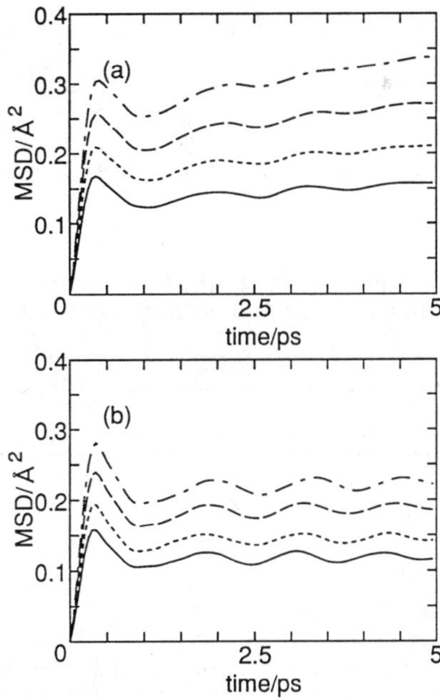

Figure 7. Mean square displacement for water at 93 K (solid line), 113 K (dotted), 133 K (dashed line) and 153 K (dash-dot line). MD simulations are started from HDA phase (a); from LDA phase (b).

$$\frac{\left(u^h - u^i + f_h^h - f_h^i\right)}{T_m} \geq \Delta s_c \tag{5}$$

and

$$\frac{\left(u^h - u^l + f_h^h - f_h^l\right)}{T_s} \geq \Delta s_c' \tag{6}$$

where we adopt the melting point for TIP4P water, $T_m \approx 240$ K (or a little higher). This predicts the upper bound of the configurational entropy of melting, 7.7 J K^{-1} mol^{-1}. The entropy of the transition from HDA to LDA is approximately 6.1 J K^{-1} mol^{-1}. It is again reasonable to assume that $\left(f_a^i - f_a^h\right)/T_m \geq \left(f_a^l - f_a^h\right)/T_s$. Thus the entropy difference between ice and LDA is very small, only 1.6 J K^{-1} mol^{-1} or less. This indicates that the number of defects in LDA is much smaller than in HDA. A further detail on the origin of small entropy for LDA is not fully understood at present time but this matches the condition that LDA undergoes first-order transition. A recent experimental study also supports such a small entropy difference (19).

Since water at room temperature belongs to fragile liquid, its potential surface should be complicated structure; a large crater includes many small potential basins. The bottom of craters can be obtained by the coarse-graining of a series of quenched structures in a similar way as in realization of V-structure (20). The coarse-graining consists of two steps, (i) the averaging the successive Q-structures over $\Delta\tau$ and (ii) subsequent quenching. The average process is performed both for translational and orientational parameters over $\Delta\tau$ (=2 ps). The coarse-graining is repeated for (*n*=) 30 times. The potential energies for Q- and coarse-grained structures are tabulated in (Table IV). The potential energy after coarse-graining is almost always lower than the simply quenched structures (*i.e.*, *n*=0). This implies that the coarse-graining procedure eliminates higher potential energy structures in the same sense that the V-structure is obtained by eliminating large distortions (21). The trajectory seems to stay inside a crater for a while to execute jumping motions among small basins.

Table IV. Total and Lennard-Jones (LJ) part of potential energy (in kJ mol^{-1}) for coarse-grained structure after *n* cycles at 0.1 MPa and 298 K

n	0	10	20	30
total	-52.20	-52.70	-52.80	-52.89
LJ	10.69	11.54	11.70	11.84

Discussion

There are two ways for assignment of the location of the second critical point. One is due to Stanley's group (4) and another is proposed by Tanaka (22, 23). The latter phase diagram is similar to the former except for a locus of the critical point: It locates at around 120 MPa according to Stanley and coworkers while it is moved into a certain negative pressure region in the latter. Despite the similarity in phase diagram, this difference is serious as to whether normal supercooled water has a thermodynamically continuous path to LDA phase. Therefore, this discrepancy must be resolved. There are many sources which make a difference in simulation results; difference in ensemble, treatment of long range interaction, simulation time, intermolecular interaction, *etc.* No justification seems to be possible as to which simulation is better to reproduce real water. A simple comparison with experimental evidence is only a way to discriminate which interpretation is more plausible. At higher pressure, no reliable data are available but no clear divergence in heat capacity is observed (2). This suggests there is no critical point at high pressure such as 200 MPa. Another evidence to support this is based on Mishima's experiment on compression and decompression of water observing LDA-HDA first order transition (8). At that pressure about 200 MPa, the experimental density difference is large, approximately 0.2 g cm^{-3}, which is to be compared with the present simulation result at 200 MPa, 0.07 g cm^{-3}. The density difference can be an order parameter to describe phase behaviors near a critical point even for LDA-HDA phase equilibrium as seen in Figure 2. The large order parameter suggests that the second critical point exists well below the transition pressure approximately 120 MPa in the experiment. The above two facts support the view that the second critical point locates in (though shallow) negative pressure region. It should be noted that water below 213 K has quite difference properties from water below 213 K, irrespective of the exact location of the critical point.

The homogeneous nucleation temperature line is almost parallel to the divergence temperature line on *T-p* plane. This suggests that there may be a common origin to the divergence and nucleation. If the divergence of density fluctuation is caused by approaching the spinodal line, it may give rise to the nucleation. Since impurities hydrogen-bonded with water are incompatible with LDA structure, no liquid-liquid phase transition and associated phenomena are expected in aqueous solutions of those solutes. This may explain why these solutions are cooled down to the glass transition temperature.

An experiment on correlation length (24) seems to be inconsistent with our interpretation and others (3,4,22,23): A small correlation length in experiment is incompatible with either view wherever the second critical point locates; below or above the atmospheric pressure. At this moment, it is impossible to account for both x-ray diffraction and thermodynamic measurement.

Acknowledgment

The author thanks Professors, I. Ohmine, M. Sasai, H.E. Stanley, H. Kanno and Drs. O. Mishima, R. Yamamoto, Y. Tamai, K. Koga for helpful discussion and I. Okabe for providing cubic ice structures. This work is supported by Grant-in-Aid from the Ministry of Education, Science and Culture, the Computer Center of Institute for Molecular Science, and Yoshida Foundation for Science and Technology.

Literature cited

1. Eisenberg D; Kauzmann, W. *The Structure and Properties of Water* ; Oxford Univ. Press: London, **1969**.
2. Angell, C.A. In *Water - A Comprehensive Treatise* ; Franks, F.; Plenum, New York, NY, **1981**, Vol. 7; pp. 1-76.
3. Speedy, R.J. *J. Phys. Chem.*. **1982**, 86, 982.
4. Poole, P.H.; Sciortino, F.; Essmann, U.; Stanley, H.E. *Nature*, **1992**, 360, 324.
5. Mishima, O.; Colvert, L.D.; Whalley, E. *Nature*, **1984**, 310, 393.
6. Johari, G.P.; Hallbrucker, A.; Mayer, E. *Nature*, **1987**, 330, 552.
7. Handa, Y.P.; Klug, D.D.; Whalley, E. J. *Chem. Phys.* **1976**, 84, 7009.
8. Mishima, O. *J. Chem. Phys*, **1994**, 100, 5910.
9. Poole, P.H.; Essmann, U.; Sciortino, F.; Stanley, H.E. *Phys. Rev.* E **1993**, 48, 4605.
10. Speedy R.J., *J. Phys. Chem.* 1992, 96, 2322.
11. Poole, P.H.; Sciortino, F.; Grande, T.; Stanley, H.E.; Angell, C.A. *Phys. Rev. Lett.* **1994**, 73, 1632.
12. Borick, S.S.; Debenedetti, P. G.; Sastry, S. *J. Phys. Chem.* **1995**, 99, 3781.
13. Angell, C.A. *Science*, **1995**, 267, 1935.
14. Stillinger, F.H. *Science*, **1995**, 267, 1924.
15. Stillinger, F.H.; Weber, T.A. *Science*, **1984**, 225, 983.
16. Nosé, S. *Mol. Phys.* **1984**,52, 255.
17. Andersen, H.C. *J. Chem. Phys.* **1980**, 72, 2384.
18. Jorgensen, W.L.; Chandrasekhar, J.; Madura, J.D.; Impey, R.W.; Klein M.L. *J. Chem. Phys.* **1983**, 79, 926.
19. Speedy, R.J.; Debenedetti, P. G.; Smith, R.S.; Haung, C.; Kay, B.D. *J. Chem. Phys.* **1996**, 105, 240.
20. Hirata, F.; Rossky, P.J. *J. Chem. Phys.* **1981**, 74, 6867.
21. Tanaka, H.; Ohmine, I. *J. Chem. Phys.* **1987**, 87, 6128.
22. Tanaka, H. *Nature*, **1996**, 380, 328.
23. Tanaka, H. *J. Chem. Phys.* **1996**, 105, 5099.
24. Xie, Y.; Ludwig, K.F., Morales, G.; Hare, D.E.; Sorensen, C.M. *Phys. Rev. Lett.* **1993**, 71, 2050.

Chapter 19

The Liquid–Liquid Critical-Point Hypothesis

H. E. Stanley[1], S. T. Harrington[1], F. Sciortino[2], P. H. Poole[3], and S. Sastry[4]

[1]Center for Polymer Studies and Physics Department, Boston University, Boston, MA 02215
[2]Dipartimento di Fisica, Università di Roma La Sapienza, Istituto Nazionale di Fisica Nucleare, Sezione di Roma I, Piazzale Aldo Moro, 00185 Rome, Italy
[3]Department of Applied Mathematics, University of Western Ontario, London, Ontario N6A 5B7, Canada
[4]Department of Chemical Engineering, Princeton University, Princeton, NJ 08544

We discuss the hypothesis that, in addition to the known critical point in water (below which two fluid phases—a lower-density gas and a higher-density liquid—coexist), there exists a "second" critical point at low temperatures (below which two liquid phases—a higher-density liquid and a lower-density liquid—can coexist). We also discuss briefly some of the evidence relating to this hypothesis. This evidence is rather tentative at the present time, and is largely based on a growing number of computer simulations using the ST2 and TIP4P intermolecular potentials. We also discuss selected experimental results that are consistent with this hypothesis.

Although water has been the topic of considerable research over the past 100 years, few would dare to claim that its peculiar properties are completely understood [1, 2]. Standard liquid theories fail to explain its dynamical and thermodynamic properties, which differ from those of most other liquids [3–6]. Many workers believe that a central role is played by the water molecule's ability to form hydrogen bonds and to create a tetrahedrally coordinated open network in both the solid and the liquid phases [7–13].

In experiments, it is generally observed that the peculiar properties of water become more pronounced when the liquid is supercooled or stretched [5, 6]. In particular, thermodynamic response functions such as the specific heat, isothermal compressibility and thermal expansivity, as well as various relaxation times, seem to exhibit a power law type singularity at a certain temperature $T_A \approx 228$K, similar to the behavior commonly encountered when approaching a second order phase transition [6, 14–16]. Accordingly, an analysis that attempts to elucidate the origin of the anomalies in water might wisely focus on the supercooled and superstretched regions in the phase diagram. Unfortunately, detailed experimental studies of deeply supercooled and stretched water remain difficult to perform due to the increasing probability of ice or gas nucleation as deeper metastable states are probed [17]. As a result, theories for the thermodynamic behavior of metastable water, such as the Speedy-Angell "stability limit conjecture" [15, 18–20], have yet to be definitively confirmed or refuted by experimental measurements. On the other hand, molecular dynamics (MD) simulations [21, 22] of supercooled and stretched liquid water are increasingly feasible due to dramatic increases in computational power.

The applicability of MD simulations as a theoretical tool to investigate the properties of liquid water has been demonstrated by many research studies. The landmark first MD simulations of water were performed in the late 1960's [23, 24]. Since then, computer simulation studies on water and aqueous solutions have immensely broadened our knowledge about this ubiquitous and unusual liquid. MD simulations are capable of mimicking a wide range of measurable properties of water, from thermodynamics [25] to structure [26] and microdynamics [27]. Larger systems (on the order of 10^5 molecules) may be simulated using new algorithms, like those found in the fast multipole methods [28].

A natural extension to our work is to consider other tetrahedrally-coordinated liquids. Examples of such systems are SiO_2 and GeO_2, known for their geological and technological importance [29, 30]. Both of these systems display features in their equations of state similar to those found in simulations of water and that can be traced to their tetrahedral configurations.

The Hypothesis

Science can progress by making hypotheses, and then testing them. Accordingly, in this talk we will examine the hypothesis that the unusual properties of liquid water are related to the existence of a "second" critical point. First we present the hypothesis, then we explain why this hypothesis is worthy of critical examination, and finally we briefly review some of the facts that are consistent with the hypothesis.

Water has a well-studied critical point C with coordinates (T_C, P_C, ρ_C), where (to one significant figure) $T_C \approx 600$K is the critical temperature, $P_C \approx 20$MPa is the critical pressure, and $\rho_C \approx 0.3$ is the critical density characterizing C. We hypothesize that there may exist a second critical point C' with coordinates $(T_{C'}, P_{C'}, \rho_{C'})$, where (also to one significant figure) $T_{C'} \approx 200$K, $P_{C'} \approx 100$MPa, and $\rho_{C'} \approx 1$.

Just as C separates a 1-fluid region (a single phase) above T_C from a 2-fluid region (two phases) below, so also C' separates a 1-liquid region above $T_{C'}$ from a 2-liquid region below $T_{C'}$. The two fluid phases that can co-exist below C—called liquid and gas—differ in volume per molecule $\bar{V}$. The specific volume difference between liquid and gas phases is a suitable "order parameter," which continuously approaches zero as one approaches C displaying a power law singularity characterized by the critical exponent β. Similarly, the two liquid phases that can co-exist below C' differ in specific volume; we shall call these two liquid phases high-density liquid (HDL) and low-density liquid (LDL). Again, this density difference, because it approaches zero as one approaches C', is an order parameter characterizing the liquid-liquid critical point.

The phases differ not only in their volume per molecule $\bar{V}$; they also differ in the entropy per molecule $\bar{S}$. Below T_C, the gas has the larger entropy, while below $T_{C'}$ the HDL phase has the larger entropy. Hence, using the Clausius-Clapeyron relation, the coexistence line in the PT phase diagram has positive slope for C but negative slope for C'.

Why Care?

One might think it senseless to invest much energy studying the critical point, as it is only a single point in an entire phase diagram. However we know that this "one little point" exerts an immense influence on properties at points in the phase diagram that are far away. For example, there are significant deviations in response functions from their corresponding values for a non-interacting fluid at temperatures more than twice the value of the critical temperature T_C.

For water, response functions are also anomalous at temperatures roughly twice the temperature $T_{C'}$. For example, the isothermal compressibility K_T is anomalously *large* at temperatures as high as 400 K (100°C); K_T actually passes through a minimum at 319K (46°C), below which it increases by almost a factor of two as temperature is further decreased. Similarly, the coefficient of thermal expansion α_P is anomalously *small* for temperatures as high as 400 K, and actually decreases to zero at 277 K (4°C), below which it continues decreases at a continually faster rate. Finally, the constant-pressure specific heat C_P is anomalously large for temperatures as high as 400 K, a fact of tremendous practical importance for using water as a coolant in, e.g., a car engine. The dynamic properties of water are also anomalous, with pronounced non-Arrhenius (often power law) behavior typically occurring at low temperature. Thus if there were a critical point at, say, 200K, then the influence of this critical point could be of relevance to the anomalies observed in water at much higher temperatures.

Why Believe?

In the remainder of this talk, we shall discuss some tentative results that are not inconsistent with the critical point hypothesis. First we present an intuitive plausibility argument, then we discuss evidence from simulations and experiments.

Plausibility Arguments. A non-interacting gas has no critical point, but a gas with arbitrarily weak attractive interactions does since at sufficiently small temperature, the ratio of the interaction to kT will become sufficiently significant to condense the liquid out of the gas. That all interacting gases display a critical point below which a distinct liquid phase appears was not always appreciated. Indeed, in the early years of this century one spoke of "permanent gases"—to describe gases that had never been liquefied. Helium is an example of what was once thought to be a permanent gas [31].

Nowadays, we understand that permanent gases cannot exist since all molecules exert some attractive interaction, and at sufficiently low temperature this attractive interaction will make a significant contribution. To make the argument more concrete, one can picture droplets of lower specific volume $\bar{V}$ forming in a single-component fluid. Once the interaction between molecules is fixed (and P is fixed at some value above P_C), then the only remaining control parameter is T; as T decreases the high-density droplets increase in number and size and eventually below T_C they coalesce as a distinct liquid phase.

Water differs from most liquids due to the presence of a line of maximum density (TMD line) in the PT phase diagram. This TMD is physically very significant, as it divides the entire PT phase diagram into two regions with remarkably different properties: the coefficient of thermal expansion—which is proportional to the thermal average ("correlation function") $\langle\delta\bar{V}\delta\bar{S}\rangle$—is *negative* on the low-temperature side of the TMD line, while it is *positive* on the high-temperature side. Here $\bar{V}$ is the volume per molecule, $\bar{S}$ the entropy per molecule, and the δX notation indicates the departure of a quantity X from its mean value.

That $\langle\delta\bar{V}\delta\bar{S}\rangle$ is negative is a thermodynamic necessity given the presence of a TMD line. What microscopic phenomenon causes it? One not implausible explanation [7] is related to the presence of local regions of the hydrogen bond network that are characterized by four "good" hydrogen bonds—and these local regions can be considered as droplets just as the high-density droplets in a gas above C. Stated more formally: the sensitivity of hydrogen bonds to the orientation of the molecules forming it encourages local regions to form that are partially ordered in the sense that if there is a region of the water network where each molecule has four "good" (strong) hydrogen bonds, then the local entropy is lower (so $\delta\bar{S} < 0$) and the local specific volume is larger (so $\delta\bar{V} > 0$), so the contribution to $\langle\delta\bar{V}\delta\bar{S}\rangle$ is negative for such regions.

As the temperature is lowered, there is no *a priori* reason why the "droplets" characterized by negative values of $\delta\bar{V}\delta\bar{S}$ should not increase in number and size, just as the droplets associated with a normal phase transition increase in number, since all water molecules exert mutual interactions on one another, and these interactions—because of their sensitivity to orientation and well as distance—favor the open clusters characterized by $\delta\bar{S}\delta\bar{V} < 0$. It is thus plausible that at sufficiently low temperature these orientation-sensitive interactions will make a larger and larger contribution, and at sufficiently low temperature (and for sufficiently low pressure), a new phase—having roughly the density of the fully hydrogen bonded network—will "condense" out of the one-fluid region.

This intuitive picture has received striking support from a recent generalization of the van der Waals theory. Specifically, Poole et al. [32] allow each water molecule to be in many bonding states, only one of which corresponds to a "good" quality hydrogen bond (with a larger number of states corresponding to "poor" quality bonds). To build in this feature, Poole et al. adopt the approach of Sastry and coworkers [33, 34] and assume that there are $\Omega \gg 1$ configurations of a weak bond, all having $\epsilon = 0$, and only *a single configuration* in which the HB is strong with $\epsilon = \epsilon_{\text{HB}}$. Thus the thermal behavior of the HBs is represented by independent $(\Omega+1)$–state systems, each described by a partition function $Z = \Omega + \exp(-\epsilon_{\text{HB}}/kT)$. Poole et al. find that for small values of the parameter ϵ_{HB}, there is no critical point (but rather a re-entrant spinodal of the form first conjectured by Speedy [19]). However for ϵ_{HB} above a threshold (about 16 kJ/mol), a critical point appears.

The possibility of a second critical point has received recent support by phenomenological analysis of Ponyatovskii and colleagues[35] and by lattice gas models[34, 36]. Also, Roberts and co-workers [37] have shown that simulation results for a microscopic "water-like" Hamiltonian confirms the presence of a second phase transition, previously deduced from approximate calculations[36].

Evidence from simulations. Simulation studies of liquid water have a rich history and have contributed greatly to our understanding of the subject. In fact, this year is the 25th anniversary of the introduction of the ST2 ("Stillinger-2") potential, in which water is represented by a central point from which emanate 4 arms—two carrying positive charges to represent the two protons associated with each water molecule, and two carrying negative charges to represent the two lone electron pairs [24]. The central points interact via a Lennard-Jones potential, while the point charges and the arms interact via a Coulomb potential. Thus every pair of waterlike particles has $4^2 + 1 = 17$ interaction terms. Corresponding to the rather "cumbersome" nature of such a potential is the fact that most studies are limited to extremely small systems—a typical number being $N = 6^3 = 216$ waterlike particles. Recently some studies have considered larger systems, but the typical size rarely exceeds $N = 12^3 = 1728$. It is hoped that by using fast multipole methods one can begin to simulate much larger systems[38].

One way to obtain less cumbersome simulations is to simplify the intermolecular potential. To this end, the simpler TIP4P potential [39] and the much simpler SPC/E potential [40] have enjoyed considerable popularity. However the opposite direction is also under active investigation: simulating more realistic potentials, such as polarizable potentials[41]. The researcher is left with the perplexing problem of which model potential to adopt!

Is it more important to steadily improve the reliability of the potential (and therefore approach, however slowly, a genuinely realistic water potential)? Or is it more important to find an extremely simple potential that encompasses the essential physics of liquid water and use this potential to study large-scale properties of the liquid? With some notable exceptions, chemists sometimes favor the former approach, and physicists the latter. Indeed, a physicist would like to discover a potential so simple that the actual connection between the various features of the potential and the resulting predictions can be understood. To this end, recent work of Buldyrev and collaborators [42] using a tunable "soft-core" potentials is promising.

With these caveats, let us very briefly summarize some recent work that might be interpreted as being consistent with (or at least not contradicting) the hypothesis that a HDL-LDL critical point C' exists. We emphasize that most of this work has not reached the stage that it can be interpreted as "evidence" favoring the hypothesis, so we also outline appropriate avenues where future work may strengthen the argumentation.

Does $1/K_T^{max}$ extrapolate to zero at $(T_{C'}, P_{C'})$? The compressibility K_T diverges at a second order critical point. Thus, we expect $1/K_T^{\max}$ to extrapolate to zero at the "new" HDL-LDL critical point C', exactly as it does for the "old" liquid-gas critical point C. Recent ST2 calculations [43] are consistent with a plausible extrapolation to a single point in the phase diagram at which $K_T^{\max} = \infty$. The caveat is that one can never know that a given quantity is approaching infinity—it could as well just be approaching a very large number. Indeed, the possibility has been raised, and seriously discussed, that there is no genuine singularity [44]; this possibility will be discussed briefly at the end of this lecture.

Is there a "kink" in the $P\rho$ isotherms for sufficiently low temperature? If there is a critical point, then we expect to find a kink in the $P\rho$ isotherms when T is below $T_{C'}$. Indeed, such a kink appears to exist for the ST2 potential, at a temperature of 235K but not at a temperature of 280 K, consistent with $T_{C'}$ somewhere between 235 K and 280 K. This finding, originally made for simulations of 216 ST2 particles [25, 45], has very recently been strikingly confirmed for a system 8 times larger [46]. An analogous kink has not been found for the TIP4P potential, but a prominent inflection occurs at the lowest temperature studied—suggesting that such a kink may be developing. Work is underway testing for inflections and possible kinks for other water potentials in three, and also in two, dimensions [42, 47].

Is there a unique structure of the liquid near the kink point? If there exists a critical point C', then we would expect a two–phase coexistence region below C'. To investigate the possible structural difference between these two phases, Sciortino et al [43] have studied the structure of the liquid at a temperature just below the estimated value of $T_{C'}$ at two values of ρ on the two sides of $\rho_{C'}$. They find that the structure of the liquid state of ST2 at $\rho = 1.05$ g/cm^3 is similar to the experimental data on high-density amorphous (HDA) solid water, while the structure of the structure at $\rho = 0.92$ g/cm^3 resembles the data on low-density amorphous (LDA) solid water. The correspondence between the HDA ice phase and ST2 water just above $\rho_{C'}$, and between the LDA phase and ST2 water just below $\rho_{C'}$ suggests that the two phases that become critical at C' in ST2 water are related to the known HDA and LDA phases of amorphous ice[48].

Does the coordination number approach four as C' is approached? Sciortino et al [43] have studied the coordination number N_{nn} of the ST2 liquid as a function of T and V, where $N_{\rm nn}$ is the average number of nearest–neighbors found in the first coordination shell of an O atom. For the high–T isotherms, their results show that a 4–coordinated "LDL"-like configuration is approached at negative P, in agreement with previous simulations of Geiger and co–workers [49]. For $T \leq$ 273 K, N_{nn} also approaches 4 at positive P. That is, if T is low enough, it appears that a 4-coordinated network can form in liquid water even for $P > 0$. This result is consistent with an experimental study of the evolution of the structure function $S(Q)$ as water is supercooled at atmospheric pressure, in which it was found that the structure tends toward that of the LDA ice [48].

Is it possible that two apparent "phases" may coexist below C'? Convincing evidence for a HDL-LDL critical point C' would be the presence of two coexisting phases below C'. This search is the focus of ongoing work. One can, e.g., partition the water molecules into two groups ("red" and "blue" molecules), those with fewer than the average number of nearest neighbors and those with more than the average and find that the red molecules and the blue molecules segregate to opposite sites of the 18 A box in which they are residing. These preliminary investigations at a temperature somewhat below $T_{C'}$ do not prove phase coexistence [46, 50], but work is underway to establish this possibility. In particular, one must first rule out the likelihood that the two "phases" are merely large fluctuations due to a large correlation length (because near a critical point there should be fluctuations of all sizes and shapes, while the sample separating into two distinct regions is rather different). Also, one must seek to find the phase separation occurring in much larger systems. To be conclusive, firstly one must demonstrate that phase separation occurs in a much larger system, and secondly one must study systematically the time dependence of $S(Q)$ as one quenches into the two-phase region from a large value of temperature.

Separate calculations of the weighted correlation function $h(r)$ for the two tentatively identified HDL and LDL phases suggest similarities with experimental results on the two amorphous solid phases HDA and LDA[50]. Additional work remains to be done to establish this point.

Do fluctuations appear on all time scales? For the ST2 potential, a histogram of hydrogen bond lifetimes reveals power law behavior over as much as two decades, with the region of "scale free behavior" extending over a larger time domain as T is decreased [51]. For the TIP4P potential, no calculations have yet been carried out, but for the SPC/E potential, non-Arrhenius behavior has also been found at high temperatures [52]. At low temperatures, it is possible that power law behavior is found [38, 46]. An important caveat in interpreting these results is that this scale free behavior is exactly what one would expect if the hydrogen bonded network were regarded as possessing defects (corresponding to molecules with fewer than four good bonds), and these defects were allowed to diffuse randomly[53, 54]. Possibly some of these ambiguities will be resolved by applying to this problem Sasai's "local structure index" that permits one to study in some detail the local dynamics[55].

Is there "critical slowing down" of a characteristic time scale? For the ST2 potential, the characteristic value of hydrogen bond lifetime, defined as the value of time at which the power law distribution of bond lifetimes is cut off by an exponential, depends sensitively on temperature and in fact is consistent with a power law divergence as T approaches $T_{C'}$ [51]. The temperature dependence of the cutoff has not been studied for other potentials.

Appearing to diverge at roughly the same temperature is a less ambiguous measure of the characteristic time—the inverse of the self-diffusion coefficient D [56]. This slowing down of the dynamics is consistent with what one expects near a critical point. Specifically, $1/D$ strongly increases as $N_{\rm nn} \to 4$. Consistent with this picture, it was found [57, 58] that additional nearest neighbors beyond 4 have a "catalytic" effect on the mobility of the central molecule, in that they lower the local energy barrier of the molecular exchanges that are the microscopic basis of diffusion, demonstrating the importance for molecular mobility of molecular environments having more than four nearest neighbors.

Because of the relation between $1/D$ and $(N_{\rm nn} - 4)$, the manner in which $N_{\rm nn} \to 4$ is also significant. At high T the decrease of $N_{\rm nn}$ with P is relatively uniform. However, as T decreases, $N_{\rm nn}$ is observed to vary more and more abruptly from a high–coordinated structure ($N_{\rm nn} > 6$) to $N_{\rm nn} \simeq 4$. It should be possible to collapse this family of curves onto a single "scaling function" if the two axes are divided by appropriate powers of $T - T_{C'}$; these tests are underway.

Is the characteristic dynamics of each "phase" different? We can identify molecules as "red"/"blue" if they are in a region of locally high/low density for a specified amount of time (say 100 ps). Looking at the mean square displacement of the red and blue "phases," we see that the red molecules (corresponding to high densities) move much further than blue molecules (corresponding to low densities)[46, 50]. The nature of transport in each phase is under active investigation, particularly in light of recent proposals for the nature of the anomalous dynamics taking place in low-temperature water [59].

Is there evidence for a HDL-LDL critical point from independent simulations? Recently, Tanaka independently found supporting evidence of a critical point by simulations for the TIP4P potential [60]. Tanaka's value of the critical temperature $T_{C'}$ agrees with the earlier estimates, but his critical pressure $P_{C'}$ occurs at roughly atmospheric pressure, or perhaps at negative pressures [60]. The resolution between the two different values of $P_{C'}$ is an open question that will hopefully be resolved shortly.

Evidence from experiments. Before discussing experiments, two very important statements must be made. The first is that the numerical values of thermodynamic variables arising from computer simulations are always rather different from the true values. It is customary to correct for this fact by comparing the values of T and P at the temperature of maximum density to the experimental values. In the case of ST2, e.g., the TMD $\rho = 1.00$ g cm^{-3} appears at T = (277+35)K and P = 82 MPa, so ST2 at these conditions overestimates the temperature by 35 K and the pressure by 82 MPa. If a correction of a similar magnitude were to hold in the supercooled region, then we would anticipate that the experimental coordinates of the HDL-LDL critical point would be approximately $T_{C'} \approx 200$ K and $P_{C'} \approx 100$ MPa. The magnitude of the correction depends upon the potential used, so simulations using the TIP4P and SPC/E potentials can be compared with ST2 or with experiment only if a correction appropriate to that potential is first made.

The second statement concerns the presence of an impenetrable "Berlin wall": the line $T_H(P)$ of homogeneous nucleation temperatures [1,4]. By careful analysis of experimental data above $T_H(P)$, Speedy and Angell pioneered the view that some sort of singular behavior is occurring in water at a temperature $T_s(P)$ some 5-10 degrees below $T_H(P)$. Regardless of the pressure, they found that the identical value of $T_s(P)$ could be used to fit the behavior of quite distinct equilibrium and dynamic quantities.

With these two caveats, one can begin to discuss relevant experimental work. Three general types of experiments have been performed, the first corresponding to a single isobar at 0.1 MPa, the second to a family of isobars up to 600 MPa, and the third to a family of isotherms below $T_{C'}$.

Density fluctuations along the P = 0.1 MPa isobar. The correlation length ξ for density fluctuations should increase close to a critical point; this quantity has recently been measured along a P=0.1 MPa isobar [61], down to quite low temperatures (239K). A gentle increase in correlation length was found, but no indication of a divergence, consistent with the possibility that the HDL-LDL critical point, if it exists, lies at a much higher pressure.

Structure along isobars up to P = 600 MPa. Bellissent-Funel and Bosio have recently undertaken a detailed structural study of D_2O using neutron scattering to study the effect of decreasing the temperature on the correlation function[48]. As paths in the PT phase diagram, they have chosen a family of isobars ranging in pressure up to 600 MPa (well above the HDL-LDL critical point of about 100 MPa). They plot the temperature dependence of the first peak position Q_o of the structure factor for each isobar. They find that for the 0.1 MPa isobar, Q_o approaches 1.7 A^{-1}—the value for LDA, low-density amorphous ice. In contrast, for the 465 and 600 MPa isobars, Q_o approaches a 30% larger value, 2.2 A^{-1}—the value for HDA, high-density amorphous ice. For the 260 MPa isobar, $Q_o \to 2.0A^{-1}$, as if the sample were a two-phase mixture of HDA and LDA.

Reversible conversion of LDA to HDA with pressure. Since the HDL-LDL critical point occurs below $T_H(P)$, it is not possible to probe the two phases experimentally. However two analogous solid amorphous phases of H_2O have been studied extensively by Mishima and co-workers[62–64]. In particular, Mishima has recently succeeded in converting the LDA phase to the HDA phase on increasing the pressure, and then reversing this conversion by lowering the pressure. The jump in density was measured for a range of temperatures from 77K to 140 K, and the density jump (when HDA is compressed to LDA) was found to occur at roughly 200 MPa. Moreover, the magnitude of the density jump decreases as the temperature is raised, just as would occur if instead of making measurements on the HDA and LDA amorphous solid phases, one were instead considering the HDL and LDL liquid phases. These results are corroborated by independently performed computer simulations using both the ST2 and TIP4P intermolecular potentials[65].

If we assume that HDA and LDA ice are the glasses formed from the two liquid phases discussed above, then the HDA–LDA transition can be interpreted in terms of an abrupt change from one microstate in the phase space of the high–density liquid, to a microstate in the phase space of the low–density liquid. The experimentally detected HDA–LDA transition line would then be the extension into the glassy regime of the line of first–order phase transitions separating the HDL and LDL phases.

Discussion

Thus far in this talk, we have presented some tentative evidence that is consistent with the hypothesis that there exists a second HDL-LDL critical point C', in addition to the well-known liquid-gas critical point C. There are also some facts, which we next present, that do not favor the existence of C'.

First, the vast experimental work of Angell and co-workers [4, 6] is equally consistent with a line of spinodal singularities or with a HLD-LDL critical point, and the generalized van der Waals theory is equally consistent with both possibilities [32].

Moreover, the observed polymorphism of amorphous *solid* water which was invoked to support the possibility of a line of liquid–liquid phase transitions terminating in a critical point supports, but certainly does not prove, the existence of a phase transition in the behavior of the supercooled *liquid.* Merely a large but finite value of K_T might be sufficient to account for abrupt density changes in the amorphous ices in the glass regime [66, 67].

Very recently, the dynamics of SPC/E water have been found to be well described by mode coupling theory, with no need to invoke the presence of a critical point[59]. For ST2 water, on the other hand, the situation may be somewhat different[46, 50, 68].

Perhaps the most compelling argument against the hypothesized existence of a HDL-LDL critical point C' is parsimony: "the correct physics is often the physics with the minimum of assumptions." Of necessity, it is not possible to experimentally verify for metastable water divergences of response functions as convincingly as one can for the liquid-gas critical point. Thus, it remains conceivable that response functions that would diverge at a critical point in fact remain finite in metastable water. In particular, Sastry and co-workers[44] have demonstrated that the low temperature increase of compressibility upon lowering temperature arises strictly as a thermodynamic consequence of the presence and negative slope of the TMD line. Without further assumptions, such increase in the compressibility results simply (and at most) in a compressibility maximum at low temperatures. Actually, the thermodynamic analysis of Sastry *et al.* lends indirect support to the HDL-LDL critical point hypothesis, for it shows that the expectation of *extrema* in the compressibility is reasonable given the type of "density anomaly" observed at higher temperatures in simulations. A thermodynamic analysis alone suffices to predict that a locus of compressibility extrema exists at low temperatures, given the locus of density maxima (TMD) as seen in simulations. What remains to be shown is that these extrema at low enough (but *finite*) temperatures are *singular* extrema, i.e., that the compressibility either diverges (as at the critical point) or becomes discontinuous (along the associated first order line) along the locus of compressibility extrema predicted by the thermodynamic analysis.

It is possible that there are other amorphous materials for which fluctuations in their local structure are enhanced due to tetrahedrality or other considerations[69]. Understanding one such material, water, may help in understanding others—whether they be other materials with tetrahedral structures (and corresponding TMD lines) such as SiO_2[70] or whether they be more complex structures like amorphous carbon which appears to display strikingly ordered local heterogeneities as it is heated toward its crystallization temperature.

Many open questions remain, and many experimental results are of potential relevance to the task of answering these questions. For example, the collective excitations that occur at low frequency and large wave vector [71] have been the object of much recent work [72, 73]. Also, the full implications of a re-entrant spinodal possibility [74–77] may not have been fully incorporated into our understanding of water. The minimum exhibited by the *adiabatic* compressibility in atmospheric pressure experiments [78] is of potential relevance since a maximum in any quantity that would be singular at a spinodal may be relevant to placing the critical point at a larger pressure than atmospheric.

Acknowledgements. We thank C. A. Angell, S. S. Borick, S. V. Buldyrev, S.-H. Chen, L. C. Cruz, P. G. Debenedetti, P. Gallo, T. Grande, P. F. McMillan, P. Ray, C. J. Roberts, R. Sadr-Lahijany, R. J. Speedy, F. W. Starr, F. H. Stillinger, H. Tanaka, P. Tartaglia, B. Widom and R. Zhang for enlightening discussions and seminal help in various aspects of this work, and NSF, BP, GNSM, and NSERC (Canada) for financial support.

References

[1] *Water: A Comprehensive Treatise*, Vol. 1–7, F. Franks, ed. (Plenum Press, NY, 1972);
Water Science Reviews, Vol. 1–4, F. Franks, ed. (Cambridge University Press, Cambridge, 1985).

[2] P. G. Debenedetti, *Metastable Liquids* (Princeton Univ Press, Princeton, 1997).

[3] G. S. Kell, "Thermodynamic and Transport Properties of Fluid Water," in *Water: A Comprehensive Treatise*, Vol. 1, F. Franks, ed. (Plenum Press, NY, 1972), pp. 363–412.

[4] C. A. Angell, "Supercooled Water," in *Water: A Comprehensive Treatise*, Vol. 7, F. Franks, ed. (Plenum Press, NY, 1980), pp. 1–81.

[5] E. W. Lang and H.-D. Lüdemann, "Anomalies of Liquid Water," *Angew. Chem. Int. Ed. Engl.* **21**, 315–329 (1982).

[6] C. A. Angell, "Supercooled Water," *Ann. Rev. Phys. Chem.* **34**, 593–630 (1983).

[7] H. E. Stanley, "A Polychromatic Correlated-Site Percolation Problem with Possible Relevance to the Unusual behavior of Supercooled H_2O and D_2O" J. Phys. A **12**, L329-L337 (1979);
H. E. Stanley and J. Teixeira, "Interpretation of the Unusual Behavior of H_2O and D_2O at Low Temperatures: Tests of a Percolation Model," *J. Chem. Phys.* **73**, 3404-3422 (1980);
H. E. Stanley, J. Teixeira, A. Geiger, and R. L. Blumberg, "Are Concepts of Percolation Relevant to the Puzzle of Liquid Water?" *Physica A* **106**, 260–277 (1981).

[8] A. Geiger, F. H. Stillinger and A. Rahman, "Aspects of the Percolation Process for Hydrogen-bond Networks in Water," *J. Chem. Phys.* **70**, 4185–4193 (1979).

[9] F. H. Stillinger, "Water Revisited," *Science* **209**, 451–457 (1980);
F. H. Stillinger and T. A. Weber, "Hidden Structure in Liquids," *Phys. Rev. A* **25** 978–989 (1982);
F. H. Stillinger and T. A. Weber, "Dynamics of Structural Transitions in Liquids," *Phys. Rev. A* **28** 2408–2416 (1983);
F. H. Stillinger and T. A. Weber, "Inherent Structure in Water," *J. Phys. Chem.* **87**, 2833–2840 (1983);
F. H. Stillinger and T. A. Weber, "Inherent Pair Correlation in Simple Liquids," *J. Chem. Phys.* **80**, 4434–4437 (1984).

[10] A. Geiger and H. E. Stanley, "Low-Density Patches in the Hydrogen-Bonded Network of Liquid Water: Evidence from Molecular Dynamics Computer Simulations," *Phys. Rev. Lett.* **49** 1749–1752 (1982);
H. E. Stanley, R. L. Blumberg, and A. Geiger, "Gelation Models of Hydrogen Bond Networks in Liquid Water," *Phys. Rev. B* **28** 1626–1629 (1983);
H. E. Stanley, R. L. Blumberg, A. Geiger, P. Mausbach and J. Teixeira, "The locally-structured transient gel model of water structure," J. de Physique **45**, C7[3]–C7[12] (1984).

[11] A. Geiger and H. E. Stanley, "Tests of universality of percolation exponents for a 3-dimensional continuum system of interacting waterlike particles" Physical Review Letters **49**, 1895-1898 (1982);
R. L. Blumberg, H. E. Stanley, A. Geiger and P. Mausbach, "Connectivity of Hydrogen Bonds in Liquid Water," *J. Chem. Phys.* **80**, 5230–5241 (1984).

[12] A. Geiger, P. Mausbach, J. Schnitker, R. L. Blumberg, and H. E. Stanley, "Structure and Dynamics of the Hydrogen Bond Network in Water by Computer Simulations," *J. Physique* **45**, C7[13]–C7[30] (1984).

[13] C. A. Angell and V. Rodgers, "Near Infrared Spectra and the Disrupted Network Model of Normal and Supercooled Water," *J. Chem. Phys.* **80**, 6245 (1984).

[14] C. A. Angell, J. Shuppert, and J. C. Tucker, "Anomalous Properties of Supercooled Water. Heat Capacity, Expansivity, and Proton Magnetic Resonance Chemical Shift from 0 to -38°," *J. Phys. Chem.* **77**, 3092–3099 (1973);
H. Kanno and C. A. Angell, "Water: Anomalous Compressibilities to 1.5 kbar and Correlation with Supercooling Limits," J. Chem. Phys. **70**, 4008 (1979);
H. Kanno and C. A. Angell, "Volumetric and Derived Thermal Characteristics of Liquid D_2O at Low Temperatures and High Pressures," J. Chem. Phys. **73**, 1940 (1980).

[15] R. J. Speedy and C. A. Angell, "Isothermal Compressibility of Supercooled Water and Evidence for a Thermodynamic Singularity," *J. Chem. Phys.* **65**, 851–858 (1976).

[16] C. A. Angell, M. Oguni, and W. J. Sichina, "Heat Capacity of Water at Extremes of Supercooling and Superheating," *J. Phys. Chem.* **86**, 998–1002 (1982).

[17] Q. Zheng, D. J. Durben, G. H. Wolf, and C. A. Angell, "Liquids at Large Negative Pressures: Water at the Homogeneous Nucleation Limit," *Science* **254**, R829–R832 (1991).

[18] R. J. Speedy, "Stability-Limit Conjecture. An Interpretation of the Properties of Water," *J. Phys. Chem.* **86**, 982–991 (1982).

[19] R. J. Speedy, "Limiting Forms of the Thermodynamic Divergences at the Conjectured Stability Limits in Superheated and Supercooled Water," *J. Phys. Chem.* **86**, 3002–3005 (1982).

[20] R. J. Speedy, "Thermodynamic Properties of Supercooled Water at 1 Atmosphere,"*J. Phys. Chem.* **91**, 3354–3358 (1987).

[21] M. P. Allen and D. J. Tildesley, *Computer Simulation of Liquids* (Oxford University Press, Oxford, 1989).

[22] M. P. Allen and D. J. Tildesley, editors, *Computer Simulations in Chemical Physics* (Kluwer, Dordrecht, 1993); J. M. Haile, *Molecular Dynamics Simulation: Elementary Methods* (Wiley, NY, 1992); E. Clementi editor, *Methods and Techniques in Computational Chemistry MOTECC–1994* (STEF, Cagliari, 1993).

[23] J. A. Barker and R. O. Watts, "Structure of Water; A Monte Carlo Calculation," *Chem. Phys. Lett.* **3**, 144–145 (1969).

[24] F. H. Stillinger and A. Rahman, "Molecular Dynamics Study of Temperature Effects on Water Structure and Kinetics," *J. Chem. Phys.* **57**, 1281–1292 (1972).

[25] P. H. Poole, F. Sciortino, U. Essmann, H. E. Stanley, "Phase Behavior of Metastable Water," *Nature* **360**, 324–328 (1992); C. Sorenson. Nature **360**, 303-304 (1992).

[26] P. G. Kusalik and I. G. Svishchev, "The Spatial Structure in Liquid Water," *Science* **265**, 1219–1221 (1994).

[27] I. Ohmine and H. Tanaka, "Fluctuation, Relaxations, and Hydration in Liquid Water. Hydrogen-Bond Rearrangement Dynamics," *Chem. Rev.* **93**, 2545–2566 (1993);
I. Ohmine, "Liquid Water Dynamics: Collective Motions, Fluctuation, and Relaxation", J. Phys. Chem. **99**, 6767-6776 (1995);
I. Ohmine and M. Sasai, "Relaxations, Fluctuations, Phase Transitions and Chemical Reactions in Liquid Water", Prog. Theor. Phys. Suppl. **103**, 61-91 (1991).

[28] L. Greengard, *The Rapid Evolution of Potential Fields in Particle Systems* (MIT Press, Cambridge, 1987);
J. Barnes and P. Hut, "A Hierarchical O(n log n) Force–Calculation Algorithm," *Nature* **324**, 446 (1986).

[29] C. A. Angell and H. Kanno, "Density Maxima in High-Pressure Supercooled Water and Liquid Silicon Dioxide," Science **193**, 1121 (1976).

[30] K. K. Mon, M. W. Ashcroft, and G. V. Chester, "Core Polarization and the Structure of Simple Metals," *Phys. Rev. B* **19**, 5103–5118 (1979).

[31] K. Mendelssohn, *The Quest for Absolute Zero: The Meaning of Low-Temperature Physics* (McGraw, New York, 1966), p. 42ff

[32] P. H. Poole, F. Sciortino, T. Grande, H. E. Stanley and C. A. Angell, "Effect of Hydrogen Bonds on the Thermodynamic Behavior of Liquid Water," *Phys. Rev. Lett.* **73**, 1632–1635 (1994);
P. H. Poole, F. Sciortino, T. Grande, H. E. Stanley and C. A. Angell, "Mean-Field Approach to Water Structure," Phys. Rev. E (preprint).

[33] S. Sastry, F. Sciortino and H.E. Stanley, "Limits of Stability of the Liquid Phase in a Lattice Model with Water-like Properties," *J. Chem. Phys.* **98**, 9863–9872 (1993).

[34] S. S. Borick, P. G. Debenedetti and S. Sastry, "A Lattice Model of Network-Forming Fluids with Orientation-Dependent Bonding: Equilibrium, Stability, and Implications from the Phase Behavior of Supercooled Water," *J. Phys. Chem.* **99**, 3781 (1995);
S. S. Borick and P. G. Debenedetti, "Equilibrium, Stability and Denisty Anomalies in a Lattice Model with Core Softening and Directional Bonding" *J. Phys. Chem.* **97**, 6292-6303 (1993).

[35] E. G. Ponyatovskii, V. V. Sinitsyn and T. A. Pozdnyakova, "Second critical point and low-temperature anomalies in the physical properties of water," *JETP Lett.* **60** 360–364 (1994).

[36] C. J. Roberts and P. G. Debenedetti, "Polyamorphism and density anomalies in network-forming fluids: Zeroth- and First-order approximations," J. Chem. Phys. **105**, 658–672 (1996).

[37] C. J. Roberts, A. Z. Panagiotopoulos and P. G. Debenedetti, "Liquid-Liquid Immiscibility in Pure Fluids: Polyamorphism in Simulations of a Network-Forming Fluid," Phys. Rev. Lett. **77**, 4386–4389 (1996).

[38] F. W. Starr et al (preprint)

[39] W. L. Jorgensen, J. Chandrasekhar, J. Madura, R. W. Impey and M. Klein, "Comparison of Simple Potential Functions for Simulating Liquid Water," *J. Chem. Phys.* **79**, 926–935 (1983).

[40] H. J. C. Berendsen, J. R. Grigera and T. P. Straatsma, "The Missing Term in Effective Pair Potentials," *J. Phys. Chem.* **91**, 6269–6271 (1987).

[41] U. Niesar, G. Corongiu, E. Clementi, G.R. Kneller, and D. Bhattacharya, *J. Phys. Chem.* **94**, 7949 (1990).

[42] S. V. Buldyrev et al (preprint)

[43] F. Sciortino, P. H. Poole, U. Essmann and H. E. Stanley, "Line of compressibility maxima in the phase diagram of supercooled water," Phys. Rev. E **55**, 727 (1997).

[44] S. Sastry, P. G. Debenedetti, F. Sciortino, and H. E. Stanley, "Singularity-free interpretation of the thermodynamics of supercooled water," *Phys. Rev. E* **53**, 6144–6154 (1996).

[45] P. H. Poole, F. Sciortino, U. Essmann and H. E. Stanley, "Spinodal of Liquid Water," *Phys. Rev. E* **48**, 3799–3817 (1993)

[46] S. T. Harrington, R. Zhang, P. H. Poole, F. Sciortino and H. E. Stanley, "Liquid-liquid phase transition: evidence from simulations," Phys. Rev. Lett. **78**, xxx–xxx (31 March 1997).

[47] L. Cruz et al. (preprint).

[48] M. C. Bellissent-Funel and L. Bosio, "A neutron scattering study of liquid D_2O under pressure and at various temperatures," *J. Chem. Phys.* **102**, 3727–3735 (1995).

[49] A. Geiger, P. Mausbach and J. Schnitker, in *Water and Aqueous Solutions*, edited by G. W. Neilson and J. E. Enderby (Adam Hilger, Bristol, 1986) p. 15.

[50] R. Zhang et al (preprint)

[51] F. Sciortino, P. H. Poole, H. E. Stanley and S. Havlin, "Lifetime of the Bond Network and Gel-like Anomalies in Supercooled Water," *Phys. Rev. Lett.* **64**, 1686–1689 (1990).

[52] A. Luzar and D. Chandler, "Effect of Environment on Hydrogen Bond Dynamics in Liquid Water," Phys. Rev. Lett. **76**, 928-931 (1996);
A. Luzar and D. Chandler, "Hydrogen-bond kinetics in liquid water," *Nature* **379**, 55–57 (1996).

[53] H. Larralde, F. Sciortino and H. E. Stanley, "Restructuring the Hydrogen Bond Network of Water" (preprint);
H. Larralde, Ph. D. Thesis (Boston University, 1993)

[54] F. Sciortino, A. Geiger and H. E. Stanley, "Isochoric Differential Scattering Functions in Liquid Water: The Fifth Neighbor as a Network Defect," *Phys. Rev. Lett.* **65**, 3452–3455 (1990).

[55] E. Shiratani and M. Sasai, "Growth and Collapse of Structural Patterns in the Hydrogen Bond Network in Liquid Water," J. Chem. Phys. **104**, 7671-7680 (1996).

[56] P. H. Poole, F. Sciortino, U. Essmann, M. Hemmati, H. E. Stanley and C. A. Angell, "Novel Features in the Equation of State of Metastable Water" in: *Hydrogen Bond Networks* Eds M.-C. Bellissent-Funel and J. C. Dore. [Proc. 1993 Cargèse Conf.] (Kluwer Academic, Dordrecht, 1994), pp. 53–60.

[57] F. Sciortino, A. Geiger and H. E. Stanley, "Effect of defects on Molecular Mobility in Liquid Water," *Nature* **354**, 218–221 (1991).

[58] F. Sciortino, A. Geiger and H. E. Stanley, "Network Defects and Molecular Mobility in Liquid Water," *J. Chem. Phys.* **96**, 3857–3865 (1992).

[59] P. Gallo, F. Sciortino, P. Tartaglia, and S.-H. Chen, "Slow Dynamics of Water Molecules in Supercooled States," Phys. Rev. Lett. **76**, 2730–2733 (1996);
F. Sciortino, P. Gallo, P. Tartaglia, and S.-H. Chen, "Supercooled water and the kinetic glass transition," Phys. Rev. E **54**, 6331–6346 (1996).

[60] H. Tanaka, "A self-consistent phase diagram for supercooled water," Nature **380**, 328–331 (1996);
H. Tanaka, "Phase behaviors of supercooled water: Reconciling a critical point of amorphous ices with spinodal instability," J. Chem. Phys. **105**, 5099–5111 (1996);
R. J. Speedy, "Two waters and no ice please," Nature **380**, 289–290 (1996).

[61] Y. Xie, K. F. Ludwig, Jr., G. Morales, D. E. Hare and C. M. Sorensen "Noncritical Behavior of Density Fluctuations in Supercooled Water," Phys. Rev.Lett. **71**, 2050-2053 (1993).

[62] O. Mishima, "Reversible first-order transition between two H_2O amorphs at -0.2 GPa and 135 K" J. Chem. Phys. **100**, 5910-5912 (1994).

[63] O. Mishima, K. Takemura and K. Aoki, "Visual Observations of the Amorphous-Amorphous Transition in H_2O Under Pressure" Science **254**, 406-408 (1991).

[64] O. Mishima, "Relationship between melting and amorphization of ice," Nature (accepted);
O. Mishima, L. D. Calvert and E. Whalley, "Melting Ice I at 77K and 10kbar: A New Method of Making Amorphous Solids," *Nature* **310**, 393-395 (1984);
O. Mishima, L. D. Calvert and E. Whalley "An Apparently First-order Transition Between Two Amorphous Phases of Ice Induced by Pressure," *Nature* **314**, 76-78 (1985).

[65] P. H. Poole, U. Essmann, F. Sciortino and H. E. Stanley, "Phase Diagram for Amorphous Solid Water," *Phys. Rev. E* **48**, 4605–4610 (1993),

[66] P. H. Poole, T. Grande, F. Sciortino, H. E. Stanley and C. A. Angell, "Amorphous Polymorphism," *J. Comp. Mat. Sci.* **4**, 373–382 (1995).

[67] F. Sciortino, U. Essmann, H. E. Stanley, M. Hemmati, J. Shao, G. H. Wolf, and C. A. Angell, "Crystal Stability Limits at Positive and Negative Pressures, and Crystal-to-Glass Transitions" Phys. Rev. E **52**, 6484–6491 (1995).

[68] D. Paschek and A. Geiger, "Simulation study on the nature of the diffusive motion in supercooled water," preprint.

[69] M. R. Sadr-Lahijany, P. Ray and H. E. Stanley, "Dispersity Induced First-Order Melting Transition in Polydisperse Solids" (preprint).

[70] P. H. Poole, M. Hemmati and C. A. Angell, "Thermodynamic Properties of Liquid Silica" Nature (accepted).

[71] J. Teixeira, M.-C. Bellissent-Funel, S.-H. Chen and B. Dorner, "Observation of New Short-Wavelength Collective Excitations in Heavy Water by Coherent Inelastic Neutron Scattering," *Phys. Rev. Lett.* **54**, 2681–2683 (1985).

[72] F. Sciortino and S. Sastry, "Sound Propagation in Liquid Water: The Puzzle Continues," *J. Chem. Phys.* **100**, 3881–3893 (1994).

[73] S. Sastry, F. Sciortino and H. E. Stanley, "Collective Excitations in Liquid Water at Low Frequency and Large Wave Vector," *J. Chem. Phys.* **95**, 7775–7776 (1991).

[74] P. G. Debenedetti and M. C. D'Antonio, "On the Nature of the Tensile Instability in Metastable Liquids and its Relationship to Density Anomalies," *J. Chem. Phys.* **84**, 3339–3345 (1986).

[75] P. G. Debenedetti and M. C. D'Antonio, "On the Entropy Changes and Fluctuations Occurring Near a Tensile Instability," *J. Chem. Phys.* **85**, 4005–4010 (1986).

[76] M. C. D'Antonio and P. G. Debenedetti, "Loss of Tensile Strength in Liquids Without Property Discontinuities: A Thermodynamic Analysis," *J. Chem. Phys.* **86**, 2229–2235 (1987).

[77] P. G. Debenedetti and M. C. D'Antonio, "Stability and Tensile Strength of Liquids Exhibiting Density Maxima," *AIChE J.* **34**, 447–455 (1988).

[78] E. Trinh and R. E. Apfel, "The Sound Velocity in Metastable Liquid Water under Atmospheric Pressure," J. Chem. Phys. **69**, 4245–4251 (1978); E. Trinh and R. E. Apfel, "Sound Velocity of Supercooled Water down to -33°C Using Acoustic Levitation," J. Chem. Phys. **72**, 6731–6735 (1980).

Chapter 20

Slow Dynamics in a Model and Real Supercooled Water

S. H. Chen[1], P. Gallo[2], F. Sciortino[2], and P. Tartaglia[2]

[1]Department of Nuclear Engineering, 24–209, Massachusetts Institute of Technology, Cambridge, MA 02139
[2]Dipartimento di Fisica and Istituto Nazionale per la Fisica della Materia, Università di Roma La Sapienza, Istituto Nazionale di Fisica Nucleare, Sezione di Roma I, Piazzale Aldo Moro, 00185 Rome, Italy

We made a molecular dynamics study of single-particle dynamics of water molecules in deeply supercooled liquid states. We find that the time evolution of various single-particle time correlation functions is characterized by a fast initial relaxation toward a plateau region, where it shows a self-similar dynamics, then followed by a slow, stretched exponential decay to zero at much later times. We interpret these results in the frame-work of a mode-coupling theory for supercooled liquids. We relate the apparent anomalies of transport coefficients in this model water, on lowering the temperature, to the formation of a long-lived cage around each water molecule and the associated slow dynamics of the cages. The experimentally observed so-called Angell temperature, which is an apparent limit of supercooling in liquid water, could thus be interpreted as a kinetic glass transition temperature predicted by the mode-coupling theory. We then discuss to what extent the experimental incoherent quasi-elastic neutron scattering data from supercooled bulk water support the idea of the existence of the slow dynamics.

PACS numbers: 61.20. Ja, 64.70.Pf

I. Introduction

The thermodynamic and transport behavior of liquid water when supercooled has been the subject of a continuing scientific debate in the last thirty years [1-3]. It has been found that there are anomalous increases of various thermodynamic quantities and apparent power-law dependences of transport coefficients on approaching a singular but experimentally inaccessible temperature T_S, of about 227 K at the ambient pressure. This discovery has stimulated a considerable amount of experimental, theoretical and computational work in an attempt to clarify the origin of the singularity. A recent review of all these works can be found in [3].

During the last five years, efforts have been directed in elucidating possible thermodynamic scenarios compatible with the trend of experimental data. The pronounced increase in isothermal compressibility, isobaric heat capacity, and the change of sign of thermal expansion coefficient of liquid water upon supercooling, can arise from three possible causes: (i) existence of a continuous, reentrant spinodal curve bounding the superheated, stretched and supercooled states of liquid water [4 - 6]; (ii) existence of a metastable, low-temperature critical point [6 - 8]; (iii) progressive increase of four hydrogen-bond (HB) coordinated water molecules, favorable due to low energy of this state, but unfavorable in terms of a high local volume and a low orientational entropy [9,10].

While the phase behavior of supercooled water has been extensively debated, not much attention has been devoted so far to the dynamics near the apparent singular temperature. For this purpose, computation of long-time behavior of dynamical quantities is needed and is now possible with modern work stations. It is thus timely to make an effort to understand the origin of the apparent divergences of transport coefficients in water on supercooling. In particular, if the dynamics could be rationalized without resorting to an underlying thermodynamic singularity, then the possibility of a singularity-free picture capable of explaining satisfactorily also the behavior of thermodynamic quantities, could be imagined [10].

Already in the late 80's. F. X. Prielmeier at al.[11] and C. A. Angell [12], based on NMR measurements and fitting of the self-diffusion constant in supercooled water, commented on a possible relationship between the power-law temperature dependences of transport coefficients in water and the prediction of Mode Coupling Theory (MCT) [13] of the so-called kinetic glass transition in supercooled liquids. More recently, translational region of Raman spectra of water has been interpreted in terms of the scaling behavior predicted by the MCT [14]. Today, due to the availability of a sufficient computing power and to an extensive development of the MCT [15], suggestion regarding the connection between the existence of a kinetic glass transition temperature and the divergence of transport coefficients in water, can be carefully examined in a model water.

In this paper we test the MCT predictions for various single-particle correlation functions, time-dependent as well as time-independent ones, in supercooled liquids with corresponding quantities calculated from Computer Molecular Dynamics (CMD) simulations of a model water, carried out for sufficiently long time, so as to allow the slow dynamics to be observed. In doing so, we try to assess to what extent the MCT, which has been proposed for describing simple liquids [16], is applicable also to the description of the single-particle dynamics of a model water, a hydrogen bonded liquid with a strong directional interactions among molecules. A detailed report on this subject has already been published[17].

This paper is organized as follows. In Section II, we briefly recall predictions of the MCT. In Section III, we discuss the states under which our CMD simulations are performed and the model potential used. Section IV is divided into subsections each presenting the numerical results of correlation functions for translational motions. Specifically, we discuss the results for the mean squared displacement, the van Hove space-time self-correlation function and the intermediate scattering function. In Section V, we present an analysis of some QENS results of supercooled water contained in pores of Vycor glass which indicates that the MCT type slow dynamics may have been observed. In Section VI, we summarize our findings.

II. The Mode-Coupling Theory For Supercooled Liquids

In this section we summarize predictions of the MCT which are relevant for interpreting our CMD data. Our discussion emphasizes results of the so-called idealized MCT [15], in which hopping effects are neglected. In contrast to a MCT for critical phenomena [18], the MCT for supercooled, dense liquids is a theory aiming at explaining fluctuation phenomena at distances comparable to the interparticle spacing. In the Fourier space, it describes correlation functions in the Q-range close to the first diffraction peak, Q_{max}, of the center-of-mass structure factor S(Q). Therefore, the Q-dependence of dynamic quantities around Q_{max} is of interest.

The MCT for supercooled liquids is a microscopic theory of liquids that focus attention on the physical phenomena in dense liquid states where the so-called cage effect is dominant. The cage effect is associated with a transient trapping of a molecule by its immediate neighbors on lowering of the temperature or on increasing the average number density. The microscopic density fluctuations of disordered, high temperature and low density fluids tend to relax rapidly with a time-scale of few picoseconds. Upon lowering the temperature or increasing the density of the liquid, there is a rapid increase in the local order surrounding a typical particle (test particle) leading to a substantial increase in the structural relaxation time. In the supercooled or dense liquid regime a trapped particle in a cage can migrate only through a cooperative rearrangement of a large number of particles surrounding it. In this sense there is a strong coupling between the single-particle motion and the density fluctuation of the liquid. According to the MCT, the spectral density of the density fluctuation (i.e. S(Q)) completely determines the long-time dynamical behavior of the liquid.

The MCT predicts the time evolution of a correlator of any local variables which has a non-zero overlap with the local density operator [19]. Examples are: the density itself, the current-density and the tagged-particle density. In the following, we will denote such a generic correlator as $\phi_Q(t)$. The evolution of $\phi_Q(t)$ is controlled by a retarded memory function, a non-linear functional of the local density. The idealized MCT predicts that on moving along a path in the pressure-temperature plane, a line $T_c(P)$ is crossed on which $\phi_Q(t)$ does not decay to zero any longer. Such a line defines the locus of the ideal ergodic to non-ergodic transition. This line separates the liquid and the glass regions. On the liquid side, the system is ergodic and $\lim_{t\to\infty}\phi_Q(t) = 0$. Close to the line, on the liquid side, $\phi_Q(t)$ has a two-step relaxation behavior displaying a fast and a slow decay. The two relaxation times become more and more separated in time scale on approaching the line. After the first (fast) process decays, $\phi_Q(t)$ reaches a plateau value f_Q^c, the so-called non-ergodcity parameter. Only after the second relaxation is completed, $\phi_Q(t)$ decays to zero. The time interval in which the correlation function is close to the plateau f_Q^c is called the β-relaxation region. The long time region after the plateau is called the α–relaxation region. The β–relaxation region is centered around a time t_σ which increases on decreasing the distance s from the critical line in the pressure-temperature plane.

The idealized MCT predicts the dependence on σ of t_σ and the functional form of the decay of $\phi_Q(t)$ in the β-relaxation region. It states: (i) that t_σ scales with $|\sigma|$ as:

$$t_\sigma = t_0|\sigma|^{-\frac{1}{2a}}, \tag{1}$$

where t_0 is a system dependent characteristic microscopic time, (ii) that the approach to the plateau is described, to the leading order in time, by a power law with an exponent a, according to

$$\phi_Q(t) - f_Q^c = B_Q|\sigma|^{1/2}\left(\frac{t_\sigma}{t}\right)^a = B_Q\left(\frac{t}{t_0}\right)^{-a} \quad (2)$$

(iii) that, on the liquid side of the phase diagram, $\phi_Q(t)$ departs from the plateau value f_Q^c according to a power-law with an exponent b

$$\phi_Q(t) - f_Q^c = -C_Q|\sigma|^{1/2}\left(\frac{t}{t_\sigma}\right)^b = -C_Q\left(\frac{t}{\tau}\right)^b \quad (3)$$

where $\tau = t_0|\sigma|^{-\frac{1}{2a}-\frac{1}{2b}}$.

Equation 3 is the leading term in the development in powers of t^b. Equation 3 is called the von Schweidler law and its region of validity is often rather limited [20].

The constants f_Q^c, B_Q and C_Q in Equations 2 and 3 are T-independent, while the exponents a and b are T and Q independent. In real space, Equations 2 - 3 express the prediction of a separation of space and time variables in a correlator. Any space-dependent correlation function $\phi(r,t)$ in the β region can be expressed as:

$$\phi(r,t) = F(r) + H(r)G(t) \quad (4)$$

where G(t) is proportional to t^{-a} or t^b depending on the time region and F(r) and H(r) are related to the Fourier transform of f_Q^c and B_Q or C_Q respectively.

The exponents a and b are related and both depend on the specific point in the transition line crossed on moving the system along the P,T plane. For example, cooling the system along different isobars will produce crossing of the transition line in different points, which will in the end imply different values for the a and b exponents. a and b are related by a transcendental equation

$$\frac{[\Gamma(1-a)]^2}{\Gamma(1-2a)} = \frac{[\Gamma(1+b)]^2}{\Gamma(1+2b)} \quad (5)$$

where $\Gamma(x)$ is the Gamma-function and $0 < a < 0.5$ and $0 < b < 1$.

Equation 3 describes the region where $\phi_Q(t)$ - f_Q^c is small. But dependence of ϕ_Q on the scaled time t/τ is predicted to be valid also in the α-relaxation regime. In other words, $\phi_Q(t)$ does not depend of T and P explicitly, but only via the P and T dependence of τ. It has been found numerically, that the decay of $\phi_Q(t)$ in the a-region is often well represented by a stretched exponential form $e^{-(t/\tau)^\beta}$. τ plays the role of the relevant time in the system and diverges on approaching the transition line with a power γ

$$\gamma = \frac{1}{2a} + \frac{1}{2b} \tag{6}$$

All characteristic times in the system are predicted to be proportional to τ. Thus, the MCT predicts that the inverse diffusion coefficient D^{-1} diverges as $|\sigma|^{-\gamma}$.

We note that Equations 5 and 6 relate the exponents a, b and γ. Thus, only one of these three exponents is truly independent. In contrast to critical phenomena, the values of a, b and γ are not universal and depend on specific point of the transition line approached, i.e. on $T_C(P)$. In particular, if we approach the line along an isobaric path, using the temperature as an external driving parameter, then the distance from the transition line is measured by $(T-T_C)$ and D goes like $(T-T_C)^{\gamma}$.

III. Molecular Dynamic Simulations

We conducted MD simulations in the N V E ensemble with 216 molecules. The effective potential used is the extended simple point charge model, SPC/E [22]. This potential treats a single water molecule as a rigid set of point masses with the OH distance 0.1 nm and the HOH angle equal to the tetrahedral angle, 109.47°. Coulomb charges are placed on each atoms. Their magnitudes are $q_H = 0.4238$ e and $q_O = -2q_H = -0.8476$ e. Only the oxygen atoms interact among themselves via a Lennard-Jones potential, with the L-J parameters, $\sigma = 0.31656$ nm and $\varepsilon = 0.64857$ kJ /Mol. Interaction between pairs of molecules is calculated explicitly when their separation is less than a cut-off distance r_c of 2.5 σ. The contribution due to Coulomb interactions beyond r_c is calculated using the reaction-field method, as described by Steinhauser [23]. Also, the contribution of Lennard-Jones interactions between pairs separated more than r_c is included in the evaluation of thermodynamic properties by assuming a uniform density beyond r_c. The MD code used here to calculate the SPC/E trajectories is the same as that used in Ref.[24] where further details are given. A heat-bath [25] has been used to allow for the heat exchange. Periodic boundary conditions are used. The time step for the integration of the molecular trajectories is 1 fs. Simulations at low T have been started from equilibrated configurations at higher T. Equilibration has been monitored via time dependence of the potential energy. In all cases the equilibration time t_{eq} was longer than the time needed to enter the diffusive regime (see Figure 6), i.e. $\langle r^2(t_{eq})\rangle$ larger than 10 Å^2.

The SPC/E potential has been explicitly parameterized to reproduce experimental value of the self-diffusion constant at ambient temperature and at density of 1 g/cm^3. It has been widely used for a number of years [26,27]. Moreover, this potential is able to reproduce a pressure-dependent temperature of maximum density (TMD) [7,24,27]. As shown in ref.[27], the SPC/E 1 bar isobar is characterized by a TMD of about 235 K and a corresponding density of 1.026 g/cm^3. The -40 MPa isobar is instead characterized by a TMD of about 250 K and a corresponding density of 1.000 g/cm^3 [24], in agreement with the experimental pressure dependence of the TMD line. We have studied the -80M Pa isobar. Seven simulations have been performed at the state points indicated in

Table I. It ranges from 35 degrees above the TMD to 45 degrees below, thus covering both the normal and supercooled states of water, for time periods ranging from a few hundreds ps at high T, to 50 ns at the lowest T.

Densities have been chosen on the basis of trial and error preliminary runs. The corresponding pressures for the chosen final densities are reported in Table I. We have preferred to work in the N V E ensemble to avoid the interference of dynamics of the pressure bath with dynamics of the system. Note also that the density dependence of the diffusion coefficient over the small range of densities studied (0.966 - 0.990 g/cm^3) is much smaller than the temperature dependence [28,29].

We also present the results of a simulation carried out for hexagonal ice at a temperature of T = 194 K with a proton disorder. We studied a box of 2.7 x 2.3 x 2.2 nm^3 containing 432 water molecules interacting via the same SPC/E potential used for simulations of liquid water. We simulated a state point along the same isobar studied for liquid water, corresponding to a density of 0.9364 gr/cm^3, an average potential energy of the system of -56.2KJ/Mol and a pressure of -77.7 M Pa.

TABLE I. Simulated State Points

T (K)	ρ_w(gr / cm^3)	E(KJ / Mol)	P(MPa)	D(in10^{-5}cm^2 / sec)
284.5	0.984	-48.1	-73±11	1.3±0.1
258.5	0.986	-50.0	-76±12	0.52±0.05
238.2	0.987	-51.6	-80±10	0.14±0.01
224.0	0.984	-52.6	-75±15	0.044±0.004
213.6	0.977	-53.4	-78±14	0.011±0.004
209.3	0.970	-53.8	-99±18	0.0051±0.0009
206.3	0.966	-54.2	-90±23	0.0018±0.0011

IV. Results

In this section we present results of the simulations. The discussion is divided into four subsections from A to D each centered around a correlation function.

A. Static Quantities.

In order to define thermodynamic states covered in our simulation we begin by tabulating the temperature dependence of the density, potential energy, and pressure in Figure 1. We note that the TMD is around (240±5)K for the SPC/E model along this isobar. In order to compare the simulation results with some real experiments in the future, we shall use the TMD as a convenient reference point to measure the temperature distance. From Figure 1 we note that the energy does not show any significant change of slope at small T, consistent with the fact that all simulations are equilibrated and the pressure is constant within the error bar.

Oxygen-oxygen radial distribution functions g(r) for some selected temperatures are shown in Figure 2. The figure shows that on cooling the system, the first peak of g(r) increases, the first minimum decreases and the second peak also increases. This illustrates the fact that, the nearest and the next nearest neighbor shells become more and more well defined. The number of nearest neighbors, calculated by integrating g(r) up to the position of the first minimum (3.2 Å), decreases from 4.2 at high T to almost 4 at the lowest temperature,

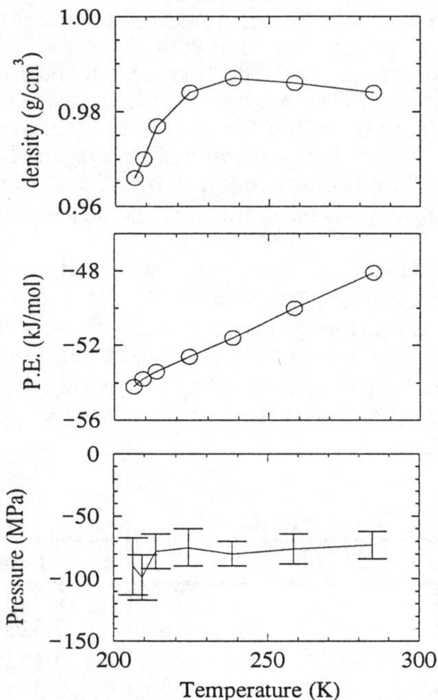

Figure 1. Temperature dependence of density (top), potential energy (P.E.) (center) and pressure (bottom). (Reprinted with permission from reference 17b. Copyright [1996] The American Physical Society. All rights reserved.)

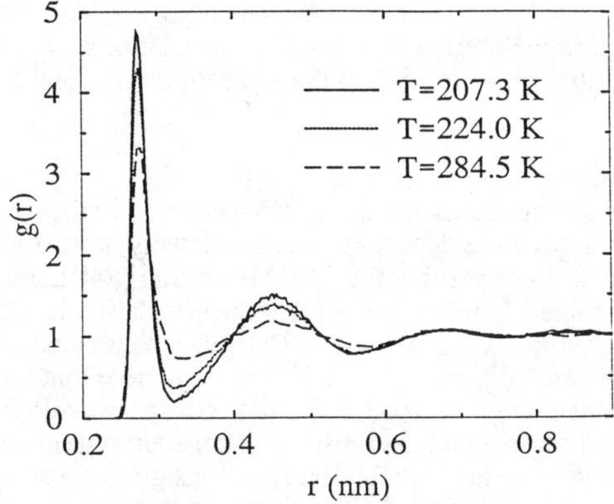

Figure 2. Radial distribution function g(r) for the oxygens for three selected temperatures. (Reprinted with permission from reference 17b. Copyright [1996] The American Physical Society. All rights reserved.)

supporting the progressive formation of a tetrahedral structure around each molecule. Still, the presence of a non-negligible population around 3.5 ∞A is indicative of the presence of five (or more) coordinated molecules, whose role in the dynamical restructuring of the HB network has been studied previously [28].

The T dependence of the oxygen-oxygen partial structure factor $S_{OO}(Q)$ is shown in Figure 3. The split first peak in water, observed in x-ray diffraction experiments [30], is recovered. As temperature decreases, the peak heights increase and peaks become better resolved. From Figure 2 and Fig. 3, we see that no dramatic changes in structure are happening on cooling the system; no sign of an increase of small wave-vector (critical) density fluctuation is observed, in agreement with previous simulations and with the basic idea of MCT. We note also that position of the first maximum Q_{max} of $S_{OO}(Q)$ does not shift significantly with T. Since T-dependence of Q_{max} is weak, we shall neglect it in the following and will compare data at different T at the same value of $Q_{max} = 1.8$ Å^{-1}.

B. Van-Hove Space-Time Self-Correlation Function.

The test-particle dynamics may be studied in a great detail by computing the space-time van-Hove self-correlation function $G_S(r,t)$ which describes the moving away of a test particle from the origin. For a system of N spherical molecules, $G_S(r,t)$ is defined as:

$$G_s(r,t) = \frac{1}{N}\left\langle \sum_{i=1}^{N} \delta(\vec{r} + \vec{r}_i(0) - \vec{r}_i(t)) \right\rangle \qquad (7)$$

Physical meaning of the van-Hove self-correlation function is that $4\pi r^2 G_S(r,t)dr$ is the probability of finding a test particle at a distance r from the origin at time t, given that the same particle was at the origin at time t = 0. In Fig. 4 (a - c) we show $4\pi r^2\, G_S(r,t)$ at the lowest T for three selected intervals of time. (5a): for t < 0.25 ps, the ballistic regime, $G_S(r,t)$ changes rapidly with time. A molecule moves out from the origin freely and explore more and more space as time increases as shown by rapid extension of the tails of $G_S(r,t)$. (5b): for times approximately between 0.25 ps and 130 ps, the cage regime, $G_S(r,t)$ changes very slowly in time. (5c): for time longer than 130 ps, the diffusive regime, tails of $G_S(r,t)$ progressively extend in space as time goes on. We note that there is no evidence for a double peak structure of $G_S(r,t)$, even at the lowest temperature, suggesting that hopping contribution to diffusion is still negligible.

The intermediate time region in which $G_S(r,t)$ is slowly varying in time corresponds to the β-relaxation region described by the MCT. In this regime, a space-time factorization is supposed to hold (see Equation 4). Figure 5 is intended to compare how this prediction of the MCT is born out in our simulated system. In Fig. 5 we show the probability that a molecule has moved less than r during time t, $n(r,t) = \int_0^r 4\pi r'^2 G_s(r',t)dr'$, as a function of time, in the β-relaxation region, for some selected r values. This representation, using the integral of G_S, is less noisy than G_S itself. We show three different r values, corresponding to distances close to, and larger than, the radius of the cage. We expect from Eq. 5 that n(r,t) can be well represented by the functional form, $f(r) - g(r)\, t^b$. From fitting n(r,t) in the b-region with such a functional form, we find that the resulting exponent b has an effective value that depends on r. It decreases on decreasing r, in apparent disagreement with the MCT. However, if we fit all curves simultaneously with an

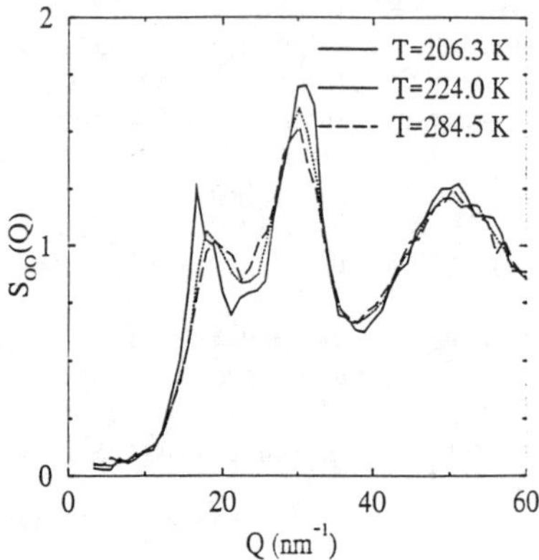

Figure 3. Structure factor $S_{oo}(Q)$ for the oxygens for three selected temperatures. (Reprinted with permission from reference 17b. Copyright [1996] The American Physical Society. All rights reserved.)

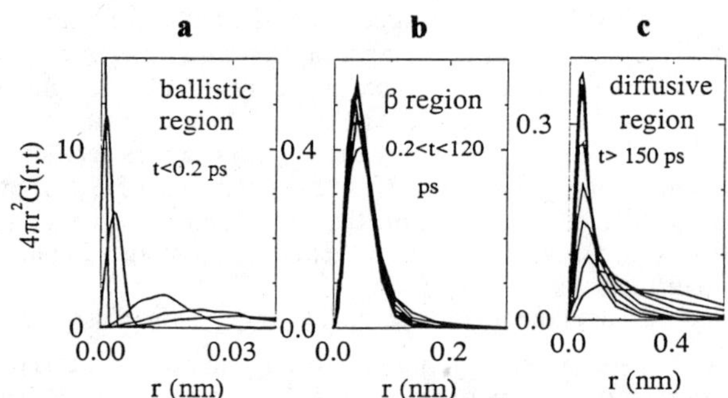

Figure 4. Self part of the van Hove space-time correlation function for three time regions of interest at the lowest studied temperatures. (a) ballistic region: t < 0.2 ps, (b) β-relaxation region: 0.2<t<120 ps, (c) α-relaxation region: t > 150 ps. (Reprinted with permission from reference 17b. Copyright [1996] The American Physical Society. All rights reserved.)

expansion in terms of t^b up to the second (or third) order, we obtain a single r-independent value $b = 0.50 \pm 0.05$. We thus suggest to use n(r) as the quantity to look at when checking Eq. 3 via simulations. The apparent dependence of the effective b on r will be discussed in more details in the following.

C. Mean square displacement and the non-Gaussian corrections.

We show next the mean square displacement (MSD) of the oxygen atoms, $\langle r^2(t)\rangle = \langle |\vec{r}(t) - \vec{r}(0)|^2 \rangle$, for all the studied temperatures, in Figure 6. This is the second spatial moment of $G_S(r,t)$. Curves have been plotted in a log-log scale in order to better display the flattening-out behavior of $<r^2>$ at intermediate times. All curves have an initial t^2 region, describing the ballistic region. At high T, the ballistic region is followed by the usual diffusive, linear dependence in t. However, as already suggested by the behavior of $G_S(r,t)$, for low T, an intermediate time region develops where $<r^2>$ remains essentially flat. A typical molecule appears trapped in a cage for a considerable amount of time before starting to diffuse away. Indeed, for the lowest temperature, $<r^2>$ becomes almost flat for three decades of time. During this time, no significant diffusion is present. The molecule vibrates and librates mainly within the cage. From the value of $<r^2>$ at the plateau, we estimate the radius of the cage to be about 0.55 Å, very slightly T dependent. Note that the onset of the cage effect appears always at the same time, 0.25 ps, regardless of temperature. This is completely analogous to the corresponding result obtained by Kob and Andersen [16] for binary mixtures of Lennard-Jones spheres close to the glass transition. While $<r^2>$ is constant, no significant structural changes happen in the system. Thus the system is frozen in a particular configuration for a long time. For longer times, onset of diffusion allows for a structural relaxation.

In Figure 7 we show the MSD for oxygen atoms in ice compared with that in liquid at three selected temperatures. As expected there is no diffusion for oxygen in ice. It is interesting to observe that there is no substantial difference in the short time dynamics of water molecules up to .25 ps in the liquid and in the solid. The plateau of the MSD of ice starts at a slightly smaller cage of radius 0.52 Å, as compared to the 0.55 Å at the lowest temperature we studied in the liquid state.

The values of D extracted from the asymptotic behavior of MSD are shown in Figure 8 together with the fitted curve to the power-law temperature dependence $D = D_0(T/T^* - 1)^{\gamma}$. We also show the values of D from [27]. For both isobars, the temperature dependence of D is well described by a power-law, as in real experiments [35]. T* and γ are pressure dependent. The difference between T* and the corresponding TMD is always about 50 K. We note also that the difference between T_S and TMD in real water is also about 50 K. Henceforth we suggest the possibility of interpreting T_S as the temperature of structural arrest. The differences in g between the two simulated isobars are consistent with the experimentally observed sensitivity of γ on pressure (see Ref. [35]) and with the MCT. Indeed, according to the MCT, γ depends on the specific point of the glass transition line $T_C(P)$ approached.

From Fig. 4 we see that $G_S(r,t)$ is apparently a Gaussian function in space only for early times. To quantify the degree of non-Gaussianity we calculate the so-called non-Gaussian parameters $\alpha_n(t)$. They are defined by

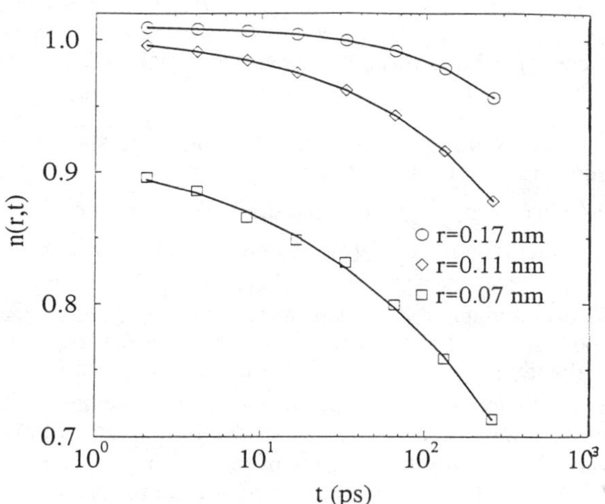

Figure 5. $n(r,t)$ as a function of time for three selected r values in the β – relaxation region. r = 1.7 Å (square), r = 1.1 Å (circle), r = 0.7 Å (diamond). Full lines are fits according to a third order expansion in powers of t^b with b = 0.5. (Reprinted with permission from reference 17b. Copyright [1996] The American Physical Society. All rights reserved.)

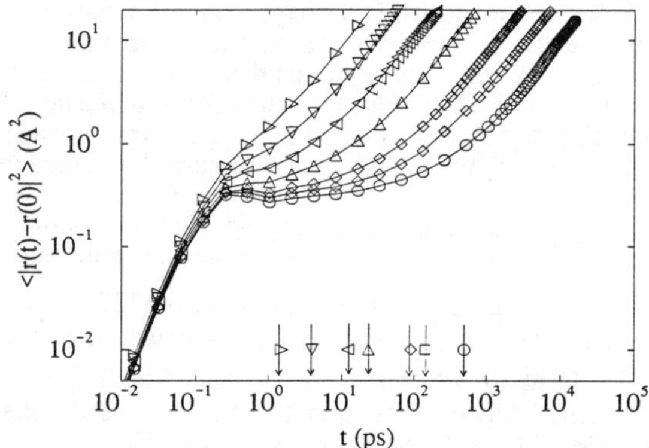

Figure 6. MSD as a function of time for: T = 206.3 K (circle), T = 209.3 K (square), T = 213.6 K (diamond), T = 224.0 K (triangle), T = 238.2 K (left-pointed triangle),T = 258.5 K (down-pointed triangle), T = 284.5 K (right-pointed triangle). The curves show the cage effect, starting at 0.25 ps, followed by an eventual diffusion of the molecule. Arrows indicate the time at which the non-Gaussian parameter $\alpha_2(t)$ is the maximum. (Reprinted with permission from reference 17b. Copyright [1996] The American Physical Society. All rights reserved.)

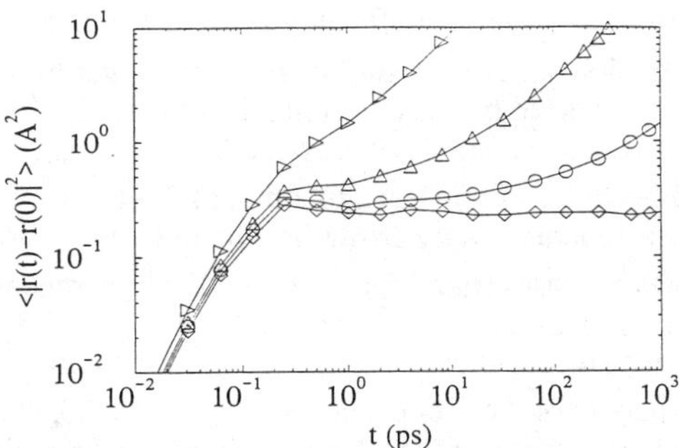

Figure 7. MSD of oxygens for hexagonal ice and liquid water. (right-pointed triangle) represents the liquid at T = 284.5 K, (triangle) the liquid at T = 224.0 K, (circle) the liquid at T = 206.3 K, (diamond) ice at T = 194 K. The lines are to guide the eyes. (Reprinted with permission from reference 17b. Copyright [1996] The American Physical Society. All rights reserved.)

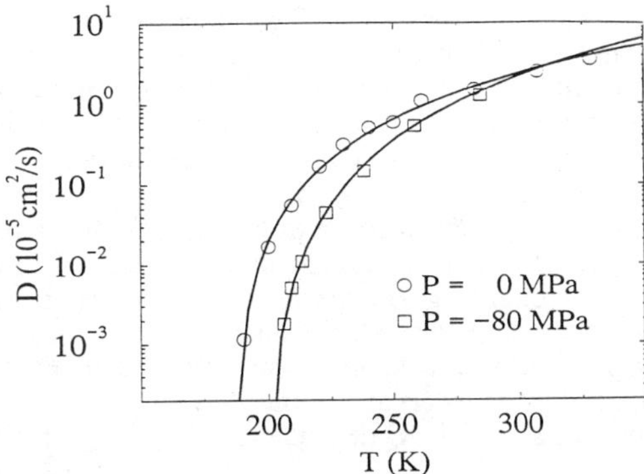

Figure 8. Temperature dependence of the diffusion coefficient D for two isobars. (square) are from Table I, (circle) from [27]. Full lines are power-law fit respectively given by

$$D = 13.93\,(T/198.7-1)^{2.73} \text{ and } D = 7.39\,(T/186.3-1)^{2.2.9}$$

where D is in cm^2/s and T is in K. (Reprinted with permission from reference 17b. Copyright [1996] The American Physical Society. All rights reserved.)

$$\alpha_n(t) = \frac{3\langle r^{2n}(t)\rangle}{c_n\langle r^2(t)\rangle^n} - 1, \qquad c_n = \frac{(2n+1)!!}{3^n} \tag{8}$$

We note that, as found in several previous simulations of liquids close to the glass transition [33,36], $\alpha_2(t)$ increases significantly in the β region and reaches its maximum when diffusion starts to be significant. For longer times, $\alpha_2(t)$ goes back to zero. This is expected because for long time $G_S(r,t)$ has to go back to a Gaussian shape [34] whose variance is controlled by the diffusion coefficient D. Figure 9 shows the behavior of $\alpha_2(t)$ as a function of T. The time at which $\alpha_2(t)$ is maximum, t_{amax}, is also indicated by arrows in Figure 6. We note that in the case of SPC/E water $\alpha_2(t)$ increases significantly in the β region, much more than it was observed in previous simulations of supercooled liquids [33,36,37]. We also note that t_{amax} has a power-law dependence on T - T* , as shown in the inset of Figure 9. The apparent exponent of the T dependence of t_{amax} is 2.5, close to the γ value of 2.7 found in the T dependence of D. A slight change in T* or the restriction of the T range to the five lowest T would allow a fit of t_{amax} with an exponent 2.7. t_{amax} could be used to locate the glass transition [38].

D. Intermediate scattering function.

We now move on to the incoherent intermediate scattering function $F_S(Q,t)$, the spatial Fourier transform of the van Hove self-correlation function $G_S(r,t)$. $F_S(Q,t)$ can be measured by an incoherent neutron scattering experiment. Previous works on the behavior of $F_S(Q,t)$ in simulated water [39 - 41], although based on rather limited simulated time scales (up to 10 ps at most) and limited temperature range, have shown that: (i) the intermediate scattering function for the center of mass, $F_S(Q,t)$, is non-Gaussian except at short times [39]; (ii) at room temperature the diffusive behavior of water molecules is not describable by a discrete jump diffusion [39 - 41]; (iii) the decay of $F_S(Q,t)$ has a fast and a slow component, the time scale of which becomes increasingly disparate upon supercooling [41].

Figure.10 shows the T dependence of $F_S(Q_{max},t)$. Note that for all T investigated, $F_S(Q_{max},t)$ decays to zero in the long time limit. This confirms that the simulations were long enough to guarantee the complete decay to zero of the test particle correlation function, i.e. that all simulations were in the liquid state and in equilibrium. $F_S(Q_{max}, t)$ also show the presence of three different regimes, the initial one characterized by a fast decay, followed by a plateau region, and by a final decay to zero. Figure 11 shows the Q dependence of $F_S(Q_{max},,t)$ at one selected temperature.

For very early times, $F_S(Q, t)$ decays following a quadratic dependence on time, characteristic of the ballistic motion. This is not surprising because we have seen already that $\alpha_2(r,t)$ is very small in this time range. Moreover, $F_S(Q, t)$ both for ballistic motion and for vibrations in a harmonic potential well, is described by a Gaussian function in space [34].

At intermediate times, $F_S(Q, t)$ is slowly varying, confirming the existence of a time region where no significant structural changes are observed. In this time region, molecule explore all the space inside the cage.

For long times, $F_S(Q,t)$ decays in a non-exponential fashion. We can fit the entire curve rather well by the following equation

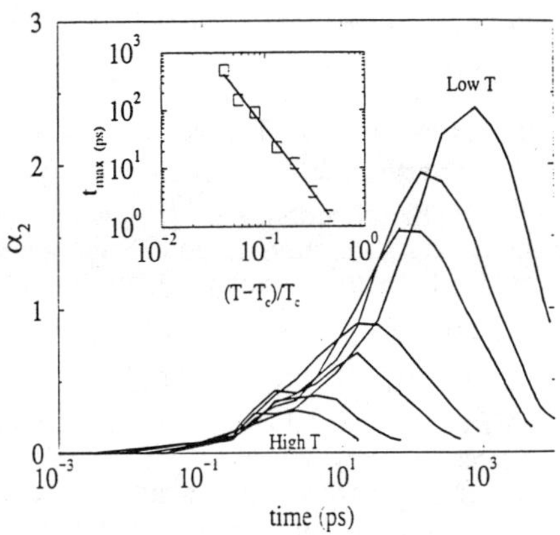

Figure 9. Non gaussian parameter $\alpha_2(t)$ as a function of time for all studied temperatures. The inset shows the T dependence of the position of the maximum t_{max} to highlight the power law dependence on $T/T_c - 1$. The full line is a power-law with exponent 2.5. (Reprinted with permission from reference 17b. Copyright [1996] The American Physical Society. All rights reserved.)

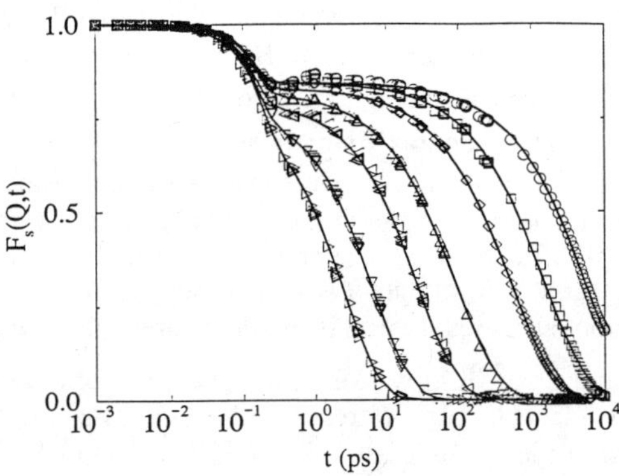

Figure 10. $F_s(Q_{max}, t)$ vs time (symbols as in Fig. 7). Solid lines are calculated according to Eq. 11. (Reprinted with permission from reference 17b. Copyright [1996] The American Physical Society. All rights reserved.)

$$F_s(Q,t) = [1 - A(Q)]e^{-(t/\tau_s)^2} + A(Q)e^{-(t/\tau_1)^\beta} \quad (9)$$

This equation models the initial decorrelation associated with the motion of the test particle inside the cage by a Gaussian term. This is followed by a term with a stretched exponential decay, describing the relaxation of the cage surrounding the test particle which results in migration of the particle. The factor A(Q) gives the level of the plateau in the intermediate times.

Full lines in Fig.10 and Figure 11 are fit to the data according to the Equation 9. From the fitting procedure we find that τ_s is rather constant and is of the order of 0.15 ps. It has a very weak T dependence as expected. A(Q), is well described by a Debye-Waller like function, $e^{-a^2Q^2/3}$, where a is (0.55 ± 0.03) Å, slightly increasing on increasing T. Figures 12 and 13 show the values of the parameters obtained from the fit, both in T and in Q, for the two most interesting parameters, β and τ_1. We note that β starts from 1 at small Q and goes to 0.5 at high Q. There is also a weaker dependence on T, suggesting that β increases on increasing T. This Q and T dependence of β is what one would expect for a phenomenon which happens on a very precise length scale. When the inverse of the magnitude of the Q vector is much larger than the cage size, the slow dynamics, manifested through the reduced values of the stretched exponent, ceases to persist. For larger distances, compared to the cage dimension, diffusion is normal and the decay of the test-particle density fluctuation goes back to the usual exponential form, e^{-DQ^2t} ($\beta = 1$). Similarly, on increasing the temperature, cages break and reform on a faster rate and the convergence of the stochastic process to a gaussian is faster. We look next to τ_1. τ_1 signifies the time it takes for the test-particle density fluctuation to die out over a length of Q^{-1}, i.e. the time it takes a molecule to diffuse over distances of the order Q^{-1}. For large distances and at all T, the leading propagation mechanism is diffusion, which implies that $\tau_1 = (DQ^2)^{-1}$. For small distances, at high T, we still observe the Q^2 behavior, in agreement with the fact that $\alpha_2(t)$ is not very large. From Figure 13 we observe that $\tau_1 DQ^2 = 1$ at low Q and high T while at high Q and low T significant deviations are present. τ_1^{-1} seems to crossover rom a Q^2 to a Q behavior. Note that a Q^2 dependence has been observed for glycerol close to its glass transition [42]. At low T, τ_1 is bigger than one would expect if diffusion where the only mechanism. This suggest that at short length scales, diffusion is lower than it would be, suggesting the presence of anomalous diffusion over small scales. It should be noted that we do not find $\tau_1 \sim Q^{-2/\beta}$ at large Q and small T, as was observed in a glass forming microemulsion and polymer melts [43,44]. This difference probably stresses the highly non-Gaussian behavior of the dynamics in SPC/E water. Indeed, the behavior $\tau_1 \approx Q^{-2/\beta}$ is expected when $G_S(r,t)$ is a Gaussian in space with a variance growing as t^β.

We now come to a more detailed study of the behavior of the correlator $F_S(Q,t)$.in the late β region. As discussed in Sec II, all correlators are supposed to decay as a power law with the von Schweidler exponent b (Eq. 3). Figure 14 shows that, for all low temperatures such that the β region is clearly established, the fit with a power-law is superior than the stretched exponential form. The Q

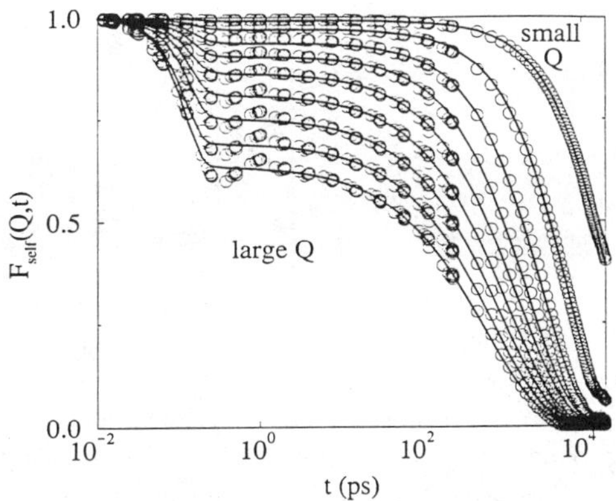

Figure 11. $F_s(Q_{max},t)$ vs time for different Q values at T = 209.3 K (from top to bottom, integer multiples of 0.33 $Å^{-1}$). Solid lines are calculated according to Eq. 11. (Reprinted with permission from reference 17b. Copyright [1996] The American Physical Society. All rights reserved.)

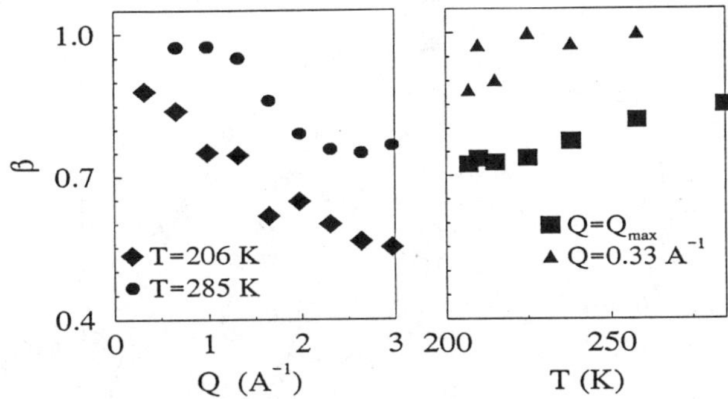

Figure 12. Results from the fitting procedure for the parameter β. Left panel: β as a function of Q at two different temperatures. Right panel: β as a function of T at $Q = Q_{max}$ and at the smallest available Q = 0.33 $Å^{-1}$. (Reprinted with permission from reference 17b. Copyright [1996] The American Physical Society. All rights reserved.)

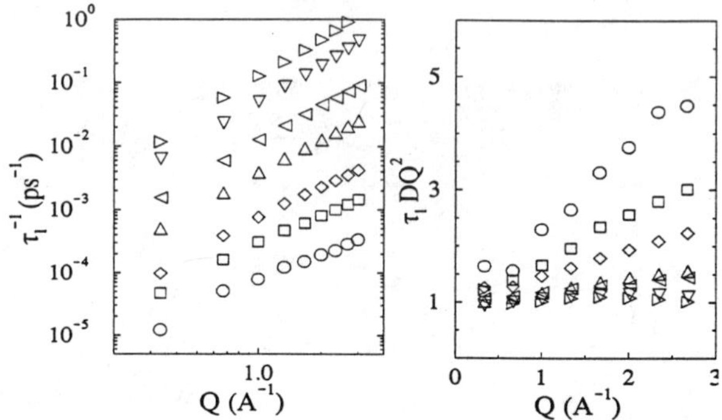

Figure 13. Results from the fitting procedure for the parameter τ_1 . The left panel shows the Q-dependence of τ_1 at all temperatures (symbols as in Figure 6). The right panel shows $DQ^2\tau_1$ as a function of Q. Note that at low T and high Q, $DQ^2\tau_1$ increases linearly, suggesting that $\tau_1 \approx Q^{-1}$. (Reprinted with permission from reference 17b. Copyright [1996] The American Physical Society. All rights reserved.)

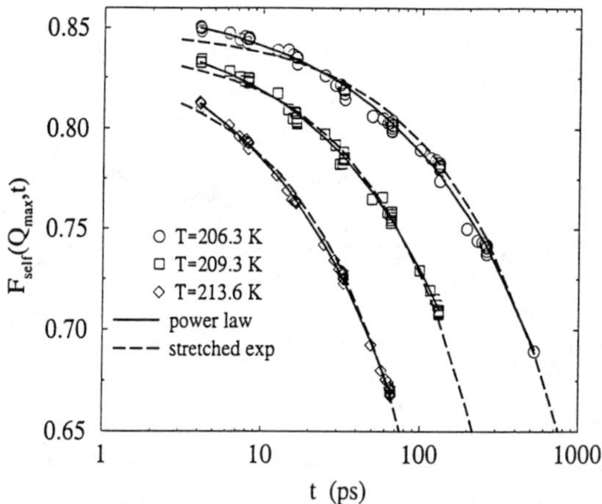

Figure 14. Comparison between the stretched exponential and the von Schweidler law in the β-region. Full curves are the comparison with $F_s(Q_{max},t)$ at the three lowest temperatures. (Reprinted with permission from reference 17b. Copyright [1996] The American Physical Society. All rights reserved.)

dependence of b is instead at odd with the predictions of MCT. We find that the apparent exponent b, as calculated from fitting the time dependence of $F_S(Q,t)$. in the β region, apparently decreases on increasing Q. It goes from the value 0.7 at the smallest Q down to about 0.3 at high Q. At Q_{max} the value of the effective b is consistently given by $b = 0.5 \pm 0.05$, independent on Q, as shown in Figure 14. The decrease of the apparent b on decreasing Q was also detected in a previous calculation of Lennard Jones mixtures close to the glass transition [16].

As discussed in Section II, the exponents b and γ are related by MCT. We independently measure both exponents, one from the t dependence of the correlators in the β region and one from the T dependence of D. The resulting (b, γ) pair obtained for SPC/E water is (0.5, 2.75). The agreement between our calculated value and the theory (Equations 5 and 6) is surprisingly good.

V. Slow Dynamics In Real Water

The most direct way of confirming the existence of a slow dynamics at supercooled temperatures as described above is to measure the self-intermediate scattering function $F_S(Q,t)$, computed in the CMD discussed in section IV D. In a QENS experiment, one measures the self-dynamic structure factor, $S_s(Q,\omega)$, which is the Fourier transform of $F_S(Q,t)$. As we have already mentioned, study of the two-step relaxation and especially the long-time alpha relaxation in time domain in the intermediate scattering function amounts to an analysis of the line shape of the self-dynamic structure factor in the frequency domain. For this purpose, we need data from a high resolution QENS experiment. The Fourier transform of the Gaussian term in Equation 9 will result in a Gaussian term in ω with a full width at half maximum equal to $4\sqrt{\ell n2} / \tau_s$. Recalling $\tau_s = 0.15$ ps, the line width is calculated to be 14.5 mev. For a typical QENS experiment, this line amounts to a flat background whose height is $\tau_s / 2\sqrt{\pi} = 0.0423$ ps, a negligible contribution compared to that from the second term in Equation 9. Furthermore, the Debye-Waller factor A(Q) has a value 0.76 at the Q_{max} and a value 0.92 at $Q = 1.0$ Å^{-1}. Thus, as far as a QENS experiment is concerned, the line shape of the dynamic structure factor is dominated by the contribution from the alpha relaxation term.

We shall re-analyze a data set taken from water contained in Vycor glass that has been published previously [50]. This QENS data set was taken with an instrument resolution of HWHM = 28 μev, which is good enough for testing the theory given by Equation 9.

Vycor brand porous glass no. 7930 is a product of Corning Glass Works. The Vycor glass has a pore volume fraction of 28% and the average pore size of 50 Å. A full hydration is defined as 0.25 g of water per g of Vycor. The QENS data set we analyzed is taken at a half hydration. For time scales of hundreds of pico-seconds and even nano-seconds, the distance that a typical water molecule travels is much smaller than the average pore size of the glass. Thus the confinement effect on diffusion of water molecules need not be considered. However, there may be some surface effects for water containing in pores, but we shall ignore it and treat the water as a bulk water in the first approximation.

Figure 15 shows some of the fits to the experimental data at T=268 K at several Q values using the model of Equation 9[51]. We fit the QENS spectra with Fourier transform of just the alpha relaxation term. In these fits the exponent β was fixed at 0.55, a value taken from the literature on a typical glass former. We are

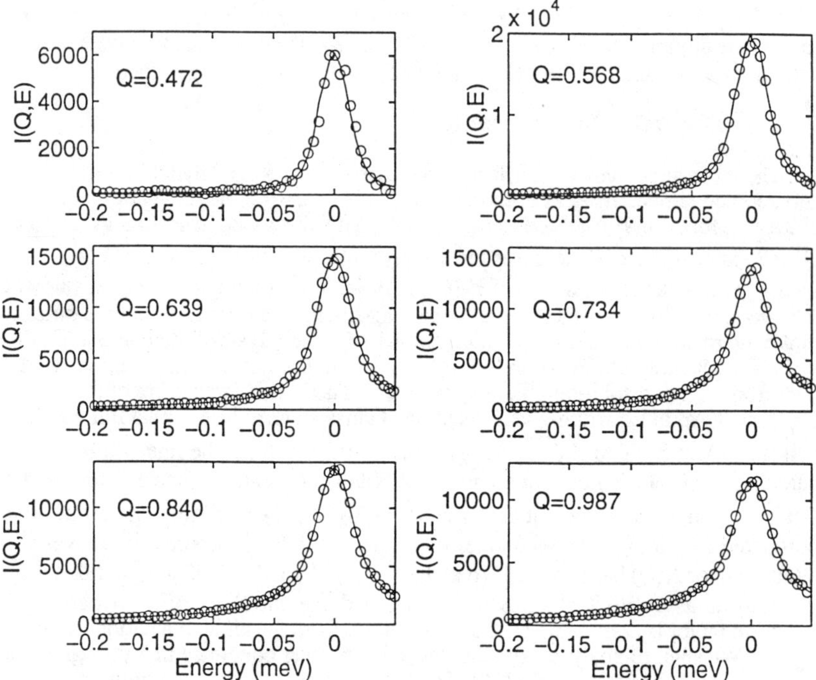

Figure 15. Analysis of QENS data for 50% hydrated Vycor at 278 K using the model of Eq.11. Note that the energy resolution of the instrument is 24 μeV FWHM. The significant point to note is the excellent fit to the wing of the quasi elastic peak by the Cole-Cole dispersion function which is an approximation to the Fourier transform of a stretched exponential function.

more interested in the magnitude of the alpha relaxation time τ , as functions of Q and temperature. As can be seen, the stretched exponential function or equivalently the Cole-Cole dispersion function fits the wing of the quasi elastic peak very well. In a previous analysis [50], this same set of data was fitted in terms of a confined diffusion model which gives an expression for the dynamic structure factor consisting of a superposition of an elastic peak and a Lorentzian quasi-elastic peak. In our present model we do not require an elastic component which is more reasonable for a liquid state [51].

Figure 16 gives the Q dependence of the relaxation rate $1/\tau$ extracted for the three temperatures indicated in the figure. It is interesting to observe that the relaxation rate $1/\tau$ does go to zero at $Q = 0$ and it is well described by a power law in Q with an exponent γ ranging from 2.51 at $T = 298$ K to 2.06 for $T=268$ K. We need to perform, in the near future, a QENS experiment with bulk water at a resolution of 10 μev for a more definite confirmation of the slow dynamics.

VI. Conclusions

We have presented evidence that SPC/E water undergoes a kinetic glass transition 50 degrees below the TMD. The CMD results are well accounted for by the idealized MCT of supercooled liquids, suggesting an interpretation of the so-called Angell temperature as the critical temperature of the MCT [17]. In this regard, the apparent power-law behavior of the temperature dependence of transport coefficients in liquid water on supercooling is traced to the formation of cages and the associated slow dynamics resulting from the presence of long-lived cages. In other words, the divergence of transport coefficients does not need to rely on the existence of a thermodynamic instability, either connected to the re-entrance of the gas-liquid spinodal or to the presence of a critical point at high pressure and low temperature.

It is important to stress that the finding of our CMD work could be tested by a careful analysis of a high-resolution QENS experiment. We found that the line shape of the slow relaxation part of the $F_S(Q,t)$ can be described very accurately by a stretched exponential decay around and above Q_{max}. Unfortunately in a previous, extensive QENS experiment of supercooled bulk water the instrumental resolution of 84 μeV was used which would not allow one to see the detailed lineshape characteristic of the stretched exponential relaxation (a Cole-Cole dispersion function) [46]. The T dependence of β at Q_{max} can also be tested, as well as the Q dependence of τ_l and its crossover from a Q^2 behavior at small Q to a Q behavior at high Q at low T. In section V we presented an analysis of a set of existing QENS data for water contained in Vycor glass. The results confirm the existence of the predicted slow dynamics as expressed by Equation 9.

The scenario described above bears a strong resemblance to the results of MD simulation for the mixed Lennard-Jones spheres carried out recently to test the MCT description of the kinetic glass transition. In this respect, the prediction of the idealized MCT seems to be robust, and able to described both a fragile liquid [47,48], a system of atoms interacting with a spherically symmetric potential, and a system with molecules interacting via a highly directional potential. It is surprising that a simple Lennard-Jones system, in which molecules are confined in cages with a large coordination number, behaves, close to its glass transition, similarly to a tethrahedrally coordinated system, like a SPC/E water, in which the cages are

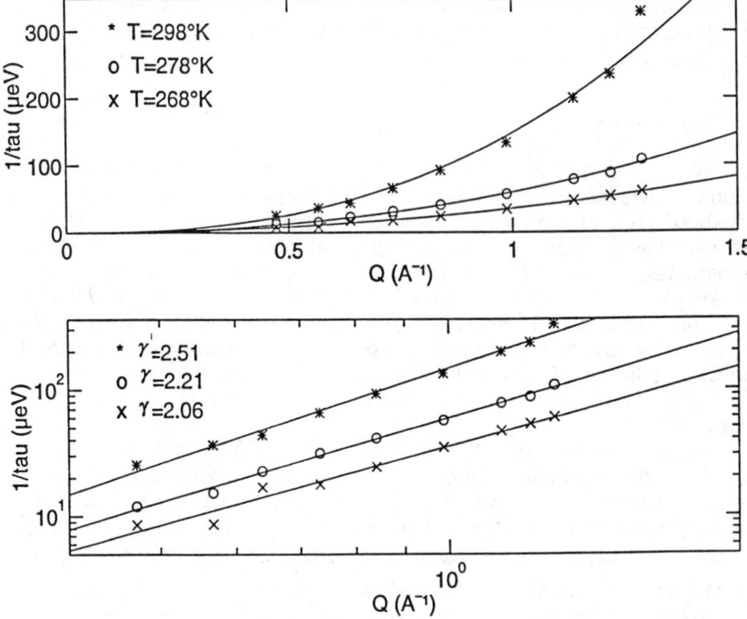

Figure 16. Q dependence of the alpha relaxation rate $1/\tau$ as a function of temperature. Note that the power law dependence on Q is well confirmed. The relaxation rate is proportional to Q to the power g indicated in the figure.

formed rather by a deep, directional hydrogen-bonding potential than by the excluded volume constraint.

The general picture we get from this study is that the system goes back to the "normal" behavior for distances much larger than the cage size. Indeed, we find that for small Q, the stretched exponential behavior tends to the usual simple exponential, diffusion becomes normal, $1/\tau_1$ scales as Q^2 and the Gaussian approximation for $F_S(Q,t)$ becomes sufficiently good. At small Q, the presence of a nearby ideal glass transition appears only via the T dependence of D and τ_1.

Acknowledgments

Research of SHC is funded by the Materials Chemistry Division of US Department of Energy. PG acknowledges financial support from the Foundation Blancleflor Boncompagni-Ludovisi. Research of FS and PT is supported by GNSM/CNR and INFM/MURST.

References

[1] C.A. Angell, *Ann. Rev. Phys. Chem.* **34**, 593 (1983).
[2] C.A. Angell, in *Water: A Comprehensive Treatise*, Ed. F. Franks (Plenum, New York, 1981), Ch. 1.
[3] P.G. Debenedetti, *Metastable Liquids* (Princeton University Press, 1997), in press.
[4] R.J. Speedy, J. Chem Phys.**86**, 982 (1982).
[5] R.J. Speedy and C.A. Angell, J. *Chem. Phys.* **65**, 851 (1976).
[6] P.H. Poole, F. Sciortino, T. Grande, H.E. Stanley, C.A. Angell, *Phys. Rev. Lett.* **73**, 1632 (1994).
[7] P.H. Poole, F. Sciortino, U. Essmann, H.E. Stanley, *Nature* **360,** 324 (1992).
[8] H. Tanaka, *Nature* **380** 328 (1996).
[9] H.E. Stanley and J. Teixeira, *J. Chem. Phys.* **73** 3404 (1980).
[10] S. Sastry, P.G. Debenedetti, F. Sciortino and H.E. Stanley, *Phys. Rev. E* **53**,6144 (1996).
[11] F.X. Prielmeir, E.W. Lang, R.J. Speedy and H.D. Ludeman, *Phys. Rev. Letts.* **59** 1128 (1987).
[12] C.A. Angell, *Nature*, **331** 206 (1988).
[13] E. Leutheusser, *Phys. Rev. A*, **29**, 2765 (2984); U. Bengtzelius, W. Götze and A. Sjölander, *J. Phys. C* **17** 5915 (1984).
[14] A.P. Sokolov, J. Hurst and D. Quitmann, *Phys. Rev. B.* **51**, 12865 (1995).
[15] W. Götze and L. Sjögren, *Rep. Prog. Phys.* **55**, 241 (1992).
[16] W. Kob and H.C. Andersen, *Phys. Rev.* E **51**, 4626 (1995) and *Phys. Rev. E* **52B**, 4134 (1995).
[17] (a) P. Gallo, F. Sciortino, P. Tartaglia, S.H. Chen *Phys. Rev. Letts.* **76,** 2730 (1996); (b) F. Sciortino, P. Gallo, P. Tartaglia and S.H. Chen, *Phys. Rev. E* **54**, 6331 (1996).
[18] K. Kawasaki, *Transport Theory and Statistical Physics* **24,** 755 (1995).
[19] J.P. Boon and S. Yip, *Molecular Hydrodynamics*, McGraw Hill, New York 1980.
[20] M. Fuchs, I. Hofacker and L. Latz, *Phys Rev.* A **45,** 898 (1992).
[21] W. Götze and A. Sjögren, *Transport Theory and Statistical Physics* **24,** 801 (1995).
[22] H.J. C. Berendsen, J. R. Grigera and T. P. Straatsma, *J. Phys. Chem.* **91**, 6269 (1987).

[23] O. Steinhauser, *Mol. Phys.* **45**, 335 (1982).
[24] P. H. Poole, F. Sciortino, U. Essmann and H.E. Stanley, *Phys. Rev. E* **48** 3799 (1993).
[25] H.J. C. Berendsen et al, *J. Chem. Phys.* **81**, 3684 (1984).
[26] Y. Guissani and B. Guillot, *J. Chem Phys.* **98** 8221 (1983) P.E. Smith and W.F. van Gunsteren, *Chem. Phys. Letts.* **215** 315 (1983).
[27] L.A. Baez and P. Clancy, *J. Chem Phys.* **101** 9837 (1984).
[28] F. Sciortino, A. Geiger, H. E. Stanley, *J. Chem. Phys.* **96** 3857 (1990);
[29] L. Vaisman, L. Perera, and M.L. Berkowitz *J. Chem. Phys.* **98**, 9859 (1993).
[30] A. H. Narten, H. A. Levy, *J. Chem. Phys.* **55,** 2263 (1971).
[31] G. F. Signorini, J. L. Barrat and M. Klien, *J. Chem. Phys.* **92** 1294 (1990).
[32] C.A. Angell *Science* **267** 1924 (1995).
[33] L. J. Lewis *Phys. Rev.* E **50**, 3865 (1994).
[34] G. H. Vineyard, *Phys. Rev A* **110**, 999 (1958).
[35] E. W. Lang and H.D. Ludemann, *Angew. Chem. Int. Ed. Engl.* **21**, 315 (1992).
[36] B. Bernu, J. P. Hansen, Y. Hiwatari and G. Pastore *Phys. Rev. A* **36**, 4891 (1987).
[37] J. P. Hansen and S. Yip, *Transport Theory and Statistical Physics* **24**, 1149 (1995).
[38] T. Odagaki and Y. Jiwatari, Phys. Rev. A **43**, 1103 (1991).
[39] A. Rahman, F. H. Stillinger, *J. Chem. Phys*, **55,** 3336 (1971); F. H. Stillinger and A. Rahman, *J. Chem. Phys*, **57**, 1281 (1972); F. H. Stillinger and A. Rahman in *Molecular Motions in Liquids*, Ed. J. Lascombe, p. 479 (D. Reidel Publishing Company, 1974).
[40] R. W. Impey, P.A. Madden, I.R. McDonald, *Mol. Phys.* **46**, 513 (1982).
[41] J.J. Ullo, *Phys. Rev.* A **36**, 816 (1987).
[42] W. Petry and J. Wuttke, *Transport Theory And Statistical Physics* **24**, 1075 (1995).
[43] E. Y. Sheu, S. H. Chen, J. S. Huang and J. C. Sung, *Phys. Rev. E* **39** 5867 (1989).
[44] J. Colmenero, A. Alegria, A. Arbe and B. Frick, *Phys. Rev. Letts* **69** 478 (1992).
[45] H. Lowen, J. P. Hansen and J. N. Roux *Phys. Rev.* **A 44,** 1169 (1991).
[46]J. Teixeria, M.C. Bellissant-Funel, S. H. Chen and A. Dianoux, *Phys. Rev.* **A31**, 1913 (1985).
[47] C. A. Angell in *Relaxations in Complex Systems*, edited by K. Ngai and G. B. Wright. (National Technical Information Service, U.S. Dept. of Commerce: Springfield, VA, 1985) p. 1;
[48] C.A. Angell, *J. Chem. Phys. Solids* **49**, 863 (1988).
[49] C. A. Angell, *J. Phys. Chem.* **97,** 6339 (1993).
[50] M.C. Bellissent-Funel, S.H. Chen and J.M. Zanotti, *Phys. Rev.* **E 51**, 4558 (1995).
[51] S.H. Chen, P. Gallo and M.C. Bellisent-Funel, "Slow dynamics of water in supercooled states in bulk and near hydrophilic surfaces", in *Non-Equillibrium Phenomena in Supercooled Fluids, Glasses and Amorphorus Materials,* Eda. M. Giordano, D.L. Leporini and M.P. Tosi, World Scientific (1996).

Chapter 21

Raman Evidence for the Clathratelike Structure of Highly Supercooled Water

G. E. Walrafen, W.-H. Yang, and Y. C. Chu

Department of Chemistry, Howard University, Washington, DC 20059

The peak frequency of the Raman OH-stretching contour from liquid water, supercooled to -33 °C at 1 atm, is only 1% higher than that of the OH-stretching peak from the solid, type II, 17:1, H_2O: tetrahydrofuran (THF) clathrate, 3190 cm^{-1} *(1)* versus 3160 cm^{-1}, respectively; and, the molal volume inferred for supercooled water at -45 °C is only 1% larger than that of a the 17:1 THF clathrate, with H_2O as the guest in all 24 voids of the 17.14 Å cubic unit cell *(2)*, 19.2 cm^3/mole *(3)* versus 18.96 cm^3/mole, respectively. Furthermore, polarized, X(ZZ)Y, Raman spectra from cold D_2O display a well-defined isosbestic point at 2430 ± 5 cm^{-1} *(4)*, which lies well within the D-bonded frequency range, thus completely separating OD-stretching components assigned to pentagonal rings, ≈2380 cm^{-1}, from those assigned to hexagonal rings, ≈2520 cm^{-1}. The foregoing observations, and many other types of data, indicate that the structure of water, toward its supercooling limit, involves a clathrate-like H-bonded network, stabilized by the strong H-bonds of pentagonal rings, by cooperative effects between pentagonal rings, and by decreased thermal disorder.

In 1968 Bernal *(5)* described a revised model of liquid water which consituted a major improvement over the 1933 model of Bernal and Fowler *(6)*. This revised model was based on the Keatite structure of silica, $P4_12_12$, which bears close similarity to the ice III structure, $P4_32_12$. Bernal's revised model involved H-bonded five- and six-membered rings, most of which were puckered. The five-membered rings of Keatite, formed by Si atoms, and of ice III, formed by O atoms, are also puckered. This Keatite-based model was found to yield the correct density, ≈1 g/cm^3, and also to agree with the known O-O pair correlation function, with respect to the first and second peak areas *(5)*.

Finney confirmed that the Keatite model was discarded after Bernal's death *(7)*, and thus we decided to develop a related model. Our model was constructed from 100 tetrahedra having angles randomized from 106 to 110 degrees. It involved flexible (bent) H-bonds (O-H···O units), as well as a range of O-O distances.

The completed model contained 44 pentagons, 30 hexagons, and 2 heptagons, most of which were puckered, and many of which shared common sides. However, 12 of the 44 pentagons were found to be planar, and measurement of the angle between two planes, which could be identified, approximately, for the remaining pentagons, yielded an average angle, over all of the 44 pentagons, of 165 degrees, i.e., 15 degrees from planarity.

Many H-bonds in our model were linear, but several deviated from linearity by as much as 20 degrees. The deviation from linearity, averaged over all H-bonds, was 6 degrees.

The number of nearest O-O neighbors in our model was found to be only slightly greater than 4, the number of second O-O neighbors was ≈18, and the density was ≈1 g/cm^3.

We believe that both Bernal' s model and our model conform most closely to cold water. Moreover, both models are devoid of periodicity, and thus they are appropriate for a liquid.

Our model building was accomplished <u>solely</u> to demonstrate that an H-bonded structure containing large numbers of pentagons and hexagons <u>is not inconsistent</u> with some known properties of water. Such models may suggest that large numbers of H-bonded pentagonal and hexagonal rings exist in the structure of water, but they definitely do not consitute a proof.

A similar suggestion comes from the simple observation that the triple point, at ≈−20 °C and ≈2kbar, involves an equilibrium between liquid water, ice III, and ice I_h. Because ice III contains H-bonded pentagonal rings, and ice I_h, H-bonded hexagonal rings, it is not unreasonable to suppose that both types of rings might also exist in liquid water at the triple point, or especially at lower pressures, e.g., 1 atm. However, this latter example raises a new possibility, namely, that supercooled liquid water may favor one type of ring, and ambient water another.

Raman data are described subsequently which indicate that pentagonal rings are favored, energetically, over hexagonal rings, and moreover that the pentagonal ring concentration increases as the temperature of the supercooled water decreases. Of course, one might arrive at the same conclusion from the simplest, geometric, bending-strain considerations, namely, that the O-O-O angle in a planar pentagon is 108 degrees, which is close to the tetrahedral value of 109.5 degrees, whereas the corresponding angle in a planar hexagon is 120 degrees, thus necessitating strain due to bending.

Apart from the preceding structural inferences, there are many thermodynamic, transport, and other data which suggest that highly supercooled water involves bulky polyhedral structures based upon H-bonded pentagons and hexagons, i.e., structures similar (except for the lack of the guest molecule) to clathrate hydrates *(8-13)*.

Molecular dynamics simulations also suggest the possibility of H-bonded pentagons and hexagons for supercooled liquid water *(14)*.

Nevertheless, expecially compelling evidence for clathrate-like structures comes from the observation that the 1 atm activation energies for shear-fluidity and self-diffusion increase to enormous values in very highly supercooled liquid water *(8)*, whereas the corresponding activation volumes are <u>negative</u> from 1 atm up to about 2 kbar, beyond which they attain the normal positive activation volumes exhibited by most other liquids. In addition, the 2 kbar self-diffusion data, when treated by the Young-Westerdahl (YW) method *(8)*, yield a constant enthalpy change of about 5 kcal/mole H_2O. This value contrasts sharply with the 1 atm values, which when treated by the same YW function *(8)*, yield increasingly large activation energies with increased supercooling.

The conclusion which seems to be forced upon us by the above facts, is that pressure disrupts the bulky, pentagon- and hexagon-based polyhedral structures, as required by le Châtelier's principle, to yield a densified water structure having properties more like those of "normal" or unassociated liquids. A pressure of 2 kbar appears to be sufficient to lower the activation energies for shear-fluidity and self-diffusion, significantly. Moreover, high-pressure produces interpenetration or structural deflation, which increases the density by filling or otherwise diminishing the size of the polyhedral voids. Such voids are the most probable source of the negative activation volumes for supercooled water.

Despite the fact that structural models, thermodynamic treatments, and molecular dynamics simulations all point to a picture involving pentagon-based polyhedral structures, it is still desirable to obtain direct experimental, microscopic evidence for such structures. Unequivocal Raman spectroscopic evidence, for example, should be in hand before the existence of such polyhedral structures can be regarded as established for supercooled liquid water. Such direct evidence is now provided in the present work by comparisons of Raman spectra from highly supercooled liquid water with those from supercooled and solid H_2O-THF, e.g., 17:1, the concentration at which the solid forms a well-defined clathrate structure, accurately measured by x-ray diffraction *(2)*.

See Sloan, Ref. *(15)* for structure, and Sloan and Long, Ref. *(16)* for Raman spectra of THF-H_2O solutions.

A digression at this point might further help to elucidate our current view of the structure of supercooled water, as contrasted to the structure of a related H_2O-filled clathrate hydrate. We refer specifically to the previously mentioned agreement between the molal volumes of supercooled water at -45 °C, and the 17:1 THF clathrate structure, where H_2O replaces THF in the 8 large voids, and also fills the 16 small voids of the unit cell *(15,17)*.

We begin with two questions.

(1). How is it possible to explain the close agreement between the above-mentioned molal volumes, which necessarily means that the H_2O molecules which fill the 24 voids in the type II unit cell are not H-bonded, with Raman observations *(1, 16, 18)* which indicate that the fraction of non-H-bonded OH-groups is decreasingly small for highly supercooled liquid water?

The fact that fully H_2O-filled types I and II clathrate structures yield molal volumes of 19.3 and $\approx$19.0 cm^3/mole *(15, 17, 19)*, respectively, in close agreement with the value of 19.2 cm^3/mole *(3)* for highly supercooled water at -45 °C suggests that such agreements are not fortuitous.

(2). How can the two above-described clathrate hole filling results be made to conform with the nearly complete H-bonding observed experimentally?

The answers to the two questions appears to be that the void-filling of the clathrate structure is equivalent, volume-wise, to structural deflation, i.e., a kind of crushing-in, probably accompanied by some tearing-apart and reconstitution, of the clathrate network. In this regard it should be noted that the solid 17:1 THF clathrate is composed of regular, planar pentagonal and hexagonal rings, whereas the above-mentioned crushing-in could be accomplished in part by ring puckering.

An analogy which might help to explain how the molal volume of the empty type II clathrate versus that of the fully H_2O-filled type II clathrate might be achieved without H-Bond breakage envisages a model of the type II unit cell constructed of fine wires, whose pentagonal and hexagonal rings are initially completely regular and planar.

We next assume an $\approx$18 % volume deflation (to give the density equivalent to filling all 24 unit cell holes with H_2O) produced by random crushing, indenting, twisting, etc., such that the pentagonal and hexagonal rings become puckered and distorted. The required rise in the global density thus corresponds to ring puckering and distortion. However, no "H-bonds" were broken in this process. Moreover, the basic clathrate network pattern remains, to wit, the connectivity does not change. This conservation of connectivity helps to clarify the term "clathrate-like" used in this chapter.

The volumes of the 16 pentagonal dodecahedral (12-sided) and 8 hexakaidecahedral (16-sided) voids in the type II lattice are about 16 X 160 $Å^3$ and 8 X 290 $Å^3$, respectively *(19)*, yielding a total void volume of 4880 $Å^3$, in a unit cell volume of 5036 $Å^3$, which amounts to ≈31 $Å^3$ void volume per H_2O molecule. The average void volume per H_2O for our previously described model is extremely difficult to determine accurately because the void shapes are very irregular. Nevertheless, we estimate that the value is roughly 25 $Å^3$ per H_2O, which is 24 % less than the value of 31, i.e., close enough to the required 18% deflation, described above. The smaller voids of our model apparently result from the many puckered and distorted rings.

An important caveat relative to the preceding conclusions is that H-bonds, and thus the pentagons and hexagons formed by H-bonded O-O nearest neighbors, are continuously breaking and reforming in supercooled water. This extremely important effect is necessarily missing from our static models.

Raman spectra from the solid 17:1 THF clathrate are presented next, followed by Raman results from highly supercooled water. Splitting of the symmetric breathing peak of the heterocyclic THF ring, upon interaction with water, is described, and this splitting is related to the OH-stretching region from a supercooled 17:1 aqueous THF solution. Further spectral comparisons are made, followed by Raman spectra from D_2O. High-pressure Raman measurements from cold equilibrium water to -9 °C are then described. The chapter concludes with a discussion of transport properties of supercooled water.

Raman Spectra From Solid, 17:1, H_2O-THF Clathrate

Tetrahydrofuran, C_4H_8O, is a cyclic ether. The oxygen atom is a member of a pentagonal heterocyclic ring whose four other members are the carbon atoms of the four CH_2 groups. (This heterocyclic pentagonal ring should not be confused with pentagonal rings of H_2O molecules in clathrate structures, also discussed here.) The two lone electron pairs of the oxygen atom, which are not engaged in bonding to carbon atoms, are available for, and certainly engage in some, H-bonding to H_2O molecules in aqueous THF solutions. However, these lone electron pairs definitely do not engage in H-bonding in the 17:1, H_2O-THF solid clathrate, as shown below.

The 17:1 H_2O-THF solid clathrate has a type II, cubic Fd3m structure whose lattice parameter is 17.141 ± 0.027 Å *(2)*. This structure contains large hexakaidecahedral cages, X, and smaller pentagonal dodecahedral cages, Y, namely, 8X·16Y·136 H_2O (*15, 17, 19*). The H_2O/THF mole ratio is 136/8 = 17/1, when the large cages are completely filled with THF.

A Raman spectrum from the solid, 17:1, H_2O-THF clathrate at 3 °C is shown in Figure 1.

The two peaks shown in Figure 1 below ≈3100 cm^{-1} arise from stretching vibrations of the CH_2 groups of THF, and hence are only of secondary importance here.

The peak near 3160 cm^{-1} and the overt shoulder near 3400 cm^{-1} arise from O-H stretching vibrations of the water in the clathrate structures. Both of these O-H stretching vibrations involve H-bonded protons.

We believe that the 3160 cm^{-1} peak is produced by H-bonded O-H groups of pentagonal rings in the clathrate structure. Such rings involve strong H-bonds which are the source of the strong proton correlation, explained next, which gives rise to the sharp, intense 3160 cm^{-1} peak *(20)*.

The term proton correlation refers to intermolecular proton motional correlation, that is, motional correlation between protons of neighboring, first or higher neighboring, H-bonded H_2O molecules.

Suppose, for example, that the symmetric OH stretches of the five H_2O molecules in an H-bonded tetrahedron (tetrahedral in the five O atoms) are all in phase. These five in-phase motions would then give rise to an additional symmetry beyond that which the symmetric stretches of the five molecules would have if they were not intermolecularly correlated. The result is that the five in-phase stretches may lower the depolarization ratio, e.g., toward the observed value of 0.04, compared to the uncorrelated case, where the value is 0.4 *(21)*.

As a second example, consider a planar, regular pentagon composed of H-bonded H_2O molecules. The totally symmetric OH stretching or ring breathing mode of such a pentagon would necessarily mean that the protons move in concert, that is, in the same phase. The depolarization ratio for such a totally symmetric mode should be zero, not too far from the value of 0.04 observed for the OH-stretching component near 3220 cm^{-1} *(21)*.

Such proton correlating mechanisms lower the depolarization ratio and yield sharp, intense Raman peaks, compared to the ordinary inhomogeneous broadening effects. This sharpening and intensification is observed for highly supercooled water, or for the 17:1 H_2O/THF solid, Figure 1.

The 3400 cm^{-1} shoulder is also produced by H-bonded O-H groups. However, this feature lacks the high degree of proton correlation of the 3160 cm^{-1} feature *(16)*. The 3400 cm^{-1} shoulder is thought to arise from H-bonded six- and higher-membered rings.

The observed high polarization (low depolarization ratio) of the 3160 cm^{-1} feature, and the much lower polarization (higher depolarization ratio) displayed near 3400 cm^{-1} *(16)* are in agreement with large and small degrees of proton correlation, respectively *(20)*. The same argument was used many years ago *(21)*, and recently *(20)*, to assign the corresponding features in liquid water.

Raman Spectra From Highly Supercooled Liquid H_2O

Hare and Sorenson (*1, 22*) have reported Raman spectra from highly supercooled water. Their spectrum at -33 °C displays an intense peak near 3190 cm^{-1}, as well as a shoulder centered near 3400 cm^{-1}. The polarization of the intense peak is large (*18, 21*) (low depolarization ratio), like that of the intense 3160 cm^{-1} peak of the 17:1 solid clathrate *(16)*. Moreover, because the frequency of the higher-temperature analogs of the 3190 cm^{-1} peak from supercooled water decline with decreasing temperature (*1, 22*), it is probable that the peak frequency from water supercooled below -33 °C will approach the frequency of the solid 17:1 clathrate peak, e.g., upon supercooling to -45°C. (Raman spectra have not yet been obtained, unfortunately, from water under such extreme supercooling.)

The close agreement between the peak frequencies of water supercooled to -33 °C, and of the solid 17:1, H_2O/THF clathrate, seems not to have been appreciated, or reported, previously. **This close agreement is of paramount importance. It provides strong evidence for the clathrate-like nature of highly supercooled liquid water. We further believe that this agreement arises**

because H-bonded pentagonal rings are common to water and solid clathrates.

"Splitting" Of The ≈920 Cm^{-1} Raman Peak From THF In Water

Raman spectra were obtained in the fundamental region from neat THF (near 1000 cm^{-1}), from aqueous solutions of THF over a wide range of concentrations, and from the 17:1 solid (none shown).

Neat THF shows an extremely sharp and narrow Raman peak near 919 cm^{-1}. This peak "splits", and develops a well-resolved, low-frequency component at about 889 to 896 cm^{-1} for aqueous THF solutions ranging from 2:1 to about 60:1, H_2O/THF. However, the Raman "splitting" disappears for the 17:1 solid THF clathrate. Only an intense, narrow peak near 919 cm^{-1} is seen, like the peak from neat THF.

The 919 cm^{-1} feature may arise from the totally symmetric stretching of the C-O-C group *(16)*, but see *(23)*. This totally symmetric stretching mode moves down in frequency when H-bonding to the lone electron pairs of the oxygen atom occurs, namely, when H_2O molecules H-bond to these lone electron pairs. The resulting interaction gives rise to the resolved low-frequency component *(14)*. However, the "splitting" disappears entirely in the 17:1 solid clathrate, because the highly-ordered cage of H_2O molecules around the THF prevents H-bonding to the lone electron pairs, possibly because of steric reasons, and/or because the H-bonds in the pentagonal rings of the enclathrating H_2O molecules may be stronger than the H-bonds to the lone electron pairs.

The new frequency near 890 cm^{-1} does not arise from removal of degeneracy. On the contrary, it arises because H-bonding to the lone electron pairs produces a separate, new mode due to the THF-H_2O hydration via the lone electron pairs of the oxygen in the THF. Klotz *(19)* has designated H-bonding to lone electron pairs of a solute molecule "disruptive H-bonding", when the clathrate H-bonding is specifically disrupted by this competing, water-proton to solute electron-pair H-bonding effect. However, the term "clathrate contravening H-bonding" seems more to the point, and is used subsequently.

The 919 cm^{-1} mode of THF probably refers solely to complete clathrate H-bonding around the THF. The THF does not engage in clathrate contravening H-bonding when it is enclathrated in the 17:1 solid. In fact, the THF does not H-bond at all in the 17:1 solid. Similarly, neat THF cannot H-bond to itself because the only protons available are the CH_2 protons. Hence, the 919 cm^{-1} mode from the THF in the 17:1 solid, and the 919 cm^{-1} mode from neat THF are essentially equivalent.

A clear discussion, using data from Glew *(24)*, of the near equality between the heat from a gas dissolving in water to yield a solution, compared with the same gas dissolving in ice to yield the clathrate hydrate, both at 0 °C, was presented by Klotz *(19)* who stated that: "This equivalence implies that the environment in both cases, in solution and in the solid hydrate, is similar with respect to coordination and spatial orientation of the water molecules in the hydration shells."

The positive excess partial molal volumes of water in solutions of non-polar molecules was also described *(19)* using data from Glew, Mak, and Roth *(25)*. Here *(19)* it is stated that: "The positive values of these molal volumes indicate that effectively water occupies a larger volume in these solutions than in the pure liquid. The expansion of the water arises from the orientation of the molecules in the hydration shell in a manner similar to that found in solid hydrates. Solute molecules occupy interstitital spaces in this hydration shell."

It is virtually certain, based upon the preceding Raman spectroscopic observations, that the above-described conclusions relative to the heat and positive excess molal volume apply equally well to the supercooled 17:1 H_2O/THF solution examined here.

OH-Stretching Raman Spectra From Supercooled, 17:1, H_2O/THF Solutions.

A Raman spectrum from the 17:1, H_2O/THF, solution supercooled to 3 °C is shown in Figure 2. (The freezing point is just above 5 °C.)

Two Raman peaks are evident from Figure 2 in the OH stretching region. The more intense peak occurs near 3250-3260 cm^{-1}, and the slightly weaker peak occurs near 3375-3395 cm^{-1}. (The CH_2 vibrations occur as a structured peak, followed by the sharp unstructured peak, both below ≈ 3100 cm^{-1}.)

The intensities of Figure 1 were multiplied by a factor, F = 33.8, to match the spectral intensity of Figure 2 at about 3370 cm^{-1}, see also Figure 3. A difference spectrum was then obtained, as shown in Figure 4.

The Figure 4 difference spectrum indicates that a pronounced sharpening and large intensity enhancement occur near 3160 cm^{-1} for the solid, compared to the supercooled solution. This difference spectrum also indicates that a sizeable decline in intensity occurs from about 3400 to 3700 cm^{-1}.

The sharpening and intensity enhancement apparent from Figure 4 near 3160 cm^{-1} indicates that the proton correlation is much greater in the solid than in the supercooled solution *(20)*. Proton correlation increases in the solid because of the large concentration of ordered pentagonal rings of H_2O molecules in the clathrate structure. The totally symmetric breathing mode of the protons in such H-bonded pentagonal rings yields increased proton correlation, as described above.

The intensity decline between 3400-3700 cm^{-1} means that broken H-bonds have been reformed upon freezing *(26)*, i.e., there is an increase of H-bonding in the solid, compared to the supercooled solution.

The intense derivative-like oscillation evident in the region below 3100 cm^{-1}, i.e., the sharp positive and negative peaks, is the result of small frequency shifts in the CH_2 peaks, obvious in Figure 3.

Comparisons Between Raman Spectra From Pure Water And Aqueous THF Solutions

When the OH-stretching contour from the supercooled 17:1 aqueous THF solution at 3 °C, Figure 2, is compared to the Raman spectrum from pure water also at 3 °C, shown next in Figure 5, it is visually evident that the order of the largest, to next largest, OH-stretching peak intensity has been inverted. The 3250-3260 cm^{-1} feature appears to be enhanced in intensity, relative to the 3375-3395 cm^{-1} feature for the supercooled aqueous THF solution, compared to the spectrum from pure water. This means that some proton corellating mechanism exists for THF, beyond that already present in water at the same low temperature *(20)*.

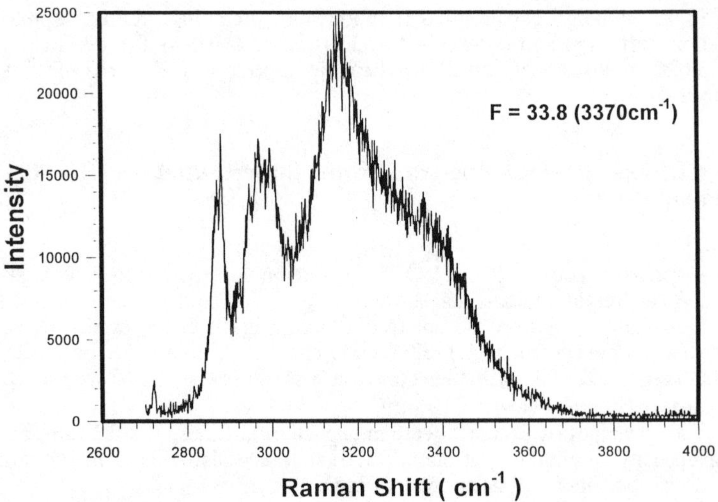

Figure 1. Raman spectrum, X(ZZ)Y polarization geometry, from 17:1, H_2O/THF solid clathrate at 3 °C. F = 33.8 refers to the amount that the intensity has been scaled up, see text.. Intensity scale in this figure, and subsequent figures, refer to photon counts per second.

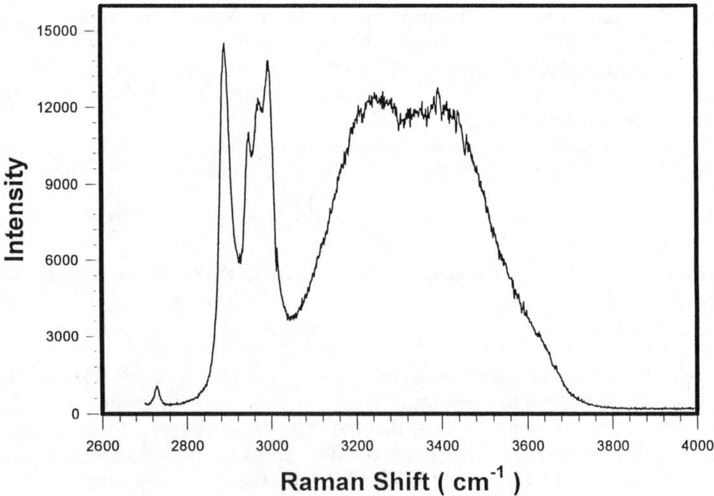

Figure 2. Raman spectrum, X(ZZ)Y polarization geometry, for the supercooled solution at 3 °C having the same composition as the solid shown in Figure 1.

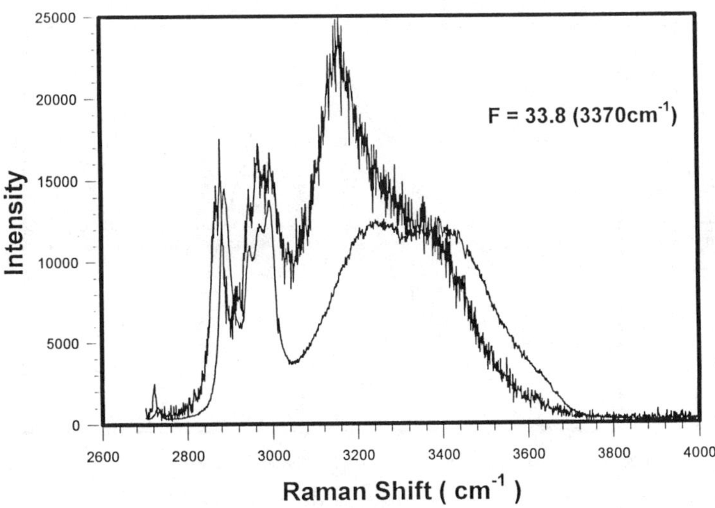

Figure 3. Raman spectra from Figures 1 and 2 normalized at 3370 cm^{-1} using the normalization factor, F = 33.8.

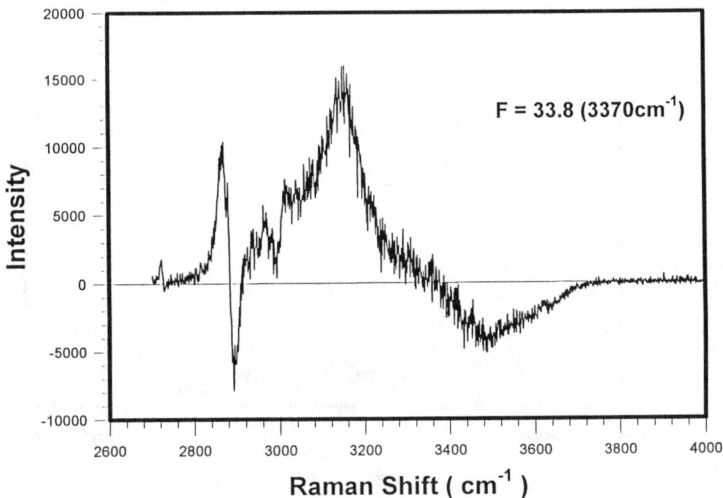

Figure 4. Raman difference spectrum corresponding to Figure 3.

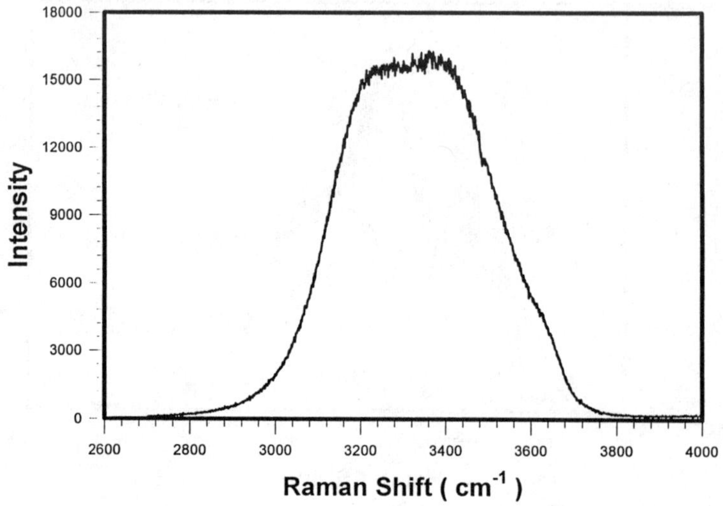

Figure 5. Raman spectrum from pure water at 3 °C, X(ZZ)Y geometry.

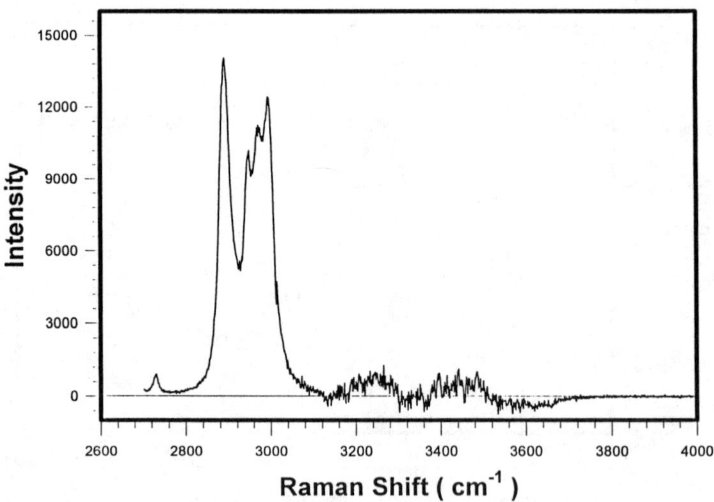

Figure 6. Raman difference spectrum obtained from Figures 2 and 5.

It should be stated that the spectra of Figures 2 and 5 were obtained under identical conditions of polarization geometry, that is, both X(ZZ)Y, with a polarization scrambler in front of the slit. Moreover, repeat determinations gave the same result. Hence, although the intensity alternation might appear to be negligible to readers not familiar with Raman spectroscopic measurements, there can be no doubt that it is real and interesting.

A difference spectrum obtained from Figures 2 and 5 is shown in Figure 6. This difference spectrum was obtained by normalizing the Raman intensities at 3370 cm^{-1} as before, see Figure 3. The OH-stretching region of Figure 6 above 3100 cm^{-1} is of particular interest.

Figure 6 gives indications of a negative difference centered just above 3600 cm^{-1}, and of two positive differences centered near 3450 cm^{-1} and 3250 cm^{-1}.

The negative difference evident in Figure 6 near 3600 cm^{-1} indicates that the broken H-bond concentration is a little lower in the THF solution than in pure water, i.e., more H-bonds in the THF solution.

The positive difference in Figure 6 near 3250 cm^{-1} indicates that the proton correlation of the THF solution has increased relative to pure water.

The positive difference of Figure 6 near 3450 cm^{-1} should be considered in conjunction with the "splitting" of the 919 cm^{-1} feature described.

H-bonding to the lone electron pairs of the oxygen atom of the THF, that is, the clathrate contravening H-bonding, gives rise to an additional peak near 890 cm^{-1}. Such an interaction should be sufficiently strong to produce a new feature in the OH-stretching region. We believe that this new OH-stretching feature gives rise to the positive difference seen in Figure 6 near 3450 cm^{-1}.

Previous Raman work with aqueous solutions also indicates that a decline of the broken H-bond population occurs when some solutes are added to water *(27, 28)*. This decline is the result of H-bonding between water and the solute (*27, 28*).

In summary, the THF-H_2O versus liquid water comparison indicates that:

(A). The amount of H-bonding rises somewhat, relative to that in pure water, when a supercooled 17:1, H_2O/THF solution is formed at the same, 3 °C, temperature.

(B). Clathrate contravening H-bonding between water molecules and the lone electron pairs of the THF also occurs. This gives rise to two new features: one near 890 cm^{-1}, from the THF ring mode, when perturbed by H_2O; and, the second near 3450 cm^{-1} (broad), due to the OH-stretching frequency, when perturbed by THF.

(C). An increase in proton correlation produces an intensity enhancement near 3250 cm^{-1}. This enhancement, compared to water at the same temperature, almost certainly arises from enclathration of the THF molecules in aqueous solution. Such enclathration is likely to be less ordered and extensive than that in the solid clathrate, that is, the frequency near 3250 cm^{-1} may be characteristic of isolated, that is, non-side-sharing, and/or disordered pentagonal rings of H_2O molecules. This proton correlation mode declines further in frequency when additional ordering due to extreme supercooling, -33 °C, 3190 cm^{-1}, or from forming the solid clathrate, 3160 cm^{-1}, occurs.

We have also considered the possibility that water molecules, when H-bonded to the lone electron pairs of THF, can further H-bond to other water molecules such that a clathrate structure is still formed around the THF. However, this situation seems implausbible for THF. H-bonds are known to form between nitrogen and the host lattice for various amine clathrate hydrates (*19*), but H-bonds do not form to the THF in its type II solid.

Raman Spectra From Hot To Supercooled D_2O

Pure anisotropic Raman spectra, X(ZX)Y, were recently reported for liquid D_2O between 95 and 22 °C (*4, 29*). These spectra, which were quantitatively intercomparable with regard to intensities, indicated an isosbestic point near 2624 ± 5 cm^{-1} (*4, 29*).

This isosbestic point separates components which are D-bonded, having components mainly below 2624 cm^{-1}, from those which are not D-bonded, having components primarily above 2624 cm^{-1}.

This isosbestic point, observed in the pure anisotropic Raman spectra from liquid D_2O, is of the ordinary type, seen many times previously in Raman spectra from water and aqueous solutions (*30-34*). However, a new kind of isosbestic point, not seen previously, was observed from the X(ZZ)Y Raman spectra from D_2O. This previously unrecognized isosbestic point is shown in Figure 7 (A) and (B).

Figure 7 (A) shows an isosbestic point at 2429 ± 5 cm^{-1} for the region from 95 to 22 °C. Essentially the same isosbestic point, about 2430 cm^{-1}, was obtained from quantitative intensity measurements at 15, 10, 5, and 0 °C, this work.

The temperature range of 15 to 0 °C is so small that the point would be obscured by the noise from all four spectra when displayed on the same figure, thus only the 5 and 0 °C spectra are shown in Figure 7 (B), where the crossing occurs at 2430 cm^{-1}.

The freezing point of pure D_2O is ≈4 °C, hence the 0 °C spectrum of Figure 7 (B) refers to the supercooled liquid.

It is certain that the X(ZZ)Y isosbestic point, 2430 cm^{-1}, separates stretching components which are only D-bonded, because this isosbestic frequency value falls solely in the D-bonded frequency region. In fact, this newly recognized type of isosbestic point, seen at 2430 cm^{-1}, is ≈200 cm^{-1} below the 2624 cm^{-1}, D-bond to broken D-bond, isosbestic frequency obtained from the pure anisotropic spectra.

The X(ZZ)Y Raman spectra from 95 to 22 °C were previously simulated with four Gaussian components, see Figure 10 (upper panels) of Ref. *(4)*. The logarithm of the ratio of the integrated intensity of a component near 2666 cm^{-1} from non-D-bonded OD stretches, to that of another component near 2475 cm^{-1} from D-bonded OD stretches, was plotted versus 1/T. This van't Hoff plot yielded a ΔH of 2.6 ± 0.2 kcal/mole for the breakage of D-bonds. The van't Hoff plot using components having the same frequencies from the X(ZX)Y spectra yielded a value of 2.8 ± 0.2 kcal/mole *(4)*, see (lower panels, Figure 10, same ref.) in satisfactory agreement with the X(ZZ)Y value. (Different Raman polarization geometries refer to the same equilibrium when the intensity ratio from the same components is measured. The values of the ratios may change with polarization geometry, but the temperature dependences do not change.)

A van't Hoff plot of the logarithm of the ratio of integrated intensities of Gaussian components centered near 2475 and 2375 cm^{-1} versus 1/T was also made, but not reported in Ref. *(4)*. Note that these components lie above, or below, the isosbestic frequency of 2430 cm^{-1} (which is almost exactly midway between the component values). The ΔH value obtained from this second van't Hoff plot was about 1.9 kcal/mole, see Figure 8.

We believe that the Raman component near 2375 cm^{-1} corresponds to D-bonded D_2O molecules in pentagonal rings, whereas the 2475 cm^{-1} component corresponds to D-bonded molecules in hexagonal, or higher-membered, rings.

The ΔH value of roughly 2 kcal/mole indicates, at least, that the energy of the pentagonal rings is lower than that of the hexagonal- and higher-membered rings.

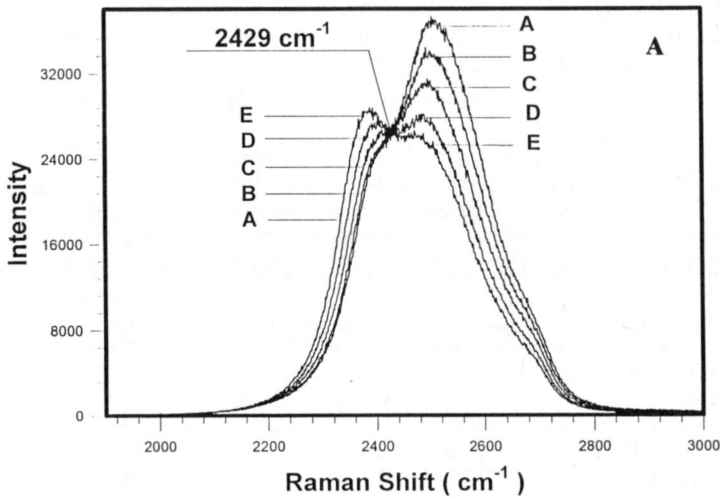

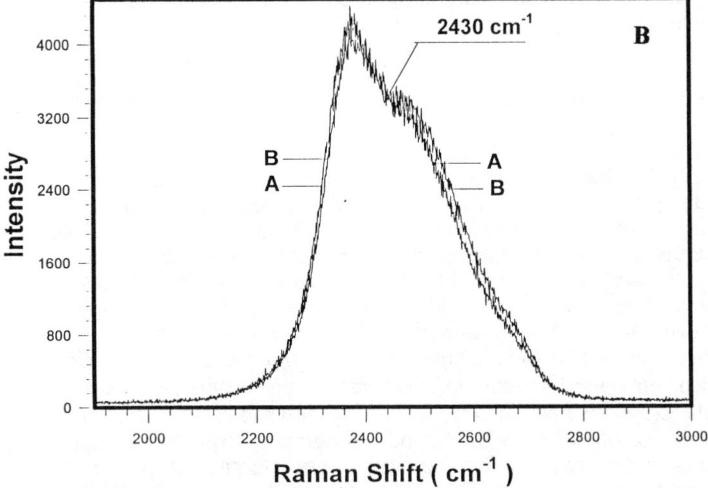

Figure 7. (A). Raman spectra from pure D_2O, quantitatively intercomparable with respect to intensities, which show an isosbestic point at 2429 cm^{-1}. A, B, C, D, and E refer to temperatures of 95, 81, 61, 41, and 22 °C, respectively. Figure 7. (B). Same type of Raman spectra as (A) obtained at a later date, hence intensity scales of (A) and (B) are not the same. Note crossing point at 2430 cm^{-1} which agrees with the (A) isosbestic point.

The above value of 1.9 kcal/mole is important, because when it is added to the previous (*30-34*) Raman value of 2.6 kcal/mole for liquid H_2O, one obtains 4.5 kcal/mole O-H···O. This means that the ΔH for the complete breakage of pentagonal H-bonds is ≈9 kcal/mole H_2O (tetrahedral network), compared to the heat of vaporization at 0 °C, which is ≈11 kcal/mole . Less energy remains to be attributed to non-H-bonded interactions than previously thought, thus resolving a long-standing problem, see Ref. *(18)*.

The ΔH of 1.9 kcal/mole, and the assignment of the 2375 and 2475 cm^{-1} Gaussian OD-stretching components to pentagonal and to hexagonal (or higher) rings, respectively, is also of key importance in the present work. The D_2O Raman measurments support the previous conclusions with regard to assignments made for liquid water, for aqueous solutions, and for the solid clathrates. Moreover, the ΔH value of 1.9 kcal/mole also supports the conclusion made in the introduction using the simplest of geometric arguments, namely, that hexagonal rings involve more bond-bending strain, and thus are high-energy structures compared to pentagonal rings. But even further agreements exist.

The isotope shift commonly observed between vibrational frequencies of ordinary and heavy water is about 1.37 (*35*). This means that the vibrational frequencies for pentagonal and hexagonal rings composed of H_2O molecules should be near 1.37 X 2375 ≈ 3250 cm^{-1} and 1.37 X 2475 ≈ 3400 cm^{-1}, respectively, in close agreement with values observed here, see Figure 2. Of course, the value of 3250 cm^{-1} calculated above refers to isolated and/or disordered pentagonal rings, the value declining to 3190 cm^{-1} for highly supercooled water, and then finally to 3160 cm^{-1} for the solid THF clathrate, for reasons given above.

High-Pressure Raman Measurements Of Cold Equilibrium Water

High-pressure Raman measurements of non-supercooled, i.e., equilibrium, water were conducted to pressures as high as 1.75 kbar and to temperatures as low as -15 °C. The same high-pressure Raman techniques were employed here, as those used previously for examining the low-frequency intermolecular spectrum of water (*36*).
Raman spectra are shown at 1.5 kbar for temperatures from 30 °C down to -9 °C in Figures 9 to 13, respectively. X(ZZ)Y geometry was used, with the sapphire windows oriented such that little or no rotation of the electric vector occurred.

Pressures of 1.50 to 1.75 kbar were found to produce an increase in the intensity contribution between ≈3300 cm^{-1} to 3000 cm^{-1}, relative to the peak intensity near 3400 cm^{-1}, as the temperature was lowered from 30 °C down to -15 °C. This effect is evident from examination of Figures 9 to 13, and it is completely obvious in Figure 14, which involves normalized spectra. However, if the same intensity comparison is made using the Hare and Sorensen spectra *(1)*, it is evident that the rise in the 3300 to 3000 cm^{-1} intensity, relative to the peak intensity, is much greater with decreasing temperature, at the low pressure of 1 atm.

In fact, the rise in intensity in the low-frequency region reported by Hare and Sorensen is so large that the peak-to-shoulder order is entirely inverted, compared to our results. Their Raman spectrum, reported at 1 atm and -33 °C, shows an intense peak near 3190 cm^{-1}, with the shoulder on the high-frequency side near 3400 cm^{-1}, just the opposite of the effect seen here in Figures 9 to 14.

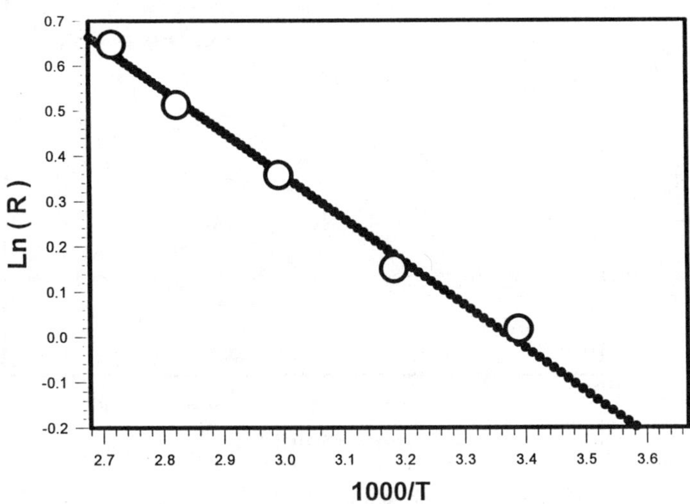

Figure 8. Van't Hoff plot of the natural logarithm of R versus 1/T, where R is the integrated Raman intensity ratio for the 2475 to 2375 cm^{-1} Gaussian components from neat D_2O between 22 and 95 °C, see Ref. *(4)*.

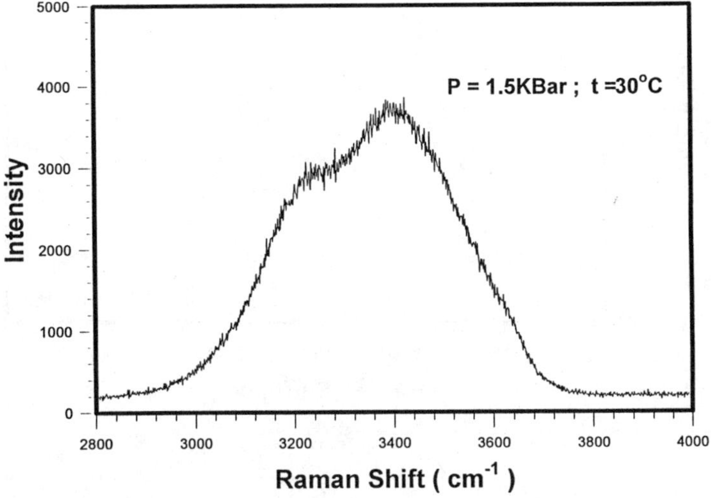

Figures 9 to 13. High-pressure, 1.5 kbar, Raman spectra from water, obtained at temperatures indicated in the figures, X(ZZ)Y polarization geometry.

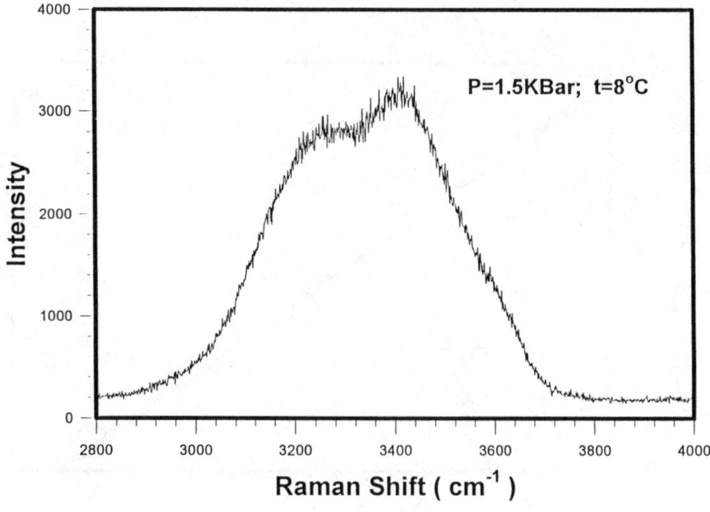

Figure 10.

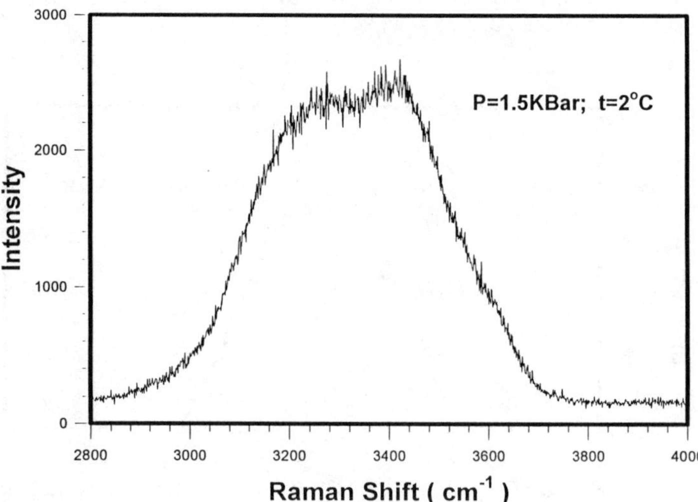

Figure 11.

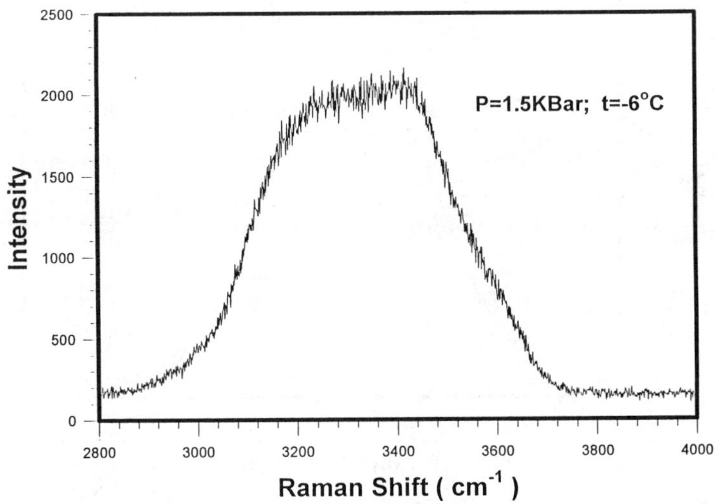

Figure 12.

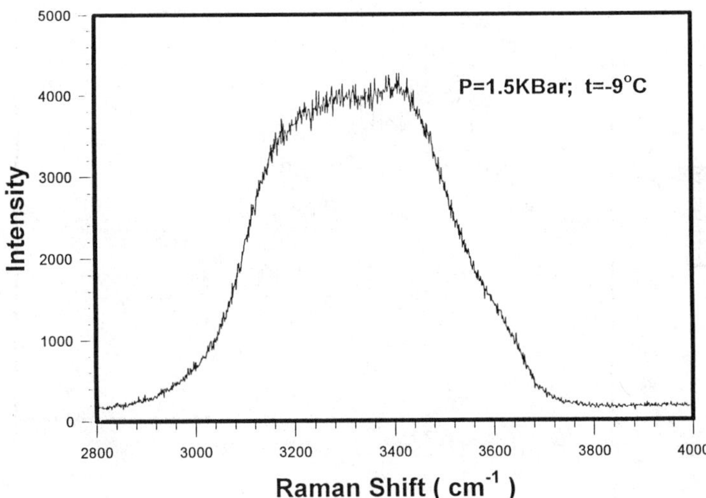

Figure 13.

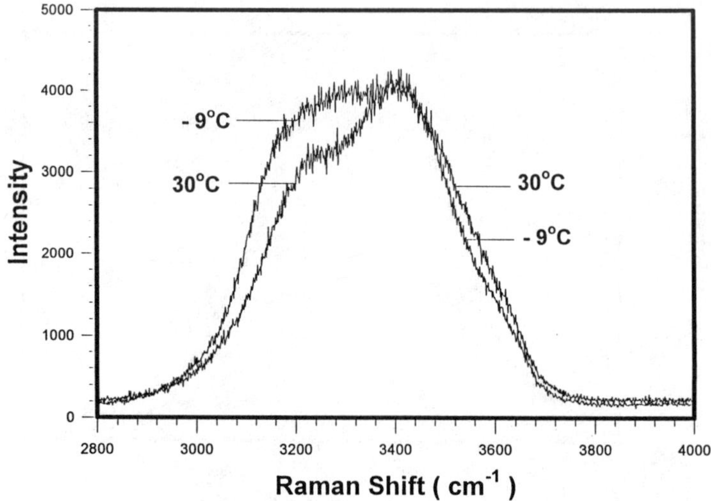

Figure 14. High-pressure Raman spectra from Figures 9 and 13 normalized at 3400 cm^{-1}.

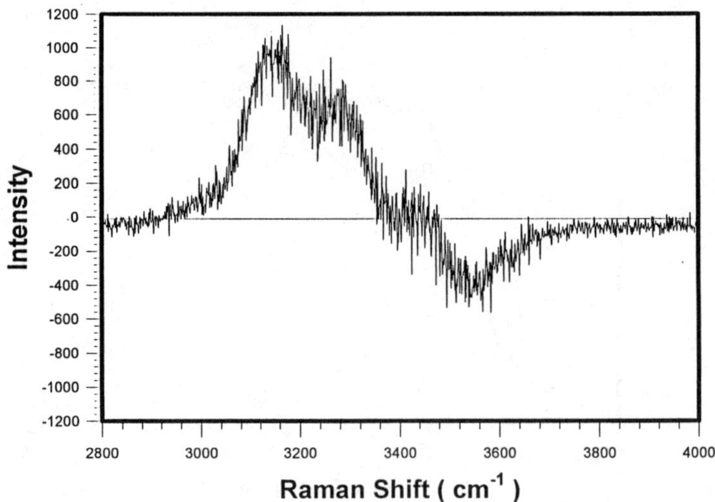

Figure 15. Difference Raman spectrum obtained from Figures 9 and 13, see Figure 14.

Direct visual comparisons between our 1.5 kbar results and the Hare and Sorensen 1 atm results at essentially the same temperature, -9 versus -10 °C, also allow a qualitative conclusion to be made, namely, that pressure rise from 1 atm to 1.5 kbar at a constant sub-zero temperature decreases the proton correlation, and more importantly decreases the concentration of H-bonded pentagon-based polyhedral clathrate-like structures. This conclusion is buttressed by Raman spectra normalized at 3400 cm^{-1}, Figure 14, and by the corresponding difference Raman spectrum, shown in Figure 15. (Normalization at the value of 3370 cm^{-1} gives virtually an identical result, e.g., for the difference spectrum.)

Figure 14 shows the 30 °C Raman spectrum normalized to meet the -9 °C spectrum at 3400 cm^{-1}. The -9 °C spectrum shows decreased intensity from 3400 to 3800 cm^{-1}, and greatly increased intensity from 3400 cm^{-1} down to 3000 cm^{-1}, compared to the 30 °C spectrum..

Figure 15 accentuates the above mentioned differences *via* a difference spectrum. The negative region from about 3400 to 3800 cm^{-1} refers to the fact that the 39 °C temperature drop has produced an increase in H-bonding. However, the more intense positive region below about 3400 cm^{-1} is of special importance to this work.

An intense peak near 3145 cm^{-1} and a weaker peak near 3280 cm^{-1} are evident from the intense low-frequency positive difference spectrum. The 3145 cm^{-1} peak is only 15 cm^{-1} lower than the peak from the 17:1, H_2O/THF solid clathrate at 3160 cm^{-1}, and the weaker peak near 3280 cm^{-1} is just above the 3250-3260 cm^{-1} value from the corresponding supercooled THF solution. (The 15 to 30 cm^{-1} differences are close to the uncertainty in peak location.)

We suggest that the 3280 cm^{-1} peak refers to H-bonded pentagons which are somewhat disordered and/or isolated, i.e., not fully bound-up into polyhedral structures. We also suggest that the 3145 cm^{-1} feature refers to pentagon-based polyhdral clathrate-like structures. However, we stress once again that the concentrations of both of the above isolated pentagonal and polyhedral clathrate-like structures are less than those at 1 atm, a conclusion which is obvious from examination of the Hare and Sorensen spectra.

We deconvoluted the positive difference of Figure 15 digitally into two Gaussian components. The central Gaussian positions which result are 3145 and 3282 cm^{-1} in agreement with the peak positions quoted above. We also obtained a difference spectrum between the -33 and -10 °C Raman spectra reported by Hare and Sorensen at 1 atm. This positive difference spectrum shows a peak at about 3130 ± 10 cm^{-1} which agrees fairly well with our 3145 cm^{-1} difference peak and indicates that this 3130 cm^{-1} difference peak, thought to be from the polyhedral clathrate-like structures, grows with increased supercooling at 1 atm.

The Hare and Sorensen results indicate that an intense peak occurs near 3190 cm^{-1} in the Raman spectrum from supercooled water at -33 °C and 1 atm, whereas only a shoulder is evident near 3400 cm^{-1}. This -33 °C spectrum is very similar in shape, except for its greater overall breadth, to the spectrum from liquid D_2O at 0 °C shown in Figure 7 (B). This peak-height to shoulder-height similarity suggests that the proton correlation is larger for D_2O than for H_2O at the same low temperature (supercooled).

An increase of proton correlation for D_2O compared to H_2O is not surprising in view of the fact that the uncertainty in the location of deuterons is much less than that of protons, as described by Chandler (*37*).

The smaller positional uncertainty of deuterons compared to protons arises essentially from the de Broglie wavelength, $\lambda_{dB} = h/p$. Because the mass of the deuteron is twice that of the proton, its de Broglie wavelength is one-half that of the proton for the same velocity. Of course, the average momentum must be employed, and when this is done one obtains the more useful relationship, namely, $\lambda_{dB} = h/\langle|p|\rangle = h/(8k_BTm/\pi)$ (*37*), which requires the deuteron's de Broglie wavelength to be ≈70 % as large as the proton's wavelength, at any fixed temperature.

Comparisons Of Results Between Raman And Transport Data

It has been known for at least 30 years that the shear viscosity, η, of water decreases with pressure rise up to about 1.5 kbar a 2.2 °C (*38*). This means that the shear fluidity, Φ, which is the reciprocal of shear viscosity increases with pressure, and hence that the activation volume for it is <u>negative</u>. The activation volume for self-diffusion, D, is also negative for cold and supercooled water up to about 1-2 kbar (*39*).

The activation volume is defined as: $V_\xi = -RT(\partial \ln\xi/\partial P)_T$, where ξ refers to self-diffusion or shear fluidity.

The interpretation of the negative activation volumes which seems most likely to be correct is that activated filling of voids in bulky clathrate-like structures occurs in the diffusional processes. Moreover, the fact that the activation volumes become positive at pressures above about 2 kbar for cold and supercooled water indicates that the bulky pentagon-based polyhedral structures are transformed into more normal, less bulky structures, e.g., by interpenetration, structural deflation, etc. Densification must result from the rise in pressure (at constant temperature), because the isothermal compressibility is always positive for any substance.

Direct Raman evidence for the breakdown of bulky, pentagon-based polyhedral structures with pressure rise was presented.

The ΔH values (activation energies) for shear fluidity and self-diffusion become enormous for highly supercooled water at 1 atm *(8)*. This conclusion is the same for Arrhenius or for Young-Westerdahl (YW) treatment of the 1 atm data *(8)*. In contrast, the YW treatment of the 2 Kbar shear fluidity and self-diffusion data gives a constant ΔH of about 5 kcal/mole H_2O *(8)*. Again the transformation of bulky, clathrate-like structures into more dense structures with pressure rise provides a satisfactory explanation of the observed effects.

A hole model involving impeded diffusion through pentagonal rings was used recently to explain the large activation energies (actually enthalpies) observed for highly supercooled water at 1 atm *(6)*. The ΔH values were shown to yield the correct, stretched O-O distance which results from forcing a water molecule through the pentagonal holes. Good agreement was obtained between the stretched O-O van der Waal's values for pentagonal rings, and that calculated from a Morse potential using experimental ΔH values *(8)*. The present Raman data, provide clear evidence for the existence of H-bonded pentagonal rings of water molecules. Moreover, the plot of Figure 8 indicates that such pentagonal rings are low-energy states, i.e., the van't Hoff ΔH for the transformation of pentagonal rings into hexagonal rings is positive and about 1.9 kcal/mole . (This van't Hoff ΔH should not be confused with the ΔH of activation discussed above.)

In conclusion, the present Raman data support the following picture of supercooled water and aqueous THF solutions:

(*i*). H-bonded pentagonal rings of H_2O molecules (or D_2O molecules) are low-energy structures, compared to hexagonal and larger rings. Lowering the temperature of water increases the pentagonal ring concentration to the point at which these rings share sides to form clathrate-like polyhedral structures, e.g., as -45 °C is approached at 1 atm.

(*j*) Substantial pressure rise, to 1-2 kbar, at sub-zero temperatures decreases the concentration of the bulky pentagon-based polyhedral clathrate-like structures in water seen from the 3000-3300 cm^{-1} versus 3400 cm^{-1} Raman intensity ratio at 1.5 kbar, compared to that at 1 atm. This decrease in the concentration of clathrate-like structures explains the effect of pressure on transport properties, e.g., the activation energies decline, and the activation volumes change sign from negative to positive at 2 kbar, for self-diffusion and shear-fluidity.

(*k*) Supercooled solutions of THF in water (17:1) give clear evidence of two apparently competitive effects: enclathration of the THF; and, H-bonding to the lone electron pairs of the O atom of THF, designated as clathrate contravening H-bonding. However, solidification of the supercooled THF solution (17:1) produces ordered enclathration, to the complete exclusion of the clathrate contravening H-bonding. The nearness of the Raman peak frequencies from the supercooled THF solution, and its solid, to the peak frequency from highly supercooled water provides direct microscopic evidence for the clathrate-like structure of supercooled water. The Raman data from THF in water are useful for understanding other nonpolar aqueous solutions, and may be of importance to the hydration of proteins.

Finally, the x-ray work of Teeter (*40*) should be mentioned, because a pentagon-based water cluster was observed in crambin which has the same structure as a sizeable portion of the type II THF clathrate hydrate cage. Three pentagonal H-bonded rings of this water cluster were found to form a cap around the CH_3 of Leu18, as expected for a hydrophobic methyl group.

Literature Cited.

1. Hare, D.E.; Sorensen, C. M. J. Chem. Phys. **1992**, 96, 13.
2. Zakrzewski, M.; Klug, D. D.; Ripmeester, J. A. J. Incl. Phenomena and Molecular Recog. in Chem. **1994**, 17, 237.
3. Speedy, R. J. J. Phys. Chem. **1987**, 91, 3354.
4. Walrafen, G. E.; Yang, W.-H.; Chu, Y. C.; Hokmabadi, M. S. J. Phys. Chem. **1996**, 100, 1381.
5. Bernal, J. D. Chapter in Liquids: Structure, Properties, Solid Interactions; Hughel, T. J.; Ed.; Elsevier: Amsterdam, **1965**.
6. Bernal, J. D.; Fowler, R. H. J. Chem. Phys. **1933**, 1, 515.
7. Finney, J. private communication, **1996**.
8. Walrafen, G. E.; Chu, Y. C. J. Phys. Chem. **1995**, 99, 10635.
9. Frank, H. S. Chapter in Structure of Water and Aqueous Solutions; Luck, W. A. P.; Ed.; Verlag Chemie Verlag Physik: Weinheim, **1974**.
10. Weingärtner, H.; Haselmeier, R.; Holz, M. J. Phys. Chem. **1996**, 100, 1303.
11. Walrafen, G. E.; Chu, Y. C. Physica A, **1994**, 206, 93.
12. Walrafen, G. E.; Chu, Y. C. J. Phys. Chem. **1992**, 96, 3840.
13. Walrafen, G. E. Chapter in Encyclopedia of Earth System Science, Volume 4; Nierenberg, W.; Ed.; Academic: Orlando, **1992**.
14. Stillinger, F. H. Science **1980**, 209, 451.
15. Sloan, E.D., Jr. Clathrate Hydrates of Natural Gases; Dekker: New York, **1990**.

16. Sloan, E.D.; Long, X. Masters thesis of the latter, Colorado School of Mines: Golden, **1993**.
17. Jeffrey, G. A. Inclusion Compounds, Atwood, J. L.; Davies, J. E. D.; MacNicol, D. D.; Eds.; Academic: New York, **1984**.
18. Walrafen, G. E.; Chu, Y. C. J. Phys. Chem. **1995**, 99, 11225.
19. Klotz, I. M., Ciba Foundation Symposium on The Frozen Cell, Walstenholme, G. E. W.; O'Connor, M.; Eds.; Churchill, J. & A., **1970**, pp. 5 to 26.
20. Walrafen, G. E.; Chu, Y. C. J. Phys. Chem. **1995**, 99, 11225.
21. Walrafen, G. E. Chapter in Structure of Water and Aqueous Solutions; Luck, W. A. P.; Ed.; Verlag Chemie Verlag Physik: Weinheim, **1974**.
22. The spectra shown in Figure 2 of Ref. *(1)* were obtained in the X(ZZ)Y orientation, designated as VV by Hare and Sorensen. The shoulders evident from 160 to 40 °C decrease in frequency from about 3250 cm^{-1}, and increase in intensity, finally changing into the intense peak at 3190 cm^{-1}, reported at -33 °C and 1 atm. If the decline in frequency, from 3250 to 3190 cm^{-1}, should continue unchanged to -45 °C, it could easily reach 3160 cm^{-1} or less.
23. The very high Raman intensity, and the extreme narrowness of the peak at ≈920 cm^{-1} suggest that its origin involves the totally symmetric breathing mode of the heterocyclic ring.
24. Glew, D. N. J. Phys. Chem. **1962**, 66, 605.
25. Glew, D. N.; Mak, H. D.; Rath, N. S. Can. J. Chem. **1967**, 45, 3059.
26. The Raman intensity of the OH-stretching region from water between 3400 and 3800 cm^{-1} increases in intensity, relative to the rest of the stretching contour, with rising temperature. This results from H-bond breakage. See Ref. *(18)*.
27. Walrafen, G. E.; Hokmabadi, M. S.; Chu, Y. C. Chapter in Hydrogen-bonded Liquids; Dore, J. C.; Teixeira, J.; Eds.; Kluwer Academic: Dordrecht, **1991**.
28. Walrafen, G. E.; Fisher, M. R. Chapter in Biomembranes, A Volume of Methods in Enzymology, Vol. 127, Packer, L.; Ed.; Academic: New York, **1986**.
29. Pure anisotropic Raman spectra obtained from D_2O, see Ref. 4, refer to the polarization geometry, X(ZX)Y. This polarization gives a Raman spectrum which refers solely to the anisotropy of the polarizability.
30. Walrafen, G. E.; Hokmabadi, M.S.; Yang, W.-H. J. Chem. Phys. **1986**, 85, 6964.
31. Worley, J.D.; Klotz, I.M. J. Chem. Phys. **1966**, 45, 2868.
32. Walrafen, G. E. J. Chem. Phys. **1967**, 47, 114.
33. Walrafen, G. E. J. Chem. Phys. **1968**, 48, 244.
34. Walrafen, G. E.; Blatz, L. A. J. Chem. Phys. **1973**, 59, 2646.
35. The square root of the moment of inertia of D_2O to that of H_2O is 1.38. The ratio of the corresponding experimental vibrational frequencies is about 1.37 ± 0.03. This ratio is difficult to determine because of the extreme breadth of the stretching contours.
36. Walrafen, G. E.; Chu, Y. C.; Piermarini, G. J. J. Phys. Chem. **1996**, 100, 10363.
37. Chandler, D. Introduction to Modern Statistical Mechanics; Oxford University Press: New York, **1987**. See pp. 194-195.
38. Bett, K. E.; Cappi, J. B. Nature **1965**, 207, 620.
39. Weingärtner, H., Z. Phys. Chem. (Frankfurt) **1982**, 132, 129.
40. Teeter, M. M. Develop. Biol. Standard, **1991**, 74, 63; Karger: Basel, **1991**.

Novel Applications: Supercooled Liquids and Protein Dynamics

Chapter 22

Cold Denaturation of Proteins

Jiri Jonas

Department of Chemistry and Beckman Institute for Advanced Science and Technology, University of Illinois, Urbana, IL 61801

By taking advantage of the phase behavior of water, high pressure can significantly lower the freezing point of an aqueous protein solution. In this way, using high-resolution, high-pressure NMR techniques, one can investigate not only pressure denaturation but also cold denaturation of proteins.

After an overview of compression effects on dynamic and hydrodynamic behavior of water and heavy water at subzero temperatures, the main part of this contribution is devoted to selected results from recent pressure and cold denaturation NMR studies of Ribonuclease A. The cold denatured state of Ribonuclease A contains partial secondary structures in contrast to its thermally denatured state which contains little or no stable hydrogen bond structures. It was interesting to find that the pattern of protection factors for the pressure and cold denatured states of Ribonuclease A obtained by hydrogen exchange experiments parallels the pattern of protection factors for the folding intermediate of Ribonuclease A reported by Udgoaonkar and Baldwin on the basis of their pulsed hydrogen experiments.

Increasing attention has recently been focused on denatured and partially folded states, since determination of their structure and stability may provide novel information for the mechanisms of protein folding (*1-3*). The native conformations of hundreds of proteins are known in great detail from structural determinations by X-ray crystallography and, more recently, NMR spectroscopy. However, a detailed knowledge of the conformations of denatured and partially folded states is lacking, and represents a serious shortcoming in current studies of protein stability and protein folding pathways (*2*).

Protein folding, the relationship between the amino acid sequence and the structure and dynamic properties of the native conformation of proteins, represents the central problem of biochemistry and biotechnology. Most studies dealing with protein denaturation have been carried out at atmospheric pressure using various physicochemical perturbations, such as temperature, pH, or denaturants, as experimental variables. Compared to varying temperature, which produces simultaneous changes in both volume and thermal energy, the use of pressure to study protein solutions perturbs the environment of the protein in a continuous, controlled way by changing only intermolecular distances. In addition, by taking advantage of the phase behavior of water high pressure can substantially lower the freezing point of an aqueous protein solution. Therefore, by applying high pressure one can investigate in detail not only pressure-denatured proteins, but also cold-denatured proteins (*4*) in aqueous solution.

Cold denaturation has been assumed to be a general property of all globular proteins (*4, 5*). However, experimental evidence for cold denaturation has been scant, owing to the fact that cold denaturation of proteins in aqueous solution is usually only observed at temperatures below 0°C at neutral pH. Different approaches have been utilized to prevent freezing of protein solutions, including the use of cryo-solvents (*6*), denaturants (*4*), emulsions in oil (*7*), supercooled aqueous solutions (*6, 8*), and site-directed mutogenesis (*5*).

High-resolution NMR spectra of complex molecules in the liquid phase usually exhibit a great deal of structure and yield a wealth of information about the molecule. Therefore, it is not surprising that the multinuclear high-resolution Fourier transform NMR spectroscopy at high pressure represents a promising technique in studies of biological systems (*9*). The information from the many advanced NMR techniques, including 2D-NMR techniques such as NOESY, COSY, and ROESY, has yet to be fully explored in high pressure NMR experiments. Recent advances in superconducting magnets make it possible to attain a high homogeneity of the magnetic field over the sample volume so that even without sample spinning one can achieve high resolution. The very first multinuclear high-resolution FT NMR experiment (*10*) on liquids at high pressure was performed in our laboratory in 1971, using an electromagnet. However, the main applications were delayed till the availability of commercial high-homogeneity superconducting magnets equipped with superconducting and room-temperature shims. The early studies by Morishima (*11*) used the capillary techniques introduced by Yamada (*12*). It is interesting to note that even though NMR is one of the most important spectroscopic tools for the investigation of biochemical systems at ambient conditions, few high-pressure NMR studies on biological molecules have been reported to date (*11-15*).

Several recent studies performed in our laboratory (*16-21*) illustrate the unique information obtained by combining high-resolution NMR techniques with high pressure.

In this contribution we report selected results of our recent cold denaturation experiments (*17, 18*) on ribonuclease A where we took advantage of the depression of the freezing point of water by high pressure to carry out unfolding of this protein at temperatures well below 0°C. For our pressure denaturation and cold denaturation experiments we have chosen ribonuclease A (RNase A) as a model system for several reasons. RNase A is a stable, well characterized protein. It is large enough to be of interest as a physiologically relevant protein, but is small enough to have a well understood structure, including a nearly complete assignment of the solution proton NMR spectrum.

RNase A is a single-domain protein, a pancreatic enzyme which catalyses the cleavage of single-stranded RNA. This protein consists of 124 amino acid residues with a molecular mass of 13.7 kDa. It has traditionally served as a model for protein folding because it is small, stable and has a well-known native structure. The ε1 protons of the four RNase A histidine residues are well-resolved from other protons in the ^{1}H NMR spectrum of the native protein in D_2O; they have been used in this work to monitor the structural changes of four distinct segments in the molecule during cold, heat and pressure denaturation processes. His12 and His119 are part of the catalytic site of native RNase A, and His 48 is at the hinge of the active site crevice. His48, His105, and His119 are in the β-sheet fold, which forms the backbone of the molecule. His12 is an α-helix near the N-terminus. The folding pathway of the protein has been extensively studied. Several studies on RNase A strongly suggest that its folding proceeds through intermediates, including an early hydrogen-bonded intermediate and a late native-like intermediate.

In order to obtain specific structural information on the pressure denatured RNase A we used the hydrogen exchange method. In this method the exchangeable amide protons (NH) of a protein are used as probes of structure and structural changes. The NH proton exchange with solvent deuterons is strongly dependent on their involvement in stable hydrogen bonds. The observation of slowed amide exchange is a powerful and site-specific probe for detecting persistent structure in proteins (*3, 22*). Two-dimensional NMR spectroscopy can be used to determine the hydrogen exchange rates of individual backbone amide protons of a protein. Schmid and Baldwin (*23*) first introduced the ^{1}H-^{3}H exchange approach in their studies on the folding of ribonuclease A, and found that at least part of the hydrogen-bonded backbone of RNase A is formed at an early stage in folding. Udgoaonkar and Baldwin (*24, 25*) used pH pulse labeling to study refolding of RNase A; their results show very fast and very slow kinetic components, and suggest that native-like secondary structures are present in early folding intermediates of RNase A.

The contribution is organized as follows. First, a brief overview of the anomalous dynamic behavior of water with compression at low temperatures illustrates the complex nature of the cold denaturation experiment at high pressure. Second, by using 1D and 2D high resolution NMR techniques including the hydrogen exchange one can show that the cold denatured and pressure denatured states contain partially folded structures in contrast to the thermally denatured state. Third, it is important to point out that the pattern of protection factors for the cold and pressure denatured states of RNase A bear a strong analogy to the pattern of protection factors reported by Udgoaonkar and Baldwin (24, 25) for the early folding intermediates of RNase A.

Experimental

The materials and experimental conditions for the various NMR experiments were discussed in detail in the original studies (*17, 18*). All NMR studies were performed using high frequency NMR system which operates at a proton Larmor frequency of 300 MHz. The system is composed of a General Electric GN-300 FT-NMR with an Oxford Instruments, Inc. wide-bore superconducting magnet (ϕ=89 mm, 7.04 T). The GN-300 is interfaced to a Tecmag Scorpio Data Acquisition System for pulse programming and experimental control, using MacNMR software.

The hydraulic pressure generation system was similar to the system described previously (*17*). As with the earlier system, carbon disulfide (CS_2) was used as the pressure transmitting fluids for proton studies.

It is appropriate at this point to mention several design considerations for high-resolution, high pressure NMR probes (*26, 27*) which are to be used for protein studies. The sample size (diameter) should definitely be greater than 5 mm., otherwise the sensitivity of the high pressure NMR probe will seriously limit the scope of problems to be studied. For a protein it should be possible to obtain good S/N for concentrations in the mM (milimolar) or lower range. Higher concentrations usually result in aggregation and even precipitation of the protein when temperature or pressure is changed. Another NMR probe performance feature decisive for successful use in studying biochemical systems is the resolution--specifically, when carrying out advanced 2D and 3D NMR experiments it is essential that the resolution is better than 5×10^{-9}.

Results and Discussion

Dynamic Structure of Water at High Pressure and Low Temperatures. In order to understand why it is not surprising to find differences in secondary structure between the cold denatured and heat denatured states of RNase A it is appropriate to review several general results obtained in NMR relaxation and transport studies of compressed liquid water and heavy water at temperatures lower than ~35 - 40°C (*28*). Simply stated water or heavy water at temperatures above 40°C is a very different solvent than H_2O or D_2O at temperatures below 0°C.

During systematic studies of density effects on the dynamic structure of liquids, we observed, for water and heavy water, that the most interesting behavior of various transport and relaxation properties occurs at temperatures between 10°C and 30°C. In several studies (*29, 30*) we took advantage of the phase diagram of water and heavy water and measured various transport properties of these liquids down to temperatures of -15°C. Nuclear magnetic resonance spin-lattice relaxation time T_1, self-diffusion coefficients, and shear viscosities were measured as a function of pressure in the temperature interval -15°C to 30°C. The pressure range was always adjusted according to the phase diagram in such a way that water always stays in the liquid phase. Before we discuss the results on water let us mention what happens with various motions in a normal liquid with compression. Simply, increased packing significantly slows down all motions in the liquid. In contrast, spin-lattice relaxation time T_1, self-diffusion coefficients, and fluidity, in water and/or heavy water, go through a maximum with initial compression and only a further increase of pressure begins to restrict motional freedom and consequently results in shorter T_1 and lower values of diffusion coefficients and fluidity.

The anomalous motional behavior of water molecules with initial compression can be qualitatively interpreted in terms of a simple physical picture based on changes in the random hydrogen-bond network. The most characteristic structural feature of liquid water is the local tetrahedral environment of each molecule beyond where there is a randomized imperfect space filling network of hydrogen-bonded molecules. As it is clear from the phase diagram of water (see Figure 1), the boundaries of the measurements were, at low pressures, ice I and, at high pressures, ice V and also ice VI. The most important structural differences (*31*) between ice I and the high-pressure ices are the distorted hydrogen-bond angles and the close approach between the non-hydrogen-bonded neighbors. By compressing liquid water in the selected temperature range, we affect the hydrogen-bond network and gradually go from optimal tetrahedral order towards a more compact packing arrangement. According to Stillinger and Rahman (*32*), there is a competition in water between the tendency of strongly directional forces to build an open, hydrogen-bond network and the tendency for external pressure to pack the molecules together more efficiently. Since the process for self-diffusion, shear viscosity, and reorientation of the water molecules necessitates the breaking and reforming of hydrogen bonds, one can expect that it is easier to break an already bent hydrogen bond than an undistorted one. Initial compression, therefore, increases motional freedom of molecules and causes diffusion, fluidity, and T_1 to go through a maximum with increasing pressure under isothermal conditions. Further compression slows down all the dynamic processes because of a more compact packing of the water molecules. Under high compression the repulsive hard-core interactions begin to compete strongly with the directional forces that are responsible for the open structure at low pressures and low temperatures.

Another result of these studies is related to the applicability of the hydrodynamic equations at the molecular level. Several studies of dielectric (*33, 34*) and NMR relaxation (*35-37*) in liquid H_2O and D_2O indicated that the Debye equation describes well the temperature dependence of the reorientation of water molecules at atmospheric pressure. However, the results of high-pressure studies (*29, 30, 38*) show convincingly that the Debye equation fails to account for the density effects on reorientation of water molecules. One has to conclude that the success of the Debye equation for reorientation of water at 1 bar is accidental. Nevertheless, it is remarkable to have such a coincidence and it remains an experimental fact that both the proton and deuteron relaxation rates in water are linearly dependent on η/T, where η is the viscosity and T temperature.

Another important hydrodynamic equation is the Stokes-Einstein equation which relates the self-diffusion coefficient to shear viscosity. Both, molecular dynamics calculations (*39*) and experiments (*28*), have shown that this equation is well obeyed in the slipping boundary condition over a wide range of densities and temperatures in simple liquids approaching the model of hard sphere fluids. In view of the strongly directional forces in water it is not surprising to find that the Stokes-Einstein equation is not valid for liquid water.

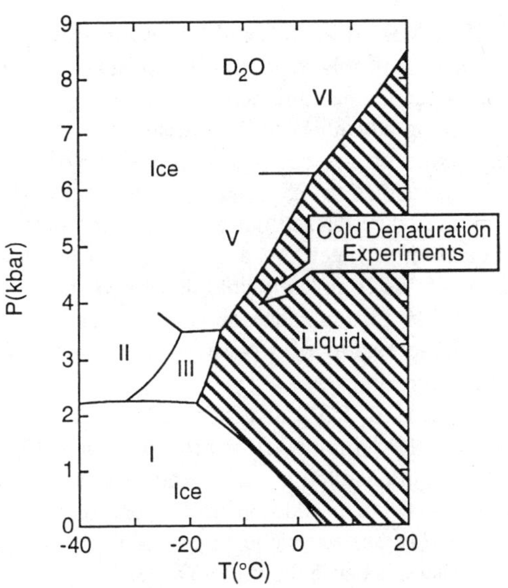

Figure 1. Phase diagram for D_2O.

One may summarize the main conclusions of the experimental studies of liquid water and heavy water at high pressures as follows:

1. All dynamic properties investigated behave anomalously with initial compression in the temperature range -15°C to ~40°C.
2. Only under higher compression does temperature and pressure have opposite effects on the transport and relaxation properties of H_2O and D_2O.
3. The Debye, and the Stokes and Einstein hydrodynamic equations are not valid for liquid H_2O and D_2O.
4. Plots of relaxation rate ($1/T_1$) vs. η/T are linear under isochoric and isobaric conditions.
5. Compression reduces the coupling between the rotational and translational motions.

High Resolution 1D NMR Study of Cold Denaturation of RNase A. It is difficult to attain the completely cold denatured state of proteins in aqueous solutions by decreasing the temperature below the freezing point of 0°C at ambient pressure. However, taking advantage of the pressure phase behavior of water, one can lower the temperature of aqueous solutions well below 0°C. At 3 kbar the RNase A solution can be cooled to -25°C without freezing. It should be noted here that the protein solution in the high salt buffer freezes well below the freezing point given in the phase diagram of water (*28*) at 3 kbar (-15°C). The phase behavior of water allows us to obtain the completely cold denatured state of RNase A with the assistance of pressure. In Figure 2, we demonstrate the behavior of the proton spectrum of the histidine region of RNase A undergoing both cold and heat denaturation at 3 kbar and pH* 2.0.

The ε^1 proton peaks of the four histidine residues are well resolved from other proton peaks in the 1D proton NMR spectrum of the native RNase A in D_2O. The protein becomes cold denatured at -22°C and heat denatured at 40°C. In the cold denatured state, one finds in addition to the composite resonance D, another resonance D′, which likely has the same origin as D′ in the pressure denatured state. This is not surprising because the protein is subjected both to a high pressure of 3 kbar and to low temperature. The linewidth broadening of resonances D and D′ is due to slower motions at low temperatures.

As discussed in detail in the original studies (*17, 18*) the percent denaturation of each histidine residue can be calculated and a pressure-temperature phase diagram of RNase can be constructed. Figure 3 shows the pressure-temperature phase diagram of RNase A. Above the phase line the protein is in the denatured state. It can be seen that the cold and heat denaturation temperatures change with pressure, and that below 2 kbar the heat denaturation temperature is not sensitive to pressure. The cold denaturation temperature data below 3 kbar are not available because the protein solution freezes before it can be completely denatured.

Since substrates or inhibitors stabilize native protein structures, it was interesting to use the histidine residues of RNase A to probe the behavior of an enzyme-inhibitor complex under cold, heat and pressure denaturation conditions. It is known that His12 and His119 are involved in the enzymatic activity. The cold,

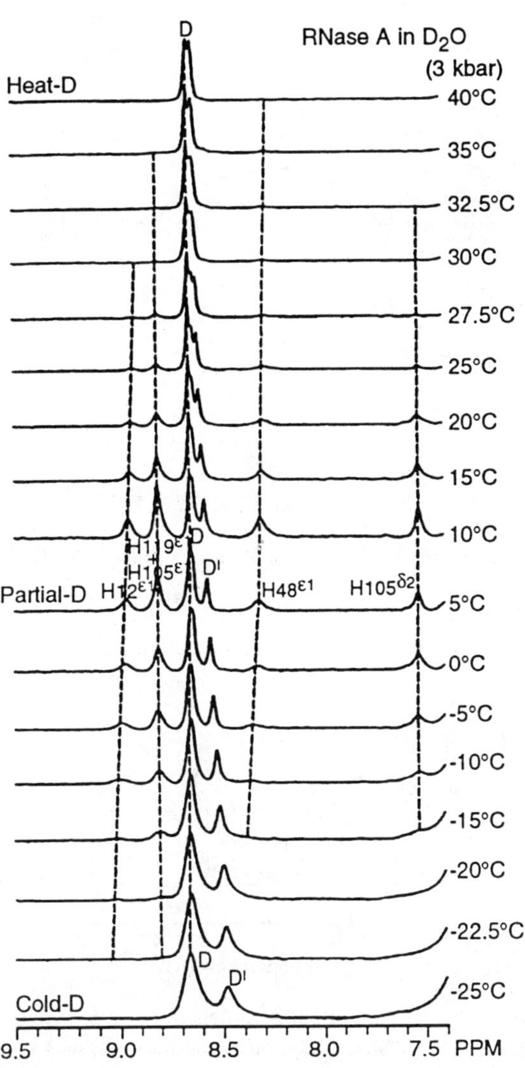

Figure 2. Histidine ε_1 ring proton NMR region of RNase in D_2O at various temperatures at 3 kbar (pH* = 2.0) (Ref. 17).

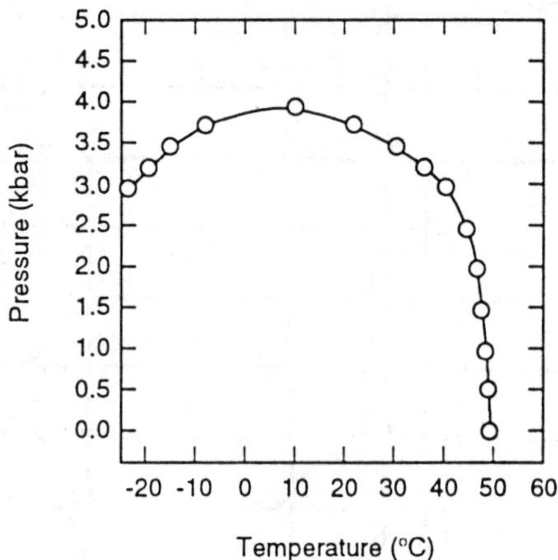

Figure 3. Phase diagram of RNase A in D_2O at pH* 2.0.

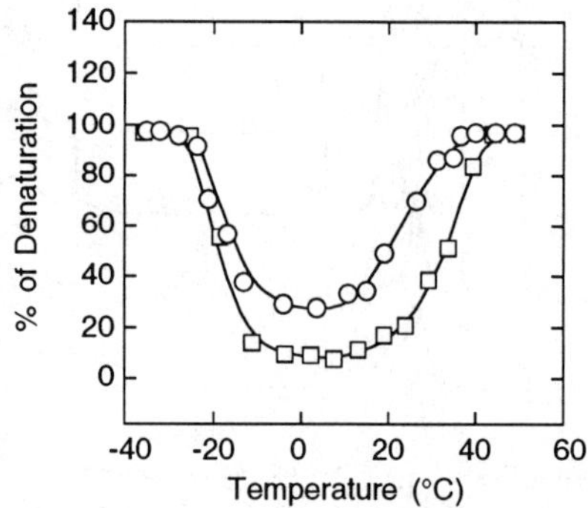

Figure 4. Comparison of the temperature denaturation curves of the RNase A protein (O) only and the RNase A-inhibitor complex (❑) at 3 kbar (pH* = 3.0) as obtained for His-48$^{\varepsilon 1}$ proton resonance (Ref. 17).

heat and pressure denaturations of the RNase A-inhibitor complex were investigated in our study (*17*) by comparison to the denaturation of RNase A without the inhibitor. The competitive inhibitor 3´-UMP(uridine-3-monophosphate) was used at a molar ratio of 3´-UMP to RNase A of 3:1.

In Figure 4, the percent denaturation of His48$^{\varepsilon 1}$ of the enzyme-inhibitor complex, and that of the enzyme without inhibitor under identical solution conditions were plotted separately as a function of temperature at 3 kbar pressure. It can be seen that the enzyme-inhibitor complex is more stable under pressure, and therefore denatures at higher temperatures than the free enzyme. From the experimental data given in Figure 4 one finds that at subzero temperatures both the free enzyme and the enzyme-inhibitor complex are cold denatured.

The appearance of a new histidine resonance D´ in the cold denatured and pressure denatured RNase A spectra, compared to the absence of this resonance in the heat denatured state, (See Figure 2) indicates that the pressure denatured state and cold denatured state contain partially folded structures, which may be similar to that of the early folding intermediate found in the temperature-jump experiment reported by Blum et al (*40*).

Hydrogen Exchange Study of Pressure and Cold Denatured States of RNase A. Since 1D NMR spectra do not allow one to arrive at unambiguous conclusions about the presence or absence of secondary structure in the pressure and/or cold denaturated states hydrogen-exchange experiments (*17, 18*) were carried out to test for the presence of partially folded structures in the pressure-denatured and cold-denatured states. The hydrogen exchange behavior contains information about the structure and conformational dynamics of proteins, resolved to the level of individual amino acid residues. The protection against exchange of the amide proton of each amino acid residue imposed by the protein structure is expressed by the protection factor, $P = k_{rc}/k_{obs}$, where k_{rc} is the hydrogen exchange rate calculated using a random coil model, and k_{obs} is the hydrogen exchange rate measured experimentally. In aqueous solutions, the rate of exchange of the peptide group NH proton with protons of the aqueous solvent is catalyzed by hydroxide and hydronium ions in pH-dependent reactions, and by water in a pH-independent way (*41*). The random coil hydrogen exchange rate can be expressed by Equation 1,

$$k_{rc} = k(acid) + k(base) + k(water)$$
$$= k_A[D^+] + k_B[OD^-] + k_W \qquad (1)$$

The random coil exchange rate is the sum of the contributions from acid, base and water catalyzed exchanges, where k_A, k_B and k_W are the respective rate constants. The rate constants k_A, k_B and k_W of individual residues of unfolded proteins are sensitive to neighboring side chains, and can be calculated from the reference rates for alanine-containing peptides by considering the additivity of nearest-neighbor blocking and inductive effects. The recently revised and extended parameters (*41*) were used to calculate the intrinsic exchange rates. To obtain the random coil exchange rates at a particular temperature and pressure, each rate constant in Equation 3 should be modified according to Equation 2,

$$k_i(T,P) = k_i(T_0,P_0)\exp(-\frac{E_a}{R}(\frac{1}{T}-\frac{1}{T_0}))\exp(-\frac{\Delta V^{\neq}(P-P_0)}{RT}) \qquad (2)$$

where i = A, B or W. E_a is the activation energy, and $\Delta V^{\neq}$ is the activation volume for hydrogen exchange. The activation energies (*41*) are: E_a (k_A) = 14 kcal/mol, E_a (k_B) = 17 kcal/mol, E_a (k_W) = 14 kcal/mol; the activation volumes (*42*) are: $\Delta V^{\neq}$ (k_A) = 0) 1 ml/mol, $\Delta V^{\neq}$ (k_B) = 6) 1 ml/mol, $\Delta V^{\neq}$ (k_w) = -20.4 ml/mol. These values were used to predict the hydrogen exchange rates of individual residues of RNase A in a random coil structure at high pressure. The experimental hydrogen exchange rate k_{obs} was obtained by fitting the measured cross-peak intensities in 2D COSY NMR spectra to the exponential function, $I = I_0 \exp(-kt)$. The data measured at eight experimental times was used to obtain the experimental hydrogen exchange rates at high pressure. The cross-peaks between the NH and $C^{\alpha}H$ hydrogens of residues of RNase A in two-dimensional COSY NMR spectra were assigned according to the literature (*43, 44*). Of 119 backbone NHs in RNase A, approximately 40 amide protons are stable to exchange with D_2O in the native state, and they can be recorded by a COSY NMR spectrum. Only 33 of the 40 amide protons show cross-peaks of sufficient intensity to allow determination of the exchange rate in the pressure experiment described in detail in the original references (*17, 18*) were used to probe the stable hydrogen-bonded structure in the pressure and cold denatured protein.

In our hydrogen exchange experiments for cold denatured state of RNase A we found 7 residues with protection factors between 5 to 10, and 22 residues with protection factors above 10. Similarly, there are 8 residues with protection factors between 5-10 and 14 residues with protection factors larger than 10 for pressure denatured state of RNase A. In contrast for the thermally denatured RNase A no significant protection was detected.

From our experimental results we conclude that in the pressure-denatured and cold denatured RNase A most of interior amide groups exchange hydrogen atoms with the solvent more rapidly than in the folded state, but more slowly than in the fully unfolded state. Clearly, at least some secondary structure, which is more or less native, seem to persist in the pressure and cold denatured states to protect the NH hydrogens from exchange with solvent molecules.

It is interesting to compare the pattern of protection factors for pressure and cold denatured RNase A with the pattern of protection factors reported by Udgoaonkar and Baldwin (*24, 25*) for the kinetic folding intermediate of RNase found in the early stage of refolding by the pulsed hydrogen exchange. Figure 5 shows the pattern for the folding intermediate with the protection factors observed both for pressure and cold denatured states of RNase A. Even if the comparison is only qualitative there is significant parallel in the pattern of the protection factors.

In summary, our experimental evidence indicates that both pressure and pressure assisted cold denatured states of proteins have more secondary structure

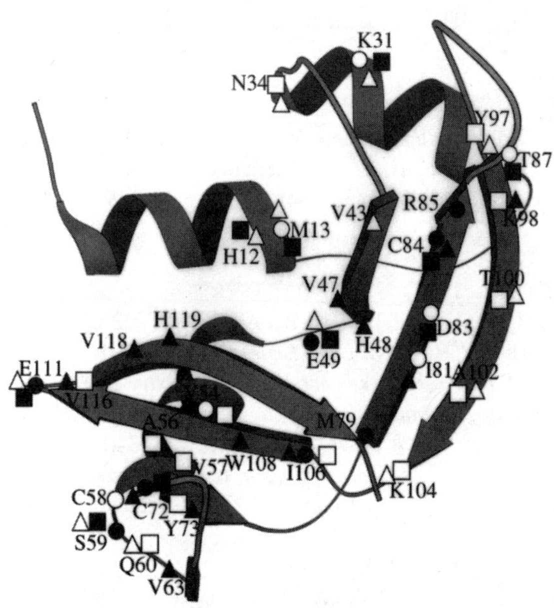

Figure 5. Comparison of the pattern of protection factors for RNase A in cold denatured state ■; ❑) pressure denatured state (●; O); folding intermediate (Δ; ▲); see ref. 40. Full symbols denote high protection, open symbols denote low protection.

than heat denatured states. The finding of parallel patterns of protection factors for pressure and cold denatured RNase A and its folding intermediate is quite important as it suggests that studies of pressure and cold denatured states of proteins may provide novel information about the folding process. There is a need for systematic studies of pressure and cold denaturation of various proteins to ascertain whether our results obtained for RNase A can be generalized.

Acknowledgments

This work was supported in part by the National Institutes of Health under grant PHS 5 R01 GM42452-08.

Literature Cited

1. Kim, P. S.; Baldwin, R. L. *Annu. Rev. Biochem.* **1990,** *59*, 631.
2. Creighton, T. E. *Proteins;* W. H. Freeman & Co.: New York, NY, 1993.
3. Matthias, B.; Radford, S. E.; Dobson, C. M. *J. Mol. Biol.* **1993,** *237*, 247.
4. Privalov, P. L. *CRC Crit. Rev. Biochem. Mol. Biol.* **1993,** *25*, 181.
5. Antonino, L. C.; Kautz, R. A.; Nakano, T.; Fox, R. O.; Fink, A. L. *Proc. Natl. Acad. Sci. U.S.A.* **1993,** *88*, 7715.
6. Hatley, R. H. M.; Franks, F. *Cryo-Letters* **1993,** *7*, 226.
7. Franks, F.; Hatley, R. H. M. *Cryo-Letters* **1993,** *6*, 171.
8. Tamura, A.; Kimura, K.; Akasaka, K. *Biochemistry* **1993,** *30*, 11313.
9. Jonas, J.; Jonas, A. *Annu. Revs. Biophys. Biomol. Structure* **1993,** *23*, 287.
10. Wilbur, D. J.; Jonas, J. *J. Chem. Phys.* **1993,** *55*, 5840.
11. Morishima, I. In *Current Perspectives in High Pressure Biology*, Janasch,H. W.; Marquis, R. E.; Zimmerman, A. M., Eds.; Academic Press: New York, NY, 1987, 315.
12. Yamada, Y. *Re. Sci. Instr.* **1974,** *45,* 640.
13. Wagner, G. *FEBS Lett.* **1980,** *112,* 280.
14. Williams, R. K.; Fyfe, C. A.; Bruck, D.; VanVeen, L. *Biopolymers* **1979,** *18,* 757.
15. Hauer, H., Ludemann, H. D. & Jaenicke, R. *Z. Naturforsch* **1982,** *37c,* 51.
16. Samarasinghe, S.; Campbell, D. M.; Jonas, A.; Jonas, J. *Biochem.* **1992,** *31,* 7773.
17. Zhang, J.; Peng, X.; Jonas, A.; Jonas, J. *Biochem.* **1995,** *34,* 8631.
18. Nash, D.; Lee, B.-S.; Jonas, J. *Biochim. Biophys. Acta* **1996,** *40,* 1297.
19. Peng, X.; Jonas, J.; Silva, J. *Proc. Natl. Acad. Sci.* **1993,** *90,* 1776.
20. Peng, X.; Jonas, J.; Silva, J. *Biochem.* **1994,** *33,* 8323.
21. Royer, C. A.; Hinck, A. P.; Loh, S. P.; Prehoda, K. E.; Peng, X.; Jonas, J.; Markely, J. L. *Biochem.* **1993,** *32*, 5222.
22. Englander, S. W.; Mayne, L. *Annu. Rev. Biophys. Biomol. Struct.* **1992,** *21,* 243.
23. Schmid, F. X.; Baldwin, R. L. *J. Mol. Biol.* **1979,** *135,* 199.
24. Udgoaonkar, J. B.; Baldwin, R. L. *Nature* **1988,** *335*, 694
25. Udgoaonkar, J. B.; Baldwin, R. L. *Proc. Natl. Acad. Sci. U.S.A.* **1990,** *87,* 8197.

26. Ballard, L.; Reiner, C.; Jonas, J. *J. Magn. Reson.* **1996,** *123,* 81.
27. Jonas, J.; Koziol, P.; Peng, X.; Reiner, C.; Campbell, D. M. *J. Magn. Reson.* **1993,** *102B,* 299.
28. Jonas, J. *Science* **1982,** *216,* 1179.
29. DeFries, T.; Jonas, J. *J. Chem. Phys.* **1977,** *66,* 896.
30. DeFries, T.; Jonas, J. *J. Chem. Phys.* **1977,** *66*, 5393.
31. Eisenberg, D. S.; Kauzmann, W. *The Structure and Properties of Water*; Oxford University Press: New York, NY 1969.
32. Stillinger, F. H.; Rahman, A. *J. Chem. Phys.* **1974,** *61,* 4973.
33. Colie, C. H.; Hasted, J. B.; Ritson, D. M. *Proc. Phys. Soc. London B* **1948,** *60,* 145.
34. Grant, E. H. *J. Chem. Phys.* **1957,** *26,* 1575.
35. Woessner, D. E. *J. Chem. Phys.* **1964,** *40,* 2341.
36. Smith, D. W. G.; Powles, J. G. *Mol. Phys.* **1966,** *10,* 451.
37. Krynicki, K. *Physica (Utrecht)* **1966,** *32,* 167.
38. Jonas, J., DeFries, T.; Wilbur, D. J. *J. Chem. Phys.* **1976,** *65,* 582.
39. Alder, B. J.; Gass, D. M.; Wainwright, T. E. *J. Chem. Phys.* **1970,** *53,* 3833.
40. Blum, A. D.; Smallcombe, S. H.; Baldwin, R. L. *J. Mol. Biol.* **1978,** *118,* 305.
41. Bai, Y.; Milne, J. S.; Mayne, L.; Englander, S. W. *Proteins: Struct. Funct. Genet.* **1993,** *17,* 75.
42. Carter, J. V.; Knox, D. G.; Rosenberg, A. *J. Biol. Chem.* **1978,** *253,* 1947.
43. Rico, M.; Bruix, M.; Santorio, J.; Gonzalez, C.; Neira, J. L.; Nieto, J. L.; Herranz, J. *Eur. J. Biochem.* **1989,** *183,* 623.
44. Robertson, A. D.; Purisma, E. O.; Eastmann, M. A. Scheraga, H. A. *Biochemistry* **1989,** *28,* 5930.

Chapter 23

Vibrational Echo Studies of Heme-Protein Dynamics

M. D. Fayer[1] and Dana D. Dlott[2]

[1]Department of Chemistry, Stanford University, Stanford, CA 94305
[2]School of Chemical Sciences, University of Illinois, 600 South Mathews Avenue, Urbana, IL 61801

The first picosecond infrared vibrational echo experiments on a protein, myoglobin-CO, are described. The experiments, performed over a temperature range of 50-300K, with native and mutant myoglobin in different solvents, provide information about protein dynamics occurring on the 10^{-12}-10^{-10} s time scales, which influence the vibrational transition of the CO ligand bound to the active site. The experiments reveal the existence of a protein energy landscape with a relatively flat distribution of energies between conformational minima, which interconvert via tunneling at lower temperature and by activated processes at ambient temperature.

Infrared (IR) spectroscopy is a valuable tool in practically all chemical applications, due in great part to the fingerprint nature of IR spectra. The application of IR to studies of macromolecular dynamics has not yet been as valuable, because macromolecular vibrational spectra are usually broad and poorly resolved, due to vibrational congestion. Nuclear magnetic resonance (NMR) spectra of macromolecules are similarly congested, but in recent years powerful multidimensional pulse techniques have been developed to overcome that difficulty. Owing to the recent explosion in the development of new techniques to generate powerful coherent pulses of infrared light (the IR spectral region can be taken to indicate the region from 2.5 to 25 μm), it ought to be possible to develop new IR techniques to relieve the vibrational congestion of macromolecular spectra. These new techniques open up exciting possibilities for exploiting nonlinear or coherent optical interactions to obtain new information about the dynamical behavior of macromolecules.

In this paper, we discuss experiments using time-resolved IR spectroscopies, especially vibrational echo techniques, to study molecular dynamics of CO bound at the active site of myoglobin (Mb), a heme protein. The vibrational echo is a time domain experiment which measures the Fourier transform of the homogeneous vibrational IR lineshape (1). With a combination of vibrational echo measurements and vibrational pump-probe measurements of the vibrational energy relaxation lifetime, it becomes possible to elucidate the various dynamical contributions to the vibrational transition. Vibrational echo experiments provide unique information about the effects of protein conformational relaxation processes on a vibrational probe molecule, the CO ligand at the protein's active site (2,4).

Previous optical coherence experiments performed on proteins studied the dephasing of electronic transitions of a Zn-porphyrin moiety (5). (Those experiments will not work on the native Fe-porphyrin found in heme proteins due to extremely fast nonradiative relaxation of Fe-porphyrins). Because of the very rapid dephasing (broad homogeneous linewidths), and the very rapid increase in the rates of dephasing processes with increasing temperature of electronic transitions, those experiments can only be performed at very low temperatures (a few degrees K). The vibrational echo experiments make it possible to use optical coherence methods to study protein dynamics at low and at physiologically relevant temperatures.

Structure and Infrared Spectroscopy of Myoglobin-CO.

Figure 1 shows the structure of the active site of Mb-CO. Myoglobin, found in muscle tissue, is used in the storage and transport of dioxygen (O_2) (6). Both dioxygen and other small ligands such as CO, NO and isocyanides bind to Mb. The active binding site is the Fe atom of the prosthetic group protoheme (Fe(II) protoporphyrin IX), which is embedded in a protein matrix consisting mainly of α-helices. A substantial literature exists that explores the structure and chemical kinetics of CO bound to the active site of Mb and its mutants (e.g. 7). One theme which continually emerges in studies of Mb-CO is the important role of histidine H64 (often termed the "distal histidine"), a polar amino acid located directly above the bound ligand in the heme pocket (6,7). The rest of Fig. 1 shows the structure of a Mb mutant, created by site-directed mutagenesis (8), where the polar H64 is replaced by a nonpolar valine residue. That mutant is termed H64V. Vibrational echo studies on H64V-CO (4) will be discussed as well.

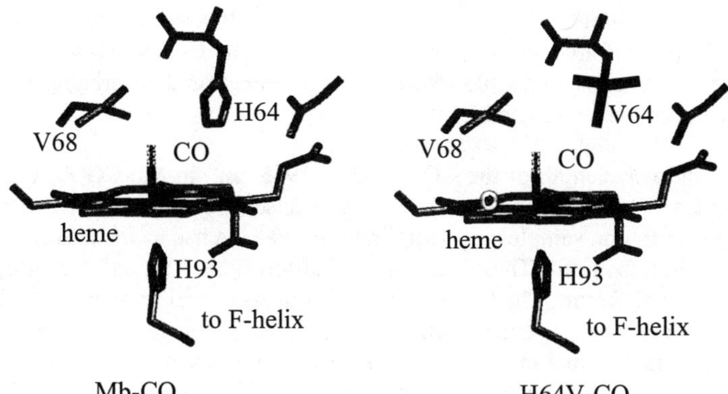

Figure 1. Structures near the active sites of native myoglobin-CO, and the H64V-CO mutant. Adapted from ref. 9.

Figure 2 shows a mid-IR spectrum of a Mb-CO sample used for time-resolved IR experiments (9). Sample preparation has been discussed previously (3,9-10). The sample consists of ~15 mM horse-heart Mb, dissolved in a pH = 7.0 buffered solution of glycerol:water (95:5 v/v) in an optical cell with CaF_2 infrared-transmitting windows, with a path length of 125 μm. We use a relatively high concentration of protein because the mid-IR transition strength is ~100 times smaller than the intense visible absorption bands of heme studied by electronic spectroscopies. We use a relatively high concentration of the cryoprotectant glycerol to insure that our samples have high optical quality at low temperature, which greatly aids our ability to detect echo signals. In Fig. 2a, the broad intense IR absorption bands are due to protein and solvent absorptions. There is a window of relative transparency in the 1800-2300 cm^{-1} region. The sharp peak in this window is due to the fundamental stretching vibration of CO bound to Mb (11). The sharp peak is the feature which allows vibrational echo techniques to bear directly upon dynamics occurring at the active site of the protein. Figure 2b is a blow up of the region of CO absorption. Careful studies (12) have shown the existence of four distinct mid-IR absorption transitions, attributed to CO absorption in different protein conformers. However in the samples we use here, only one of these conformers, denoted A_1, is dominant. It appears as an IR absorption peak with a spectral width of ~13 cm^{-1}.

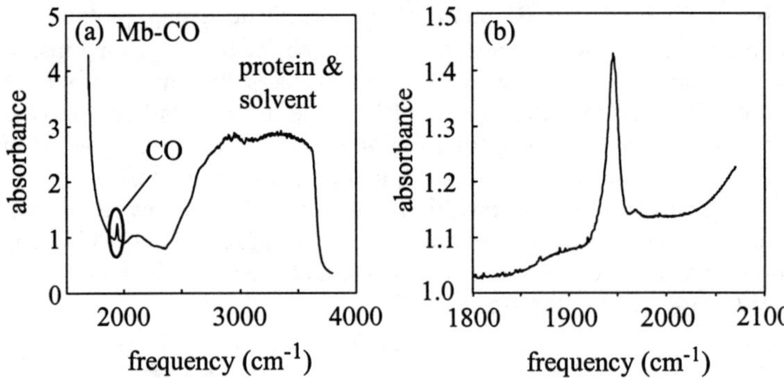

Figure 2. (a) Infrared spectrum of Mb-CO. (b) The CO stretch vibrational transition on an expanded scale. Note the large background absorbance due to protein and solvent. Adapted from ref. 9.

At the wavelength of the CO stretch (~1945 cm^{-1} in Mb-CO A_1 conformer), the solvent and protein provide a background absorption which even in our high protein concentration samples is about three times as intense as the absorption of the CO stretching transition. These two types of absorptions (CO and background) are fundamentally different. The CO absorption is due to a smaller number of absorbers (one for each Mb-CO unit) with a larger absorption cross-section, and the background arises from a much larger number of absorbers, e.g. water, glycerol, C-C and amide stretches of the protein backbone, etc., all of which have much smaller absorption cross-sections. Conventional IR spectroscopy does not make a distinction between these different situations because it measures the sample absorbance, which is the product of the cross-section and concentration (9).

Infrared Vibrational Echo and Pump-probe Spectroscopies

Unlike conventional IR spectroscopy, in echo and pump-probe experiments, the strength of the signal observed depends sensitively on the magnitude of the absorption cross-section (9). When performing experiments on Mb-CO with mid-IR pulses tuned to 1945 cm^{-1}, the signals which are observed are almost entirely dominated by the CO vibrational transition, and contributions from the much larger absorption background are almost entirely absent. In fact, theory allows for the possibility of using intense mid-IR pulses to extract dynamical information from the vibrational transitions of small ligands in protein solutions, even when the ligand's absorption transitions are totally buried in a broad background of weaker absorbers.

All the experiments described here were performed using a free-electron laser (FEL) located at the Stanford FEL Center. The use of FELs to study protein dynamics is quite new, and a detailed discussion of protein spectroscopy with FELs, and the relative merits of FELs versus other techniques such as optical parametric generation for mid-IR pulse generation, which can also be used for vibrational echo spectroscopy, are discussed in more detail in ref. 9. The Stanford FEL is an extremely stable source of tunable pulses in the mid-IR. The mid-IR pulses used to study Mb-CO had Gaussian temporal envelopes 1.7 ps FWHM, and transform-limited spectra 8.6 cm^{-1} FWHM.

In principle, information about vibrational dynamics might be obtained from the Fourier transform of the absorption spectrum. The loss of vibrational energy from an oscillator, as well as the loss of phase memory (vibrational dephasing) due to time-dependent perturbations of the transition, lead to homogeneous broadening of the spectrum (13). The Fourier transform of the spectrum is proportional to the time-correlation function of the normal coordinate of the vibrational oscillator, and the dynamical processes which broaden the spectrum cause the correlation function to decay in time. However, in a disordered condensed matter system such as Mb-CO, even a well-resolved vibrational spectrum will not provide dynamical information (14). The dynamics of interest are washed out by inhomogeneous broadening. In the Mb-CO system, inhomogeneous broadening is caused by a static or quasistatic (see below for the meaning of quasistatic) distribution of protein conformations, that results in a distribution of protein-CO interactions.

The use of pump-probe and vibrational spectroscopies to elucidate the dynamics of CO bound to Mb is shown in schematic form in Fig. 3. The ordinary IR absorption lineshape is dominated by inhomogeneous broadening. The vibrational echo experiment, which is the vibrational analog of the spin echo of NMR and the photon echo of visible and UV spectroscopy (1), eliminates inhomogeneous broadening to reveal the underlying homogeneous lineshape. In our experiments, the vibrational echo signal decays exponentially with time, (echo signal) α $\exp(-4t/T_2)$, where T_2 is the time constant for vibrational dephasing. When the echo decay is exponential, the homogeneously broadened spectrum has a Lorentzian lineshape with spectral width $\Delta\nu = 1/\pi T_2$. There are two distinct processes which contribute to the vibrational dephasing time constant T_2. These are vibrational relaxation (VR), which denotes the irreversible loss of vibrational energy from the CO oscillator to the rest of the system (protoheme, protein and solvent), and pure dephasing, which denotes the irreversible loss of phase memory (without losing vibrational excitation energy).

The VR process is characterized by time constant T_1, often called the vibrational lifetime. Pump-probe experiments on Mb-CO (9,10,15), which are techniques for measuring the rate of recovery of absorption saturation induced by an intense pulse, give a signal which also decays exponentially with time, (pump-probe signal) α exp $(-t/T_1)$. The observation of a vibrational lifetime T_1 indicates that the VR contribution to the homogeneous lineshape is a Lorentzian of width $1/2\pi T_1$. Pure dephasing is characterized by a time constant denoted T_2*. In our experiments, T_2* is determined from the results of echo and pump-probe experiments, using the well-known relation,

$$\frac{1}{\pi T_2} = \frac{1}{\pi T_2^*} + \frac{1}{2\pi T_1}. \tag{1}$$

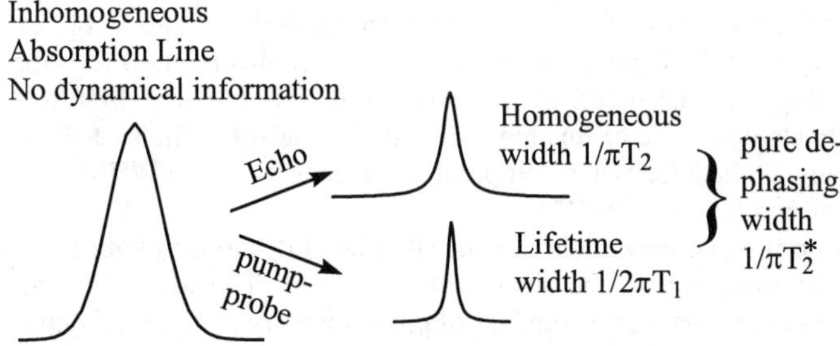

Figure 3. Contributions to the vibrational lineshape. The homogeneous linewidth from echo experiments, and the lifetime-broadened width from pump-probe experiments are combined to give the pure dephasing time constant T_2*.

The experimental apparatus used for pump-probe and echo experiments is diagrammed in Fig. 4. Individual picosecond duration mid-IR pulses are selected from a burst of pulses emitted by the FEL using an acousto-optic modulator (AOM). A fraction of the pulse (50% for the echo and 10% for the pump-probe) is picked off using a beam splitter and sent directly to the sample. The remainder of the beam is passed through a second AOM, which chops the beam to enable the detection electronics to suppress noise due to scattered light and intensity fluctuations. This beam is sent down a computer-controlled delay line and then to the sample. The two IR beams are focused to a diameter of 100 μm on the Mb-CO sample, which is held in an optical cryostat. Echo and pump pulse energies of 150-300 nJ are typically used, with probe pulse energies of ~20 nJ. In the pump-probe experiment, the signal detected is the modulation, by the more intense pump pulse, of the weaker probe pulse transmitted through the sample. In the echo experiment, the signal being detected is the echo pulse emitted after the sample is subjected to two intense incident pulses, a pump and a rephasing pulse. Fig. 4 shows the echo pulse is detected by being spatially separated from the two incident pulses.

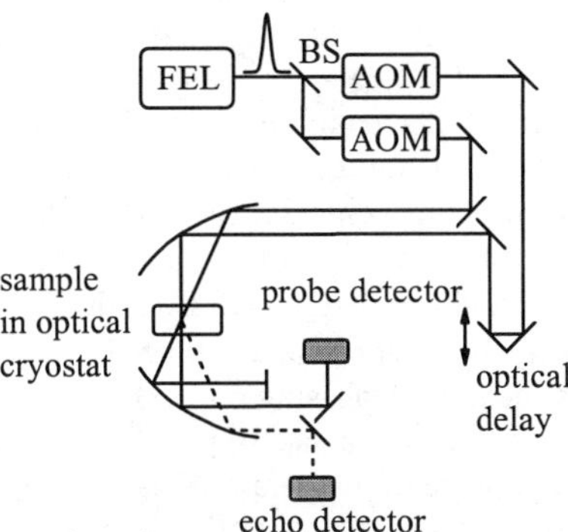

Figure 4. Experimental schematic (3). FEL denotes free-electron laser. AOM denotes acousto-optic modulator.

Results from Vibrational Echo Experiments

Figure 5 shows some representative echo data (2), obtained from an Mb-CO sample at 80K in glycerol:water 95:5 (v/v). The inset log plot shows the echo signal decay is exponential in time. At 80K, the echo data yield $T_2 = 26.5 \pm 0.6$ ps, corresponding to a homogeneous linewidth of 0.40 cm^{-1}. At this same temperature, the width of the absorption spectrum is 12 cm^{-1}. Therefore the spectrum is massively inhomogeneously broadened, with the two widths differing by a factor of 30. At room temperature, the homogeneous linewidth determined from echo measurements is 2.7 ± 0.5 cm^{-1}, which is still approximately 5 times narrower than the 13 cm^{-1} width of the room temperature absorption spectrum. The observation of inhomogeneous broadening at room temperature allows us to conclude that on the time scale of at least a few hundred ps (see below), the A_1 conformer of Mb-CO actually exists in many different substates, characterized by slightly different transition frequencies of the CO stretch.

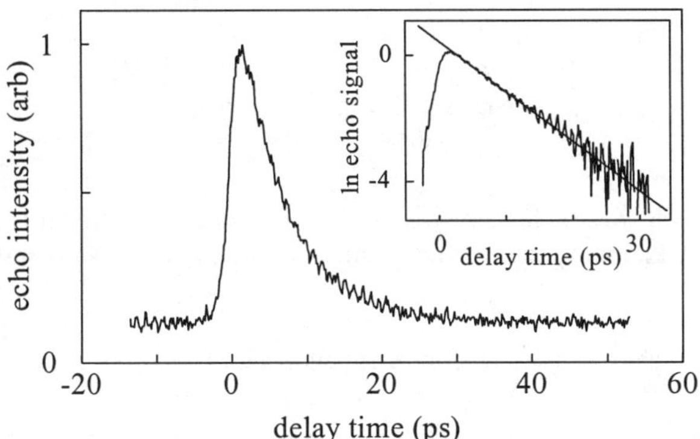

Figure 5. Vibrational echo data (3) on Mb-CO at 80K.

Pump probe and echo measurements were made from 50K to 300K. The temperature dependent T_1 and T_2 data, as well as the pure dephasing times T_2^* obtained with equation 1 are displayed in Fig. 6. In this paper we will not discuss the T_1 measurements in detail. Both the temperature dependence of T_1 and the dependence of T_1 on the structure, elucidated by studying many different proteins and synthetic porphyrin compounds, have been discussed previously (9-10,15). In our experiments, T_1 and T_2 could be determined from the data within ±3% error. The error in the derived quantity T_2^* is approximately ±5%.

Figure 7 displays the pure dephasing rate $1/T_2^*$ temperature dependence on a logarithmic plot. There is a break in the slope of the temperature dependence at ~185K, which is within the range of temperatures associated with the glass transition temperature of the glycerol:water solvent (2). Below this transition, the data fall on a straight line, indicating power law behavior of the form aT^α, where $\alpha = 1.3 \pm 0.05$. This fit is indicated by the dashed line in Fig. 7. Above the solvent glass transition, the Mb-CO dephasing dynamics exhibit a deviation from the low temperature power law behavior. The higher temperature data can be fit by an activation term, $b\exp(-\Delta E/kT)$, where ΔE is a barrier height. Thus (2,3),

$$\frac{1}{T_2^*} = aT^\alpha + be^{-\Delta E/kT}. \tag{2}$$

The smooth curve through the data points in Fig. 7 is a fit to equation 2 where the activation term is fit by a barrier height $\Delta E = 1250 \pm 200\ \text{cm}^{-1}$.

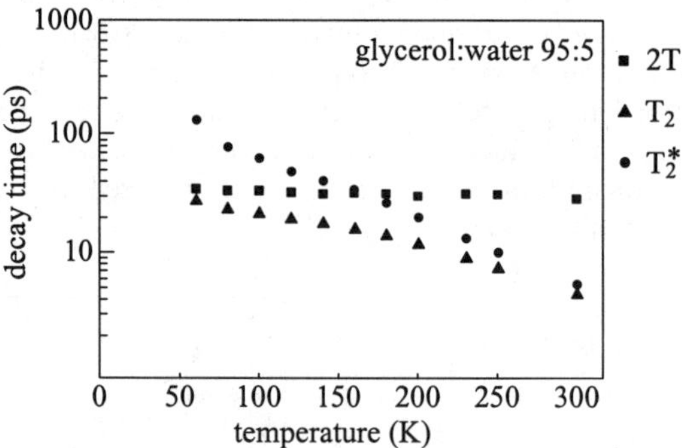

Figure 6. Temperature dependence (3) for Mb-CO of vibrational echo decay time constant T_2, two times the vibrational lifetime T_1, and T_2^* computed from equation 1.

Figure 8 shows the temperature dependence of the pure dephasing rate $1/T_2^*$ for Mb-CO in different solvents. These were glycerol:water 95:5 (v/v) as in Fig. 7, ethylene glycol:water 50:50 (v/v), and buffered water. The ethylene glycol data are similar to the glycerol data, and in fact can be fit with the same values of ΔE and α, but there is one dramatic difference. The break in slope in the ethylene glycol data occurs at a lower temperature of ~140K, which is in the range associated with the glass transition temperature of that solvent. Data in water were obtained only at ambient temperature. Figure 8 shows that at ambient temperature, the pure dephasing rate decreases in the order water > ethylene glycol > glycerol. That is the same ordering as the viscosity of these solvents, which at 300K are roughly 1 cP, 100 cP, and 1000 cP.

Figure 9 compares the temperature dependence of $1/T_2^*$ for the mutant protein H64V-CO (4) to the native form of Mb-CO, under the same conditions (pH, glycerol:water solvent, etc.). The H64V-CO data are strikingly similar to the Mb-CO data, and again can be fit by the same values of ΔE and α, but there is again one dramatic difference. At every temperature, the pure dephasing rate of H64V-CO is always 20% *slower* than in Mb-CO.

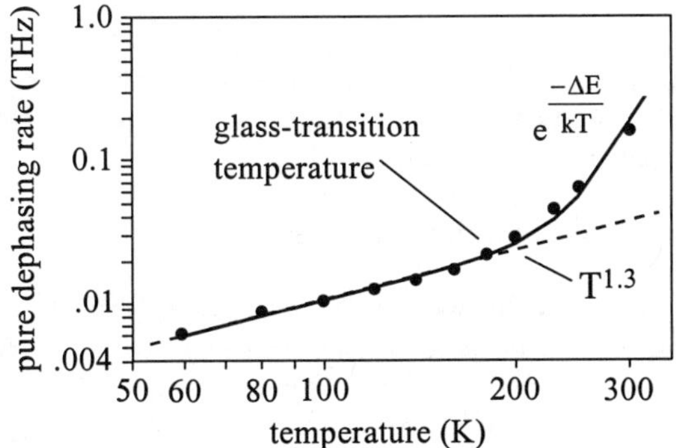

Figure 7. Logarithmic plot (3) of pure dephasing rate $1/T_2^*$ versus temperature for Mb-CO in glycerol:water is fit by equation 2.

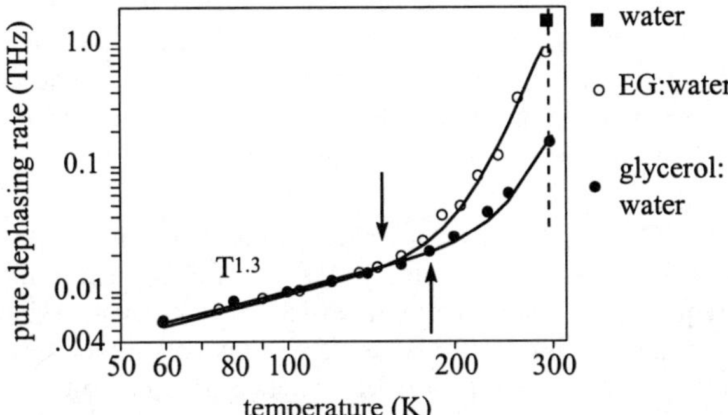

Figure 8. Logarithmic plot of pure dephasing rate versus temperature for Mb-CO in glycerol:water (95:5), ethylene glycol:water (50:50) and buffered water. The top and bottom arrows indicate respectively the temperatures of the glycerol:water and ethylene glycol (EG): water solvent glass transitions.

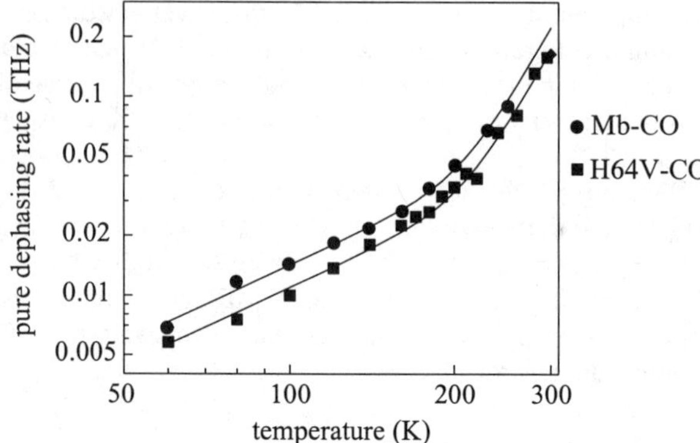

Figure 9. Logarithmic plot of pure dephasing rate versus temperature for H64V-CO in glycerol:water (95:5). Also shown are the data for native Mb-CO in the same solvent. Adapted from ref. 4.

Interpretation of Pure Dephasing Measurements

The following somewhat oversimplified discussion follows the treatment of magnetic resonance experiments described by Kubo (16). The more complicated theory needed for a detailed quantitative understanding of vibrational echo experiments in disordered condensed media is discussed in ref. (17). In our echo experiments, an intense mid-IR pulse causes an ensemble of CO ligands to begin oscillating in phase. The pure dephasing process characterized by T_2^* is caused by interactions between CO and its surroundings. We term the driven oscillator the "system", and the surroundings consisting of protoheme, protein and solvent the "medium" or "bath" (9). An interaction which causes pure dephasing induces a phase shift $\Delta\phi = \Delta\Omega\tau_c$, where $\Delta\Omega$ is the magnitude of the vibrational transition frequency shift induced by the interaction, and τ_c is a characteristic correlation time for the interaction. Interactions between the system and medium which do not induce a frequency shift $\Delta\Omega$ will not induce pure dephasing.

The frequency shift $\Delta\Omega$. Using sophisticated molecular mechanics programs, it might be possible to calculate the values of $\Delta\Omega$ for various specific interactions. For example the frequency shift induced by a particular amino acid residue in the heme pocket (Fig. 1) might be computed. Here we will take a simpler approach based on the data in Table I. Let us take as a reference state protoheme-CO with no protein (Ph-CO in D_2O solvent). The data in the Table show the following: (1) the total CO frequency shift $\Delta\Omega$ induced by replacing the D_2O solvent in the reference state with a myoglobin protein is on the order of tens of cm^{-1}; (2) replacing certain amino acid residues in the heme pocket can change the protein-induced shift by tens of cm^{-1}; and (3) the solvent surrounding the protein has essentially no effect on CO frequency shift. Also important to note is that Fig. 2 shows there is a dispersion ($\pm 6\ cm^{-1}$) in the protein-induced frequency shift, as seen from the inhomogeneous width of the A_1

spectrum, which indicates that the substates of the A_1 conformer have slightly different CO vibrational frequencies.

Table I. CO vibrational frequencies in heme systems[a]

system and solvent	*frequency (cm^{-1})*
1. Protoheme-CO in D_2O	1977
2. Myoglobin-CO A_1 state in glycerol:water	1945 (±6)
3. Myoglobin-CO A_1 state in *poly*-(vinyl alcohol)	1946
4. Myoglobin-CO A_1 state in water	1944
5. Myoglobin-CO A_1 state in trehalose	1945
6. Myoglobin-CO A_1 state in crystalline Mb	1946
7. Myoglobin-CO H64V mutant in glycerol:water	1967
8. Myoglobin-CO V68N mutant in glycerol:water	1917

[a]Adapted from data in ref. 3.

The correlation time τ_c. When $\Delta\Omega\tau_c >> 1$, interactions are said to be in the *slow* limit (16). In the slow limit, frequency-shifting interactions change slowly or not at all on the time scale of the vibrational excitation lifetime T_1. This is the meaning of the term quasistatic interactions used above. Interactions in the slow limit cause inhomogeneous broadening which dominates the ordinary IR spectrum, but which is eliminated by the echo pulse sequence. When $\Delta\Omega\tau_c << 1$, interactions are in the *fast* limit (16). In the fast limit, frequency-shifting interactions change so rapidly with time that the system sees only a time averaged effect. This motional averaging process does not cause pure dephasing. The important (14,16) case for understanding vibrational echo experiments in disordered media occurs when $\Delta\Omega\tau_c \approx 1$, termed the *intermediate* limit. In that limit, interactions produce substantial frequency shifts during the vibrational lifetime, which result in efficient pure dephasing.

Mechanism of pure dephasing. For values of $\Delta\Omega$ indicated by Table I, interactions with τ_c faster than ~100 fs are too fast to cause pure dephasing. Effects of interactions far slower than the ~20 ps vibrational lifetime (say a few hundred ps) are eliminated by the echo pulse sequence. Therefore the pure dephasing time constant is a measure of interactions between the medium and the CO oscillator which (1) induce a frequency shift $\Delta\Omega$, and (2) which have characteristic correlation times τ_c over the range of a few tenths of ps to a few hundred ps. Pure dephasing is caused by dynamics of the protein matrix, which occur on the approximate time scale of 10^{-12}-10^{-10} s. However, only a subset of those dynamical motions are observed by the echo experiment, namely those motions which induce a shift of the CO resonance frequency. Motions of the solvent have no direct effect on the phase relaxation of the CO oscillator, since the solvent does not cause a significant frequency shift. The solvent does, of course, play a role in providing a thermal bath for the heme protein (2-3).

The effects of heme structure on CO stretching frequency have been studied in much detail. The heme structure induced CO frequency shift is in most part due to different electrostatic interactions in the heme pocket, which affect the extent of back donation from the heme d_π-p_π orbitals to the antibonding π^* orbitals of CO (back-bonding). Increased back-bonding weakens the CO bond and red shifts the vibrational frequency. It seems likely, therefore, that pure dephasing is caused by fluctuations in protein structure, which induce fluctuations in the extent of back-bonding. It is thought the most significant structural fluctuations are in large part those which affect the locations of the amino acid residues near the heme pocket shown in Fig. 1, but certainly motions of other components of the Mb-CO system must be involved as well. The picture that emerges is that heme acts as an antenna that receives and communicates protein fluctuations to the CO stretching vibration, and that pure dephasing of the CO vibration is sensitive to a wide sampling of protein structural fluctuations.

Low Temperature Protein Dynamics Revealed by Vibrational Echoes

Figures 7-9 show pure dephasing of the CO oscillator in heme proteins has a low temperature power-law phase and a higher temperature activated phase, which depends on solvent viscosity. The $T^{1.3}$ low temperature phase is reminiscent of the temperature dependence that has been observed for pure dephasing of electronic transitions of molecules in low temperature glasses (14), except the $T^{1.3}$ dependence in glasses persists only to a few degrees K, whereas in proteins it is seen at much higher temperatures. Several experiments, including ligand rebinding (18) and pressure relaxation studies (19), suggest that protein behavior is similar to glasses. The most successful theoretical treatment of glass dynamics has been the tunneling two-level system (TLS) model (20). Thus we try to understand the lower temperature phase in terms of a protein two-level system (PTLS) model (2,3). Other possible models for understanding low temperature dephasing are mentioned below. In the PTLS model, some molecular groups of the protein can reside in either of two minima of the local potential surface. Each side of the double well potential represents a distinct local configuration of the protein. The bulk protein sample contains an ensemble of these PTLS systems, characterized by a broad distribution of tunnel splittings and tunneling parameters. Transitions between PTLS can occur by tunneling through the potential barriers or at sufficiently high temperatures by activation over these barriers. Thus the complex potential surface of the protein is modeled as a collection of double well potentials.

The low temperature pure dephasing rates can be computed using the TLS uncorrelated sudden-jump model (14,17). It is found the rate is a power-law function of temperature T^α, where α is determined by the probability distribution P(E) for the PTLS energy difference E. In general, for P(E) α E^μ, $\alpha = 1 + \mu$. Thus applying the

PTLS model to our data, we find the distribution of PTLS energy differences E $P(E)_{Mb\text{-}CO} \propto E^{0.3}$.

The PTLS model of the Mb protein indicates it has an energy landscape on which the distribution of energy differences between conformations is broad and almost flat (2,3). Figs. 7-8 show this tunneling landscape is not controlled by the nature of the solvent, since $T^{1.3}$ behavior is seen in both glycerol and ethylene glycol solvents. Figure 9 shows the tunneling landscape is not affected by a single point substitution of valine for the distal histidine. The energies involved are much greater than those invoked to explain dynamics in low temperature glasses (14). Our echo results suggest there are higher barriers in Mb that play an important role at temperatures on the order of 100K. While the PTLS model successfully explains our results, it is possible that a less restrictive set of assumptions which still involve an energy landscape with a broad distribution of barriers might explain the data as well (2,3). For example, thermal activation over barriers with a broad distribution of barrier heights ΔE, where $P(\Delta E) \propto (\Delta E)^{0.3}$, can also give a power-law temperature dependence. In other words, the power-law temperature dependence can be explained as well by a PTLS model, which postulates a distribution of tunnel splittings E, or an activated process model which postulates *the same* distribution of barrier heights ΔE. In either case, the protein potential energy surface is seen to be a complex landscape with a very broad spread of energy parameters. Regardless of the specific mechanism of low-temperature dephasing, these protein measurements are in stark contrast to dephasing in glasses, where the power-law temperature dependence is superseded by an exponential temperature dependence due to thermal phonon populations, at just a few degrees K.

High Temperature Protein Dynamics Revealed by Vibrational Echoes

In the high temperature phase, above the solvent glass transition temperature, the pure dephasing process appears to be exponentially activated. That change can likely be attributed to the softening of the boundary condition placed on protein motions by the solvent (2,3). When the solvent is rigid, certain conformational changes will have very high barriers which cannot be surmounted. In fact the only conformational changes likely to occur at low temperature are those which preserve the essential shape of the protein surface in contact with the rigid solvent. But when the solvent becomes fluid, new conformational changes become possible via activation over barriers which become lower when the range of protein motions are no longer restricted by a rigid solvent. The observation in Fig. 8 that the rates of the activated processes at 300K increase with decreasing solvent viscosity strengthen the interpretation that the activated processes responsible for ambient temperature pure dephasing involve large scale motions of the protein, which are sensitive to the viscosity of the surrounding solvent. One point is worth stressing concerning the activation barrier value of ~1250 cm^{-1} used to fit the data in Figs. 7 and 8. This number is uncertain to several hundred cm^{-1}, and since the activated temperature dependence is observed over a relatively narrow range of temperature, the single activation energy cited here might well represent a distribution of ΔE values with an average in the range of 1250 cm^{-1} (2,3).

The activated pure dephasing process seen at ambient temperature involves relatively large scale, viscosity dependent protein motions which have a direct effect on the frequency of the CO oscillator. It is a tantalizing possibility which needs further investigation, that these processes might be closely related to the protein rearrangements needed to allow CO ligands to escape from the heme pocket, or enter the heme pocket from the solvent (21).

Vibrational Echo Studies of Heme Mutants

The data obtained on H64V mutant proteins (4) indicate the global protein dynamics are not affected by this mutation, since the functional form of the temperature dependence of the pure dephasing seen in Fig. 9 is unchanged. However the strength of the coupling of the protein fluctuations to the CO are evidently reduced (by a factor of ~20%) because the distal histidine, one of the closest amino acid residues, whose polar nature can have a substantial effect on the CO frequency (Table I), is removed (7).

In the absence of the distal histidine in H64V-CO, pure dephasing still occurs because the dynamical interconversion between different protein conformations is still communicated to the CO vibrational oscillator by the other amino acid residues of the proteins (4). Further use of site-directed mutagenesis techniques will likely reveal the extent of coupling between these other residues and CO at the active site of the protein. Perhaps certain mutations will, unlike the H64V substitution, have an effect on the overall global dynamics of the protein.

Summary

Vibrational echo experiments are a new method of examining protein dynamics, and have already yielded intriguing insights into how protein dynamics are transmitted to the active site of myoglobin, at low temperatures and perhaps more significantly at physiologically relevant temperatures. The vibrational echo is the vibrational analog of the spin echo of NMR, which ushered in a new dimension in magnetic resonance spectroscopy. Without coherent pulse sequences, NMR would not be the powerful probe of structure and dynamics it is today. As the spin echo did for NMR, the vibrational echo is expanding the scope of IR vibrational spectroscopy.

Acknowledgment

This research was supported by Office of Naval Research contracts N00014-95-1-0259 (D. D. D.) and N00014-94-1-1024 (M. D. F.), and National Science Foundation, Division of Materials Research grants DMR 94-04806 (D. D. D.) and DMR 93-22504 (M. D. F.). The experiments described here were the result of an effort involving many valued collaborators, including Prof. H. A. Schwettman, T. I. Smith, C. W. Rella and Alfred Kwok of the Stanford Free Electron Center, Kirk Rector of the Department of Chemistry, Stanford University, and Dr. Jeffrey R. Hill, Prof. Stephen G. Sligar and Dr. Ellen Y. T. Chien of the School of Chemical Sciences of the University of Illinois.

Literature Cited

1. Zimdars, D.; Tokmakoff, A.; Chen, S.; Greenfield, S. R.; Fayer, M. D.; Smith, T. I., Schwettman, H. A., *Phys. Rev. Lett.* **1993**, 70, 2718. Tokmakoff, A.; Zimdars, D.; Urdahl, R. S.; Francis, R. S.; Kwok, A. S.; Fayer, M. D., *J. Phys. Chem.* **1995**, 99, 13310.
2. Rella, C. W.; Kwok, A.; Rector, K. D.; Hill, J. R.; Schwettmann, H. A.; Dlott, D. D.; Fayer, M. D. *Phys. Rev. Lett.* **1996**, 77, 1648.
3. Rella, C. W.; Rector, K.; Kwok, A.; Hill, J. R.; Schwettman, H. A.; Dlott, D. D.; Fayer, M. D. *J. Phys. Chem.* **1996**, 100, 15620.
4. Rector, K. D.; Rella, C. W.; Hill, J. R.; Kwok, A. S.; Sligar, S. G., Chien, E. Y. T.; Dlott, D. D.; Fayer, M. D. *J. Phys. Chem.* **1997**, 101, xxxx (in press).
5. Leeson, D. T.; Wiersma, D. A. *Phys. Rev. Lett.* **1995**, *74*, 2138.
6. Antonini, E.; Brunori, M. *Hemoglobin and Myoglobin in their Reactions with Ligands*; North Holland: Amsterdam, 1971.
7. Springer, B. A.; Sligar, S. G.; Olson, J. S.; Phillips, G. N., Jr. *Chem. Rev.* **1994**, *94*, 699.
8. Springer, B. A.; Sligar, S. G. *Proc. Natl. Acad. Sci. USA* **1987**, *84*, 8961.
9. Hill, J. R.; Dlott, D. D.; Rella, C. W.; Smith, T. I.; Schwettman, H. A.; Peterson, K. A.; Kwok, A.; Rector, K.; Fayer, M. D., *Biospectroscopy* **1996**, 2, 277.
10. Hill, J. R.; Tokmakoff, A.; Peterson, K. A.; Sauter, B.; Zimdars, D.; Dlott, D. D.; Fayer, M. D. *J. Phys. Chem.* **1994**, 98, 11213.
11. Alben, J. O.; Caughy, W. S. *Biochem.* **1968**, *7*, 175.
12. Ansari, A.; Beredzen, J.; Braunstein, D.; Cowen, B. R.; Frauenfelder, H.; Hong, M. K.; Iben, I. E. T.; Johnson, J. B.; Ormos, P.; Sauke, T.; Schroll, R.; Schulte, A.; Steinbach, P. J.; Vittitow, J.; Young, R. D. *Biophys. Chem.* **1987**, *26*, 337.
13. Gordon, R. G. *J. Chem. Phys.* **1965**, *43*, 1307.
14. Narasimhan, L. R.; Littau, K. A.; Pack, D. W.; Bai, Y. S.; Elschner, A.; Fayer, M. D., *Chem. Rev.* **1990**, 90, 439.
15. Hill, J. R., Dlott, D. D.; Rella, C. W.; Peterson, K. A.; Decatur, S. M.; Boxer, S. G., Fayer, M. D., *J. Phys. Chem.* **1996**, 100, 12100. Hill, J. R.; Ziegler, C. J.; Suslick, K. S., Dlott, D. D.; Rella, C. W.; Fayer, M. D., *J. Phys. Chem.* **1996**, 100, 18023.
16. Kubo, R., *Adv. Chem. Phys.* **1969**, 15, 101.
17. Bai, Y. S.; Fayer, M. D., *Phys. Rev. B* **1989**, 39, 11066.
18. Austin, R. H.; Beeson, K. W.; Eisenstein, L.; Frauenfelder, H.; Gunsalus, I. C. *Biochem. 1975*, 14, 5355.
19. Iben, I. E. T.; Braunstein, D.; Doster, W.; Frauenfelder, H.; Hong, M. K.; Johnson, J. B.; Luck, S.; Ormos, P.; Schulte, A.; Steinbach, P. J.; Xie, A.; Young, R. D. *Phys. Rev. Lett.* **1989**, *62*, 1916.
20. Anderson, P. W.; Halperin, B. I.; Varma, C. M., *Philos. Mag.* **1972**, 25, 1. Phillips, W. A., *J. Low Temp. Phys.* **1972**, 7, 351.
21. Case D. A.; Karplus, M. *J. Mol. Biol.* **1979**, 132, 343.

INDEXES

Author Index

Affiliation Index

Subject Index

Highlights from ACS Books